D0742445

ESPRIT Basic Research Series

Edited in cooperation with
the Commission of the European Communities, DG XIII

Editors: P. Aigrain   F. Aldana   H. G. Danielmeyer
O. Faugeras   H. Gallaire   R. A. Kowalski   J. M. Lehn
G. Levi   G. Metakides   B. Oakley   J. Rasmussen   J. Tribolet
D. Tsichritzis   R. Van Overstraeten   G. Wrixon

G. A. Orban   H.-H. Nagel (Eds.)

Artificial and Biological Vision Systems

Springer-Verlag

Berlin Heidelberg New York
London Paris Tokyo
Hong Kong Barcelona
Budapest

.05207897

Volume Editors

Guy A. Orban   **OPTOMETRY**
Katholieke Universiteit Leuven, Faculteit Geneeskunde
Laboratorium voor Neuro - en Psychofysiologie, Campus
Gasthuisberg
Herestraat, B-3000 Leuven, Belgium

Hans-Hellmut Nagel
Institut für Algorithmen und Kognitive Systeme
Fakultät für Informatik der Universität Karlsruhe (T.H.)
and
Fraunhofer-Institut für Informations- und Datenverarbeitung IITB
Fraunhoferstraße 1, W-7500 Karlsruhe, FRG

ISBN 3-540-56012-2 Springer-Verlag Berlin Heidelberg New York
ISBN 0-387-56012-2 Springer-Verlag New York Berlin Heidelberg

Publication No. EUR 14600 EN of the Commission of the European Communities,
Scientific and Technical Communication Unit, Directorate-General Telecommunications, Information Industries and Innovation, Luxembourg
Neither the Commission of the European Communities nor any person acting on behalf
of the Commission is responsible for the use which might be made of the following
information.

Typesetting: Camera ready by authors
45/3140 - 5 4 3 2 1 0 - Printed on acid-free paper

Foreword

As is true with most areas of Artificial Intelligence, there is real need for a symbiotic relationship between the biological and artificial – a need for problems to be viewed from many different angles, and particularly so in the study of vision. The INSIGHT consortium is taking steps in this direction. In a traditional sense, the papers in this volume are represented by the areas of neuroscience, psychophysics and traditional computer vision. However, to gain deeper insight into vision processes, it is the interaction of scientific ideas from these areas that is essential.

The scope of the topics discussed has a definite interdisciplinary flavour: at one end of the spectrum we have experiments performed and direct measurement of the responses of neurons to visual stimuli; at the other end we have the mathematical and computational aspects of optical flow (the relative motion between observer and object) and approaches of tackling vision through binocular disparities (stereopsis). Traditional edge detection (essential for the initial classification of shape) is also covered as is the study of natural texture patterns that occur on object surfaces.

A fundamental aim of the Basic Research part of the ESPRIT programme is the production and maintenance of a pool of research expertise in Europe, from which both further research and industry can draw. As the authors state in their preface, this project has not only succesfully merged the talents of senior researchers from different backgrounds, but also brought many young ones along.

This volume comes at a watershed moment when a number of new ESPRIT Basic Research Projects are starting. We feel sure that INSIGHT-2 will build upon the successes of its predecessor.

George Metakides

Preface

This volume presents an overview of achievements attained in an interdisciplinary study of vision systems. In order to put these achievements into perspective, it may be useful to provide some background on this 30-month study which has been made possible by a Basic Research Action (BRA) supported by the European Community in the framework of the European Specific Programme for Research in Information Technology (ESPRIT). When this INSIGHT project was planned, three basic ideas guided its conception.

First, the groups involved were convinced that given the progress in neuroscience and psychophysics as well as the technological and methodological facilities nowadays available for computer vision, a fruitful interdisciplinary interaction could be set up, linking these three areas of expertise in vision systems. Indeed, progress on the biological side was such that not only could the brain and in particular the visual system of primates be considered as an existence proof – an argument frequently discussed by many familiar with these disciplines – but that the time appeared to have come to unravel biological solutions to hard vision problems and to compare these solutions in detail with ideas explored by computer vision approaches. Not only was computer vision to benefit from such a collaboration, but also the biological side. Computer vision scientists no longer explore machine vision problems by heuristic algorithms only, but they attempt to formulate their ideas in a manner which facilitates an analysis of the consequences of their assumptions. We expected that this tendency could become a rich source of experimental questions to be tested not only by machine vision groups, but also by researchers in biology. Given the complexity of biological vision systems and the problems encountered in isolating and measuring particular processing steps, neuroscience naturally is heavily preoccupied with attempts to understand the functioning of primate vision by an inductive approach. Computer vision, on the other hand, can always inspect all details of the information processing designed into its system. Thus, since the processing power has recently become available to avoid cutting every corner in order to obtain results within a reasonable processing interval, computer vision has begun to reorient itself towards a more deductive approach. Machine vision systems begin to be synthesized on the basis of increasingly precise specifications of what they should do. It thus becomes possible to test whether hypothetical solutions deduced from biological experiments really perform as expected.

The second idea underlying INSIGHT postulated that different levels of processing had to be tackled simultaneously. Traditionally, 'early vision' or 'low level vision' is considered to comprise transformations close to the level where the irradiance distribution impinging on the retina is transduced into signals to be processed by the vision system. Eventually, subsequent processing steps will result in descriptions at a symbolic or even conceptual level. These latter levels of processing are usually addressed as 'higher levels of vision'. It is obvious that research on such 'higher vision levels' benefits from a precise description of the results which can be made available by the initial processing steps. On the other hand, unraveling of the details of initial processing steps will be supported by precise expectations about what their results will be used for. Although we deliberately did not exclude any level of vision, emphasis was clearly given to intermediate levels of description. This comprises higher order, local operators which are expected to bridge the gap between the initial local operations immediately following the signal transduction and those processing steps where concepts about the depicted three-dimensional world begin to influence the type of information to be represented. The integration of different types of cues and the representation of surface patches then naturally follow. Cue integration has been studied both regarding the recovery of depth and regarding the determination of boundaries. Each of these topics has been approached by all three disciplines involved, although it is fair to say that our work in neuroscience mainly contributed to research on higher order, local operators.

A third idea has influenced our activities in a less explicit manner. It appears natural to assume that artificial vision systems which are designed according to a better understanding of the building principles of human vision would be easier to use by humans. In view of the broad range of applications for robust machine vision systems, the interface between artificial and biological vision systems becomes very important. Having both kinds of systems functioning according to similar rules is expected to facilitate the interaction between them. This consideration inspired the full title of our project: *Vision Systems for a Human, Natural Environment*. The word 'natural' stressed the general purpose scope of anticipated machine vision systems – in analogy with the capabilities of biological vision. The word 'human' in the title was meant to capture three aspects: that the human vision system is expected to be a major source of inspiration, that it is part of the environment in which artificial systems have to operate, and that artificial vision systems will have to be used by humans.

The interdisciplinary nature of the research reported in this volume is demonstrated by the fact that the contributions are about equally divided between investigations into biological vision systems – both from a psychophysical and neurophysiological point of view – and into computer vision systems. The authors have been particularly asked to deemphasize the technical vocabulary in an attempt to make their contributions accessible to members of the other community. The reader will realize that such an attempt presents its own challenge if it is combined with a quest for scientific rigor in the presentation of results.

Ideally, a project such as INSIGHT should be based on a theory of vision applicable both to biological and artificial vision systems. Although we feel we have made considerable progress in understanding the problems facing the other disciplines as well as their methodological approaches, we have not yet reached this stage of a unifying theory. The order in which the contributions are presented in this volume should convey, however, our 'vision' about how this goal may be approached.

We start with the chapter by Koenderink et al. on *Local Operations: The Embodiment of Geometry*. The research by Koenderink and co-workers has inspired many participants of INSIGHT. The next group of chapters explores this same topic from many different aspects, most of them related by joint efforts to clarify the definition, estimation, and use of optical flow. *The Analysis of Motion Signals and the Question of the Nature of Processing in the Primate Visual System* by Orban sets the stage from the biological point of view. The next chapter by Hoffmann on *Motion Perception and Eye Movement Control* extends this discussion towards the problem of how optical flow estimates may be used by the brain.

The subsequent three chapters complement this description from the computer vision side. The chapter *Computational Aspects of Motion Perception in Natural and Artificial*

Vision Systems by Verri et al. presents a computer vision approach towards the estimation of optical flow and integrates this with attempts to model some of the biological data obtained by the Leuven group of Orban and co-workers. The contribution *Four Applications of Differential Geometry to Computer Vision* by Deriche et al. covers not only optical flow estimates, but treats its use – in combination with other approaches – for the extraction of surface descriptions. This research – again ranging from the investigation of initial processing steps towards the use of their results in levels addressing already the description in terms of scene attributes – is taken as a node which provides a natural link to two additional investigations. Demazure et al. analyze *Geometry of Vision* in their quest to exploit the complicated interaction of these factors for the estimation of surface structure from images. The essay by Longuet-Higgins on *A Method of Obtaining the Relative Position of 4 Points from 3 Perspective Projections* illustrates the difficulties of inferring structure precisely from a minimal number of observations. The chapter by Nagel on *Direct Estimation of Optical Flow and Its Derivates* leads back from the preceding excursion into three-dimensional aspects of vision to the intricacies of understanding optical flow. The subsequent chapter by Proesmans and Oosterlinck on *The Cracking Plate and Its Parallel Implementation* illustrates the application of numerical techniques to the solution of problems which exhibit a structure similar to the estimation of optical flow.

A second very important topic for INSIGHT, in addition to optical flow, has been the interactions between stereo and texture. Rogers emphasizes in his contribution on *The Perception and Representation of Depth and Slant in Stereoscopic Surfaces* the similarity between optical flow and stereo and thus naturally introduces this second part. The following chapter by Frisby and Buckley on *Experiments on Stereo and Texture Cue Combination in Human Vision Using Quasi-Natural Vision* reports on the complicated interactions between stereo and texture cues. It thus represents an obvious link to the subsequent chapter on *The Analysis of Natural Texture Patterns* by Watt which is followed by *Segmenting Textures of Curved Line Elements* by Simmons and Foster. Eklundh et al. nicely illustrated the interaction between computer vision and psychophysical investigations in the final chapter on *Extraction of Shape Features and Experiments on Cue Integration*.

The interaction between many research groups from different disciplines as documented by these investigations not only stimulated the senior researchers involved. It simultaneously provided a challenging opportunity for young people, in particular doctoral students and postdoctoral researchers participating in this endeavor, to extend their horizon – sometimes taking consolation from the visible efforts of their supervisors not to disappear in the yawning canyons between the apparent secure grounds of each discipline.

We hope that this volume will convey some of the excitement we had in carrying out this interdisciplinary research. All participants gratefully acknowledge the support of the European Community, and in particular ESPRIT Basic Research, without which it would not have been possible to assemble such a broad range of expertise, bridging boundaries both across disciplines as well as across languages and nationalities. Our thanks are also due to the referees who accompanied this project and our meetings with their advice, recommendations and encouragement.

	Guy A. Orban	Hans-Hellmut Nagel
August 1992	Leuven	Karlsruhe

Table of Contents

Local Operations: The Embodiment of Geometry

Jan-Johan Koenderink, Astrid Kappers and Andrea van Doorn

Buys Ballot Laboratorium, Universiteit te Utrecht

We consider the default structure of the visual front end based on very general symmetry considerations and invariance principles. It turns out that such very general principles constrain the possible sampling structures greatly, and that the resulting default structure is not unlike the actual structure of the primate front end visual system and also not unlike the structures as they have "evolved" in computer vision. The general formalism suggests various useful relations that are an immediate consequence of the front end structure but are generally being rediscovered in ad hoc ways in image processing. The formalism allows a novel interpretation of the concept of a "feature", an otherwise elusive concept in organic and computer vision alike. This interpretation may well turn out to be useful in image understanding.

1 The "Front End Visual System"

Vision is best defined as "optically guided behavior". It is sustained by a dense, hierarchically nested and heterarchically juxtaposed tangle of cyclical processes. On the coarsest scale of description these processes comprise both causal connections in the environment of the agent ("ecological optics"), as well as somatic processes. In this chapter we focus upon the *interface* between the light field and those parts of the brain nearest to the transduction stage. We call this the "visual front end". This interface is an obvious bottleneck of optically guided behavior since all optically specified information has to be passed to higher centers by way of the front end. Thus a thorough analysis of this stage is a prerequisite for the description of higher order processes. Of course, the exact limits of the interface are essentially *arbitrary*, but nevertheless the notion of such an interface is valuable.

We suggest that the demarcation be drawn according to the following considerations (which may be applied with greater or lesser severity according to whim or necessity):

- the front end is a "machine" in the sense of a *syntactical transformer* (or "signal processor");

- there is no semantics (reference to the environment of the agent). The front end merely processes *structure*;

- the front end is *precategorical*, thus – in a way – the front end does not compute anything;

- the front end operates in a *bottom up* fashion. Top down commands based upon semantical interpretations are not considered to be part of the front end proper;

- the front end is a deterministic machine, *i.e.*, it doesn't "hallucinate" or "dream"; but all output depends causally on the (total) input from the immediate past (the "specious moment" as defined in the psychology of time perception).

This roughly outlines the meaning of the term "visual front end". The notion is similar in spirit (though more technical) to Orban's distinction between literal and interpretative representations. (This book.)

The *task* of the front end is to transform, encode and distribute the transduced spatiotemporal irradiance distribution in such a way as to enable efficacious visual processing. That means: processing subserving efficacious optically guided behavior of the organism as a whole. What is not explicitly encoded by the front end is irretrievably lost. Thus the front end should be universal (undedicated) and yet should provide explicit data structures (in order to sustain fast processing past the front end) without sacrificing completeness (everything of potential importance to the survival of the agent has to be represented somehow). Clearly there are some incompatible objectives here: for instance, how can one encode explicit data structures without commitment?

The visual front end is being studied in physiology and psychophysics. We aim at a *general formal framework*, as the only way to proceed from mere factcollecting to natural philosophy. Mathematical structures are used to reorder existing facts. Certain mathematical structures can be used to *represent* given physical structures, for instance geometrical objects represent the local behavior of extended structures. The emphasis will be on these structures themselves, rather than on formal manipulation or representation, *i.e.*, on *geometrical* aspects.

Typically "geometrical objects" are defined as equivalence classes of other objects. For instance, a "tangent vector" is an equivalence class of curves. All tangent vectors at a point span "tangent space", which is a local picture of the space itself. This is the preferred level of description from a physicist's point of view: one concentrates on the essential structure without bothering too much about accidental representation. (*E.g.*, technicalities such as coordinate representation.) The same position is a natural one for the description of the visual system: for instance, every point of the visual field carries a copy of tangent space. Indeed the cortical hypercolumns may be interpreted as their embodiments. The language of "fiber bundles" thus very aptly describes the structure of the visual front end.

The geometrical language provides a universal language and a uniform format in which to describe front end structures. The language is abstract and basically devoid of any *meaning* (*i.e.*, it is pure syntax). Indeed, meaning is due to interpretation, a top down action of the organism. A top down query to the front end entails a "logical format" that bestows meaning on the front end structure. (Thus the same structure may acquire many different meanings.) It's like the format of a "read" command in many computer languages. The same *datum* may be treated as an ASCII-character, a memory address, or an integer number, depending on the format statement associated with the read command. The meaning is not (only) in the structure of the datum, but in the read action performed upon the datum. Of course it doesn't necessarily make sense to interpret a given datum in any old way: it is the responsibility of the process that issues the command to make it a sensible one, otherwise gibberish results. (An unfortunate effect that is not unfamiliar to computer users!) In the neurosciences this problem is known historically as the problem of "local sign", or of the "homunculus". We will not address such important problems in this chapter.

The front end cannot represent everything. There has to occur some process of *selection*. This is a very serious matter, because what's not represented doesn't so much as *exist* for the agent, in the sense that it cannot codetermine efficacious action of the agent.

By "representing" we mean something like "segregation of quality", putting things in distinct, addressable pigeonholes. This "segregation" implies a parallelism, that is a minimization of lateral connections. The primitive notions of "continuity" and "coherence" make that one particularly useful "segregation of quality" is a division in terms of *locality*. In a local representation one can do without extensive (that is spatial, or geometrical) properties and represent everything in terms of intensive properties. This obviates the need for explicit geometrical expertise. The local representation of geometry is the typical tool of differential geometry. For instance, a *vector* summarizes a *bilocal* property (vector as "arrow with tip at $\mathcal{B}$, tail at $\mathcal{A}$") in a purely *local* manner. The columnar organization of representation in primate visual cortex suggests exactly such a structure. Local sign has to be attached to the (hyper-)columns, whereas the activity within such a (hyper-)column can be *local* in the sense that only intensive, rather than extensive, operations need to be performed.

1.1 Consequences of Non-Commitment

Non-commitment means not making choices: no place is *a priori* different from any other, no orientation special, no level of resolution more important than any other. "Not seeing the wood for the trees" is a serious blindness, but one doesn't want to loose sight of the trees either. Any firm commitment of the front end may allow especially powerful processing of some of the structure, but also necessarily entails a limit on the agent's repertoire of efficacious behavior. In practice extreme dedication of the front end is only found in species for which ecological niches exist that allow them to get away with diminished abilities in other areas. *Homo sapiens* may be the most universal animal around; if so, then its front end must be the least dedicated.

In this paper we explore "ideal", that is completely undedicated front ends. Every real species falls short of this, sometimes for good reasons. Exploration of the ideal limit is a help in understanding real systems nevertheless: it is the only prototype available against which to judge the merit of real systems.

The consequences of non-commitment are that the front end must implement a structure that is invariant under certain basic symmetry groups. The collective symmetries characterize the front end. They are like axioms of a formal system. You can't *deduce* the symmetries from first principles, you *postulate* them. They sum up the common sense notion of non-commitment: the following are a set of symmetries that appear very basic indeed:

homogeneity no location or moment is special. This entails translational symmetry in space and time;

scale invariance no spatial or temporal scale is special (for the eternal eye the eagle's perspective is no more important than the mole's). This entails scale invariance (or self-similarity) in space and time;

isotropy no spatial orientation is singled out. (The vertical is special to us, but how about astronauts?) This entails rotational invariance;

separability various dimensions are independent. This entails separability, *e.g.*, what happens in time doesn't depend on what happens in space, *e.g.*, on whether I happen to look through a telescope or a microscope;

linearity if inputs are superimposed I expect the responses to be likewise superimposed. This symmetry is called "linearity". Nonlinearities always imply distinguished parameter ranges, *i.e.*, dedication to specific phenomena;

semigroup property of scaling scale transformations should combine gracefully: whether I change scale in one go or via a number of stages shouldn't matter at all. However, one can't require that scale change can be undone, thus we require only the "semigroup" property;

contrast invariance TV movies look essentially the same on different TV sets, irrespective the fact that no two sets have exactly the same (nonlinear) transfer, or "gamma" and "brightness" settings. Hence it appears prudent to require that the front end be invariant with respect to contrast transformations, where luminance is scaled by an arbitrary power law.

Clearly one could either add to this list or curtail it. One then obtains front ends of various degrees of generality. The list proposed here is extensive enough to constrain the front end structure considerably, yet it yields a structure that is still more universal than any known biological system. Any *real* system will of course be dedicated to the generic environment and lifestyle in which the species evolved: thus there can be *no general theory* of vision in the strict sense, only theories of specific instances.

The present approach attempts to build a starting platform for such more specific theories. Moreover, we don't even attempt a "theory of everything", but build in essential limitations right from the start: For instance we don't address spectral discrimination or binocular information in this chapter. Our approach is very much akin to that of the physicist: for example, the theory of the "ideal gas" is extremely important in physics, despite the fact that no such a thing exists in nature at all and every real gas is an exception! Moreover, the ideal gas does not even properly "condense" to the fluid state, thus it can't be part of any "complete" theory of material constitution. The scientific advance made possible by such fortunate abstractions is evident enough.

The items discussed above are far more intricate than might appear at first blush. Most of them cannot simply be tested empirically, but their value can only be assessed in a much later stage of development of the theory. We lack the space to develop such important considerations here. Just a simple example: one might strike out the linearity assumption by pointing at the neurophysiological literature which indicates that all neural processes are of a highly nonlinear nature. However, it would be fairly easy to implement our constructs in such a way that it would look likewise highly nonlinear to the superficial eye, yet in no way jeopardize the value of the analysis: Such a simple "test" is not decisive at all. (In order to see at least the possibility of such a state of affairs you may think of a linear problem programmed on a digital computer. To the user the system is linear, though all the digital gates and processes that "implement" the system are of an essential nonlinear nature.) Similar considerations apply to the other items.

1.2 Remapping of Dimensions

The representation of any dimension typically includes taking ratios with some fiducial object, singling out an "origin", *etc.* By assumption all origins are equivalent.

For the spatial domain we typically pick (any) fiducial location and call it "the origin". Distances between locations are given as the ratios to the length of some (arbitrary) fiducial "yardstick". Often the natural thing to do is to pick the resolution as a yardstick. Then distances become pure numbers, whereas only the integer parts of these numbers are relevant. (This is loosely speaking of course: we don't imply that you should truncate coordinates. The problem of discretization is an important one that will not be taken into account here.)

Resolution (or "inner scale") is a dimension on its own right. The resolution is the minimum length over which significant changes may be expected. Thus it is a positive number, expressed in terms of the fiducial yardstick. It is convenient to use the highest available resolution as the fiducial value and treat this scale as uniform. There still is a problem: we don't have self-similarity. Self-similarity implies that you take the logarithm of the ratio of the actual resolution to the fiducial one. We call this the "natural resolution parameter". This parameter ranges from minus to plus infinity. The origin of the scale depends on the fiducial yardstick. Since no yardstick is singled out we regard all origins as equivalent. If you express the spatial distances and the resolution in the indicated manner you can no longer find out whether you look through a microscope or a telescope. Thus scale invariance has been arrived at. This trick should be familiar to the neuroscientist as akin to the "Weber-Fechner law" in various sensory domains. It is also familiar in statistics: the only way to express total ignorance for a parameter that may assume all positive real values is to assume a logarithmically uniform prior probability distribution.

Time is a more complicated dimension than space is. We assume that different observers may agree on simultaneity of events, but we don't assume everyone has the same clock. (With "observers" we indicate parts of the front end here. Different parts may use different temporal scales.) We do assume that all clocks are regular though, they only differ in rate ("unit of time") and epoch. (Days since the birth of Christ or lunar cycles since the battle of Hastings will do equally well.) "Regularity" is taken to mean that if two observers record any three events A, B, C (say), then the ratio $(t_B - t_A)/(t_C - t_A)$ will agree for both observers. Let N ("now") denote the present moment. Suppose we point out two fiducial events P, Q (Q later than P) to all observers. We ask all observers to report the time of occurrence of some event E as the number $\tau = (t_E - t_N)/(t_P - t_Q)$. Now all observers agree (you easily check that the numbers τ reported are equal). These numbers are always positive since all events (including the fiducial one) exist in the past. Self-similarity in time is obtained if we take the logarithm and treat this scale as uniform. Then the origin is delayed $t_P - t_Q$, whereas the present ("now") maps to minus infinity, the infinite past to plus infinity. All choices of origin are to be considered equivalent.

Finally we regard the irradiance domain. Irradiance is always positive. Complications are due to the fact that different observers may use different units (such as lux, or Watt per meter squared) and that we require invariance with respect to contrast transformations. A simple way to handle the problem is the following: we designate two points in the input that are at different irradiances I_A, I_B with $I_B > I_A$ (say). We report the numbers $\log(I_C/I_A)/\log(I_B/I_A)$ for the irradiance at any point C (say). These numbers agree even for different photometers (lux or Watt per meter squared) and even under arbitrary contrast transformations. (That is for transformations of the form $I^* = \alpha I^\beta$, with $\alpha, \beta \in (0, \infty)$.)

The invariance is obtained only if the contrast transformation also affects the fiducial irradiances at A, B. An apt choice for the fiducial anchor points I_A, I_B are the 25% and 75% quartiles of the pixel intensities. An automatic gain control following a logarithmic transducer function (like we find in the visual system) will perform essentially the same task.

In most cases the scale transformations are mere formal devices that are most convenient for the description, and there is no obvious need to implement them, often it is not even clear that such has any meaning. (*E.g.*, in the case of a shift of the origin.) In some cases the transformation is easily incorporated in the structure of the receptive fields. (*E.g.*, in the case of the temporal transfer.) However, in the case of the irradiance we have to insist on a hardware implementation at a very early stage because the non-linear transformation doesn't commute with the linear transformations implemented by the receptive fields. Then we may as well forget about this transform in describing the receptive field structure. The only datum needed at the interpretation stage is the fact that irradiance *ratios* map to internal *differences*.

1.3   Local versus Multilocal Operations

Suppose you want to find the value of some scalar field at some location. All you can ever come up with is the average over an area s that is forced upon you by the measuring apparatus. Different pieces of apparatus may allow you to vary s, thus you obtain a one-parameter family of values. Think of the apparatus as an operator that can be characterized by its location and its inner scale (s). The operator takes a bite of a field and spews out a number, the average value over the area s at the location of the operator. Such an operator might well be called a "point operator". Although the point has a size, it "has no parts" as Euclid would have it. For the special linear fields $X(x, y) = x$, $Y(x, y) = y$, the operator yields the numbers x and y say, which are its "Cartesian coordinates".

A "point" is obviously a "local" entity. However, notice that the inner scale is essentially arbitrary, so the average may well extend over the whole image! Points lack internal structure, but they do have sizes.

A "vector operator" also spews out a number when you feed it a field. The number is the average slope or the slope of the smoothed field (the order makes no difference because of linearity) over an area s in the direction of the vector times the modulus of the vector. Again this is a purely local operation. Contrast this with the following method of finding the result of a vector operation: take two points A, B (A denotes the tail, B the tip of the vector) and let the points operate on the field. Subtract the results and divide by the distance of the points. This is a *multilocal* method, and it is fraught with problems. Not only do you need to be able to point out a location ("find back a point", *i.e.*, Lotze's "local sign"(Koenderink 1990; Lotze 1884)), but you need topological expertise (A and B have to be close but not coincident) and even a metric (the distance $\| \vec{AB} \|$).

We do not assume that the front end is *that* sophisticated. Hence we consider only local operations. Multilocal operations have to be implemented at further stages of processing. We don't loose much, because *e.g.*, all the operations from differential calculus and geometry are of a purely local nature. It is perhaps not superfluous to stress the fact that being *local* in the technical sense has nothing to do with "small size" or "simple structure": thus "local processing" in no way rules out the existence of very large receptive fields with complicated internal structures.

We assume all operators to be centered on the same location. Without loss of generality we may take this location to be the origin.

1.4   Scale Transformations

The symmetries of the front end put a very strong constraint upon the possible implementations of point operators. In fact, it is well known that the only admissible structure is a linear operator with Gaussian profile. Thus the action of a point operator $\mathcal{P}$ on a field $f(\mathbf{r})$ defined on the plane $\mathcal{R}^2$ is explained as follows:

$$\mathcal{P}\langle f \rangle = f(\mathbf{r}_\mathcal{P}, s_\mathcal{P})$$

with

$$f(\mathbf{r}_\mathcal{P}, s_\mathcal{P}) = \int_{\mathcal{R}^2} G(\mathbf{r}, s_\mathcal{P}) f(\mathbf{r}) \, d\mathbf{r},$$

where the kernel G describes the "size" and position of the point $\mathcal{P}$:

$$G(\mathbf{r}, s_\mathcal{P}) = \frac{e^{-\frac{(\mathbf{r}-\mathbf{r}_\mathcal{P})\cdot(\mathbf{r}-\mathbf{r}_\mathcal{P})}{4 s_\mathcal{P}}}}{4\pi s_\mathcal{P}}.$$

For points in *any* position we note that the convolution $G \otimes f$ predicts the output in any case.

The operation of a *vector* can be explained in various ways, all basically equivalent. If you conceive of a vector as of a "bilocal object" (arrow with tip and tail), then you may define its action on a field as the difference of the field's values at tip and tail of the arrow. If you conceive of a vector as of a *rate*, then you may think of its action in terms of the rate of change of the field as a moving point experiences it when it moves at a velocity given by the vector. If you are more formally minded, then you may conceive of a vector as of a *directional derivative*. Thus the unit vector in the x-direction operates on the field $F(x, y)$ to yield $F(x + 1, y) - F(x, y)$ (vector as bilocal object, tip at $x = 1$, $y = 0$), as $\partial F(x+t, y)/\partial t$ (vector as rate of change for the orbit $(x+t, y)$), or as $\partial F(x, y)/\partial x$ (vector as directional derivative in the x-direction). In the limit all these views merge.

Thus the operator $\mathbf{e}_x$ (the unit vector in the x-direction) is explained through its action:

$$\mathbf{e}_x \langle f \rangle = \frac{\partial f}{\partial x}(\mathbf{r}_{\mathbf{e}_x}, s_{\mathbf{e}_x}),$$

where $\mathbf{r}_{\mathbf{e}_x}$ is the position of the tail of $\mathbf{e}_x$, whereas $s_{\mathbf{e}_x}$ is its "size" (as distinct from its modulus, which is unity). The corresponding weighing function is simply

$$G_{10}(\mathbf{r}, s) = \frac{\partial G(\mathbf{r}, s)}{\partial x},$$

and again the convolution $G_{10} \otimes f$ predicts the output in any case. Because of the linearity of the convolution operation you have formally $\partial(G \otimes f) = \partial G \otimes f = G \otimes \partial f$, *i.e.*, convolution and differentiation commute. This simple observation enables us to construct arbitrary differentiation operators. In figure 1 we depict such differential operators of orders zero, one and two for dimension one.

The formalism presented here is based upon the extreme assumptions that the image is the Euclidean plane and that the resolution may range from $s = 0$ (infinite acuity!) to $s = \infty$ (no detail whatsoever resolved!). Of course real life is different. The typical picture has a finite extent, its *scope*, and is given at a finite resolution, its *grain size*. We run

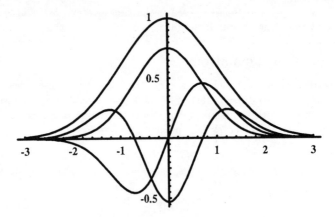

Figure 1: *The differential operators for $s = \frac{1}{2}$ in dimension 1. The operators G_0, G_1 and G_2 are depicted. For very high orders the operators become very wiggly. (The number of zeros equals the order.) Asymptotically you obtain "Gabor functions", i.e., trigonometric functions modulated with a Gaussian envelope. The wide Gaussian in this figure depicts this asymptotic envelope. In a later section we introduce a characteristic function A that exactly equals this asymptotic envelope.*

algorithms that see a limited region of interest (ROI) at a time, the *outer scale*, and use a limited resolution, the *inner scale*. (See figure 2.) If the outer scale approaches the scope you get into the *boundary problem*. If the inner scale approaches the grain size you run into the problem of *insufficient resolution*. Our formalism works fine in cases where resolution and boundary don't pose problems. If one runs into such problems one should strive to increase the scope and/or decrease the grain size, rather than construct fudge "solutions" that will never work well anyway. In the primate visual system algorithms (apparently frozen in pieces of hardware) run in the reasonable regimes (Bijl and Koenderink 1989; Koenderink and van Doorn 1978). When the animal runs against the limits it changes its behavior and strives for more reasonable input (explorative vision). In machine vision many algorithms have been designed to run into the resolution limit. A consequence is the myth that higher order operators are virtually impossible ("not robust"(Horn 1986)). In the primate visual system operators of order four and more are not at all rare and apparently do well (Young 1985; Koenderink and van Doorn 1987; Koenderink 1988).

2   The Blob Hierarchy

Intuitively, the irradiance distribution is a deeply nested set of light and dark regions. It may be compared with a landscape in which hills mimic the light regions, dales the dark regions. (This metaphor is useful if you try to forget the polarization of the typical landscape: Since water runs downwards and eventually has to go somewhere, there are hardly any pot-valleys, whereas isolated hills are numerous. Here we consider "symmetrical" landscapes.) One possible formalization of the hills and dales structure was pioneered by Cayley and later Maxwell. It divides the landscape into "natural districts" by way of the watersheds or the water courses. Thus you obtain a parcellation in either hills, *or* dales. These parcellations interlock and are in many respects dual.

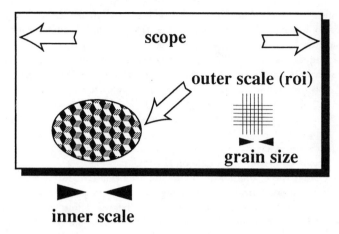

Figure 2: *Definition of cardinal extents. The extent of the whole picture is its scope, the sampling density (e.g., pixel separation) defines the picture's "grain size". Algorithms need input from a region of interest (r.o.i.) that defines the "outer scale", with a resolution that defines the "inner scale". Notice that scope and grain size pertain to the picture, but inner and outer scale to algorithms you may wish to run on the picture. The reasonable regime is that where the grain size is much finer than the inner scale, whereas the scope far exceeds the outer scale.*

For our purposes we need a parcellation into hills *and* dales. The natural way to do this is via the "isophotes", *i.e.*, the loci of equal irradiance. If the landscape is smooth (as we will assume), the isophotes are smooth curves that are either closed, or otherwise end on the boundary. We don't treat the boundary problem here, it is essentially trivial. For a finite number of very special heights, some isophotes may have a double point: this happens exactly if the height is that of a pass. At a pass two hills or two valleys meet. It is also possible that a hill meets a valley, as on the top of a volcano. If the height is taken between two successive pass heights, then the patterns of nearby isophotes are qualitatively identical. (The naïve reader will find these observations substantiated on perusal of a topographic map of some mountainous region containing the lines of constant height above sea level.)

When we say "hill" (or "light blob") we indicate one connected component of a level set, *i.e.*, that part of the landscape that is above a certain (arbitrary) fiducial height. (Similar for the dales or dark blobs.) A hill might well have internal structure (hills and dales), but these are all above the fiducial height. This is the only definition of a "light blob" that makes operational sense. Attempts to define "blobs" in terms of extrema are bound to fail, because of the fact that they are highly sensitive to noise. Even an infinitesimal amount of noise on a hilltop is bound to create any number of spurious hilltops! Thus the only reasonable way to define blobs is via the finite height ranges for which the fiducial isoheight curve landscape contains no critical points (extrema or saddles), but only finite slopes. (Eklundh's chapter in this book develops the notion of "blobs" in more detail.)

For every fiducial height you obtain one possible set of hills and dales. These are juxtaposed and nested to arbitrary depth. The structure is that of a tree (in the sense of computer data structures). For a different fiducial height you obtain a different structure. When two fiducial heights are divided by just one pass height, then the patterns differ only by a single "blob merge" or "blob split". If they are further apart the structures can be very different indeed. You may characterize the complete landscape in a qualitative fashion if you pick a series of fiducial heights such that each pair of successive heights is divided by a single pass height. Then you may set up a geneological tree for each blob, specifying its parents and offspring as the fiducial level is changed. When you raise the fiducial level hills divide into two or vanish and dales suddenly appear or merge (Possibly with themselves, like a snake biting itself in the tail. Think of a lake that grows from a round basin into a circular one with an island in the middle.) When you lower the level the opposite happens.

Note that you never create a hill out of the blue (that is: creation occurs only via splitting) on raising the level, nor a dale on lowering it.

When you change the *resolution* the landscape itself changes. Intuitively, the landscape should *simplify* when you turn on the blurring. This is very useful because you obtain a "generalized", "abstracted" view of the landscape that will allow you to pick out major features without being bothered with detail. You'd be unpleasantly surprised if blurring would actually introduce detail. Yet this is what happens when you use just any old blurring method (Koenderink 1984). It has been shown that the only way to blur correctly is to let the landscape "diffuse" in the technical sense. (You apply the diffusion equation.) This guarantees that summits decrease and immits (a term introduced by Maxwell in the 19th century as the opposite of a summit) increase on blurring, thus the landscape truly "erodes". The diffusion equation is

$$\Delta I(x, y, s) = \frac{\partial I}{\partial s},$$

where $\Delta I(x, y, s)$ denotes the Laplacean of the illuminance.

This very nice property of scale space is so important that it has given rise to a considerable literature, including reports that "causality" might be violated, even for the diffusion process. Such reports arise from careless interpretations of causality. First of all the very structure of the diffusion equation expresses the fact that summits decrease and immits increase under blurring. This means that for any fiducial level you only loose hills or dales upon blurring, they never arise out of the blue. This condition can never be violated. Please note that this does in no way imply that the number of hills can't increase on blurring. Because splitting processes are not at all rare, it often happens that a hill will give rise to two hills. Such splittings are intuitively reasonable if you think of dumpbell shaped hills (or dales). A saddle-extremum point may also be generated "out of the blue", but this process will *never* introduce new blobs. Progressive blurring will eventually reduce *any* image to a featureless one, thus extrema generated through blurring must vanish again on even further blurring.

The intuitive notion of causality in scale space is that of the natural "erosion" of a landscape: hills wear away and become progressively lower, whereas (pot-)valleys fill up and their bottoms become progressively higher. Examples of "noncausal" behavior would be catastrophic events such as the genesis of a mountain (volcanic eruption), or of a hole (as when a subterranean cavity collapses). The mathematical scale space structure essentially vetoes such events. For our present purposes "causility of scale space" is perhaps best defined as "evolution according to the diffusion equation". However, it would be foolish to forget the intuitive notion altogether.

3   Differential Image Structure

The standard way to study any entity in the neighborhood of a point is to differentiate it. The paradigmatic example is the "first derivative", which is the best linear approximation for arbitrarily small neighborhoods. This immediately generalizes into Taylor's expansion, which expresses the local structure in polynomial form. Can one do anything like that at a finite level of resolution? This is necessary because one can't make operational sense out of derivatives given real (observed) signals.

The gamut of "neighborhood operators" from image processing comes near to the answer, but the standard approach fails miserably on many counts. The reason is that the apparatus used tends to be arbitrary, constrained only by various types of *ad hoc* conditions. If you care to proceed in a principled manner, then there is very little leeway in picking the operators.

One cue on how to progress is to observe that blurring and differentiation commute. As observed above you have formally

$$\partial(G \otimes I) = \partial G \otimes I = G \otimes \partial I.$$

Thus the differential operators are just the derivatives of the blurring kernel. Because the Gaussian is the unique blurring kernel that complies with the basic front end symmetries, it follows that only the derivatives of Gaussian operators respect these basic symmetries. If you use difference operators (difference of Gaussians, Sobel operators, Canny edge finders,...*ad infinitum*), you tie yourself to a specific position, orientation, or scale, and you have lost the universal viewpoint. Thus the common view that the type of edge finder is essentially arbitrary and you may apply further constraints in order to obtain desirable properties (good localization, optimum noise rejection, *etc.*) is just nonsense. There is one exception: linear combinations of derivatives of Gaussians are also appropriate, *e.g.*, the Laplacean operator $\Delta G = G_{xx} + G_{yy}$. Such linear combinations may assume many unexpected guises, and it makes sense to try to find a general characterization of their structure.

What is needed is a way to formalize the structure of allowable operators in a general manner. Then we may use the formalism to derive general rules. It is not obvious how to proceed. We have found that a very simple, but illuminating way to handle this problem is to start by defining the local region through an "aperture". This aperture may be interpreted as a weighing function, or "window function", with weights that are everywhere very small, except in the fiducial region. A judicious choice of the aperture leads to a simple formalism. An obvious choice is to pick an aperture that transforms in a simple manner under diffusion.

Somehow the operators should have a spatially limited support, so let us define this support by way of the aperture $A(\mathbf{r}, s)$. If you have any image $I(\mathbf{r})$, then $A(\mathbf{r} - \mathbf{r_0}, s) \cdot I(\mathbf{r})$ picks out a small *subimage* (of diameter proportional to $\sqrt{s}$) of the image, centered at $\mathbf{r_0}$. One way to pick A is to watch the behavior of derivatives of Gaussians: for a high order of differentiation they are like Gabor functions with envelope $\exp(-\mathbf{r} \cdot \mathbf{r}/8s)/8\pi s$. So let us fix A to this envelope. (See figure 1 and its legend.) The choice is the more a particularly nice one because blurring the window function A clearly leaves its shape invariant. The support of A is conveniently measured by its effective radius, which equals $\sqrt{8s}$.

The next step is to observe that the operators should be scale invariant and that the blurring should be a "causal" one in the above defined sense. Hence the operators should satisfy the diffusion equation. Moreover, they should respect the front end symmetries.

At this point it is convenient to introduce a notion of "natural coordinates". Notice that the factor $\sqrt{4s}$ is a length and can be interpreted as the *equivalent radius*, *i.e.*, the radius of a "pillbox" operator with the same integrated weight and same maximum amplitude as the Gaussian corresponding to the point operator $\mathcal{P}$. We denote this "equivalent radius" with $r_E = \sqrt{4s}$. The natural coordinates are defined as $\xi = \mathbf{r}/r_E$. Please notice that the *point operator* is modulated with the exponential factor $\exp{-\xi \cdot \xi}$, whereas the *aperture* is modulated with the exponential factor $\exp{-\xi \cdot \xi/2}$. This probably appears confusing at first blush, however, it is simply a reflection of the fact that the asymptotic envelope of a high order derivative of a Gaussian is *broader* than the width of that Gaussian itself. (The reader who doubts this is in for a somewhat extensive calculation that would be rather out of place here.)

That the operators should respect the front end symmetries means among more that the space dependence must be some function of the natural coordinates, say $\Phi(\xi)\,A(\xi)$. The operator in terms of the normal coordinates and the inner scale will be designated $\Psi(\mathbf{r}, s)$.

Finally, we note that the operators may have various orders. For instance, an "edge detector" is like a first derivative and must have order one. An operator of order n will have the dimension of length to the power minus n. This is illustrated by the derivatives of Gaussians, whose amplitudes vary with scale as r_E^{-n}.

Thus we are let to the *Ansatz*

$$\Psi_n(\mathbf{r}, s) = \left(\frac{1}{\sqrt{4s}}\right)^n \Phi_n\left(\frac{\mathbf{r}}{\sqrt{4s}}\right) A(\mathbf{r}, s) = r_E^{-n-2}\Phi_n(\xi)\frac{\exp{\frac{-\xi \cdot \xi}{2}}}{2\pi},$$

with the understanding that Ψ_n satisfies the diffusion equation. Since the derivatives of Gaussians can be written in exactly this manner, we may rest assured that solutions exist. The mathematically inclined reader may want to find a few examples of the Φ_n at this point, starting from derivatives of Gaussians as the Ψ_n. A few of the Φ_n are illustrated in figure 4.

Substitution of the expression Ψ_n in the diffusion equation yields a very simple PDE (partial differential equation) for the form factor Φ in terms of the natural coordinates, namely

$$\Delta\Phi_n + ((2n+1) - \xi \cdot \xi)\Phi_n = 0.$$

The exact form of the equation is of little importance (except for the fact that it happens to be a famous equation for mathematical physics (Powell 1961), so the solutions are well known), the important fact is that we have arrived at a PDE in natural coordinates at all. There is a rich theory of PDE's, thus we are in a position to formulate general statements concerning the set of allowable operators at large. This is remarkable and important in view of the fact that no such a general theory of neighborhood operators exists today, and in fact the set of operators in practical use is an odd lot with hardly any internal consistency. We now see that all linear neighborhood operators that respect the front end symmetries are local solutions (*i.e.* vanishing at infinity) of one and the same PDE. Some very important facts we get "for free" from the theory of PDE's are due to the fact that the PDE has a set of solutions that is complete and allows an orthonormal base. This has some important consequences:

- any local image structure $(A \cdot I(\mathbf{r}))$ can be represented completely via linear combinations of the Φ_n.

- the coefficients are the result of applying the Ψ_n to the image $I(\mathbf{r})$.

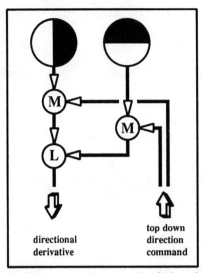

Figure 3: *A simple network that can simulate an edge finder of arbitrary orientation. The network is based upon two orthogonal edge finders, these have to be conceived of as coincident. (They have not been superimposed in the figure for the sake of clarity.) A command input sets the desired orientation. The elements designated "M" are multiplicative nodes, the element marked "L" is a linear (additive) node.*

- when one truncates the representation to some finite order, one obtains the best approximation to the local image structure in the least squares sense. Note that the Φ_n are orthogonal, whereas the Ψ_n are not. Usually one will normalize the Φ_n and thus obtain a convenient orthonormal basis.

- solution of the PDE in various coordinate systems in which the PDE is separable yields families of operators with specific symmetry properties. For instance, we may construct solutions with radial or azimuthal symmetry.

- rotations of the coordinate systems can easily be accommodated via orthogonal transformations of the observations per order. This means, for instance, that you need only two edge finders: any other can be obtained by rotation, that is linear combination. (See figure 3.) Likewise you need only three line finders, *etc.*

- transformations between representations (*e.g.*, polar to Cartesian or vice versa) can be done through linear orthogonal transformations per order.

- The PDE allows us to construct a complete taxonomy of receptive fields.

This immediately yields a tremendous increase in power of the neighborhood operator approach.

When the PDE is solved in Cartesian coordinates we obtain simply the mixed partial derivatives of the Gaussian as solutions. (See figure 4.) Many of such solutions are reminiscent of receptive field profiles as reported in the primate visual system (Jones and Palmer 1987a, 1987b; Young 1985), though by no means all of the possibilities appear to have been reported. When you solve the PDE in other coordinate systems you obtain families of operators with specific types of symmetry properties. For instance, the simplest

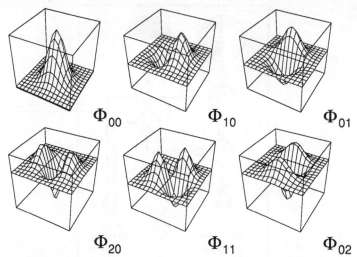

Φ_{00} $\qquad$ Φ_{10} $\qquad$ Φ_{01}

Φ_{20} $\qquad$ Φ_{11} $\qquad$ Φ_{02}

Figure 4: *The solutions of the partial differential equation that governs the allowable operators for orders up to the second in dimension two. The PDE has been separated and solved in Cartesian coordinates. In this simple case the operators are just the mixed spatial derivatives of the zeroth order Gaussian kernel.*

types of center-surround organized receptive fields appear as lowest order radial derivatives of the polar family. The structures of simple symmetry appear to have obvious application in image processing and are perhaps worth a search in neurophysiology: notice how the theory permits a "prediction" of local structure reminiscent of the prediction of elementary particles in high energy physics.

Application of a Cartesian operator, *e.g.*, of order (n, m) yields the exact (!) mixed partial derivative of order (n, m) of the image at the resolution level of the operator. This puts us in a position to implement any formula for differential geometry in terms of these operators: every derivative is the output of an operator, the structure of the formula then specifies how these outputs should be combined. Since the operators are pieces of hardware, we obtain a wiring scheme for a (nonlinear) network that implements the differential geometrical formula.

A simple example is "boundary curvature". This may be taken to mean the curvature of the local isophote. The formula from differential geometry is

$$\kappa(x, y) = \frac{-I_y^2 I_{xx} + 2 I_x I_y I_{xy} - I_x^2 I_{yy}}{(I_x^2 + I_y^2)^{\frac{3}{2}}}.$$

(Consider an equation like this to be taken "straight from the book". The reader may want to check, but the derivation is essentially irrelevant to the present discussion.) It is straightforward to "compile" the formula into a little nonlinear network (Koenderink and Richards 1988) (see figure 5), by implementing the I_{xx}, *etc.*, with the Cartesian operators, and addition, multiplication and exponentiation via hardware adders, multipliers and non-linear transfer elements. However, in practice you would compute the pair of numbers $(-I_y^2 I_{xx} + 2 I_x I_y I_{xy} - I_x^2 I_{yy})$ and $(I_x^2 + I_y^2)^{\frac{3}{2}}$. At some higher stage we have to make sure that there *exists* an edge in the first place (*e.g.* that $\| \nabla I \| = (I_x^2 + I_y^2)^{1/2}$ is locally very high) and then the curvature follows immediately. The front end might just compute the

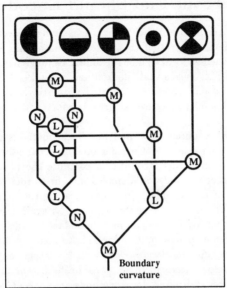

Figure 5: *A network that computes edge curvature. Because the output is the curvature of the local isophote the network is insensitive to arbitrary (non-degenerated) intensity transformations. The input elements are of orders one and two. They must be conceived of as coincident. (They have not been superimposed in the figure for the sake of clarity.) Notice that we picked input operators of polar form. This makes no difference because the polar operators are just linear combinations of the Cartesian ones. The elements marked "M" and "L" have the same meaning as in figure 3, the elements marked "N" are nonlinear transducers.*

two numbers at each point of the image and pass them on as a local description of the illuminance distribution.

The example is of course a trivial one. However, it is easy enough to construct more complicated geometrical invariants with various desirable properties. Such invariants may lead to very complicated constructions indeed. In order to illustrate this we give an almost arbitrary example that is far more complicated than anything being attempted in image processing today: in a (suitably rotated) coordinate system such that $I_x = 0$ and $I_y > 0$, the expression

$$\text{sgn}(I_{xx})\,(-9\,I_{xy}^2\,I_{xx}^2 + 9\,I_{yy}\,I_{xx}^3 - 18\,I_y\,I_{xx}^2\,I_{xxy}+$$

$$18\,I_y\,I_{xy}\,I_{xx}\,I_{xxx} - 5\,I_y^2\,I_{xxx}^2 + 3\,I_y^2\,I_{xx}\,I_{xxxx})/(9\,I_y^{4/3}\,I_{xx}^{8/3})$$

is a curvature measure that is invariant against arbitrary area preserving affinities. (The so called "affine curvature", look for it in any book on advanced differential geometry of the plane. This too is just an example: no need to check if one is not so disposed.) Again, it is straightforward to compile the formula into a network. The resulting network has a very intricate nonlinear structure, and it is *a priori* highly unlikely that standard electrophysiological methods would suffice to unravel its machinery. (This remark may perhaps help to cure the common but mistaken notion that linear operators must lead to trivial results.)

4   What the Front End Should Encode

The front end should encode local image structure allowing the relevant local properties to be computed from intensive, rather than extensive data, thus obviating the need for geometrical expertise (such as local sign or the ability to judge parallellity of orientations at different locations). This means that the front end should encode up to an order that will allow such computations. More complicated properties will have to be computed multilocally, which is bound to require far greater effort and may well turn out to be impossible.

The difference between purely local and multilocal methods perhaps needs further elaboration: we provide an example. Consider boundary curvature again. The method explained above uses second order structure to compute the boundary curvature from purely *local* data. Another method would employ only first order structure. This suffices to find the orientation of the boundary (first order) along the boundary (zeroth order). The rate of change of orientation along the boundary is again the boundary curvature. Notice that this method allows us to compute boundary curvature if we: 1. Can compare orientation at different places, and 2. Are able to measure finite distances between distinct locations. Thus this *multilocal method* asks for rather sophisticated geometrical expertise. (For instance, one needs methods to calibrate distances, transport orientation along curves, *etc.*). In general purely local methods are conceptually much simpler and much more plausible from the viewpoint of physiological implementation. If the brain indeed prefers the simple, local algorithms, then this has important consequences for the repertoire of the front end.

One requirement that has not been introduced so far is that the front end should encode in a *coordinate free* manner. This is again a consequence of the general requirement of non-commitment. For instance, although two edge finders (in x and y-direction say) suffice for the first order, one should not commit oneself to a specific choice of x-y-axis.

In some cases the requirement can be met fairly easily. For instance, in the even orders there exist linear combinations of operators that are isotropic and thus require no special choice of coordinate system. An example is the Laplacean operator. A general way to deal with the problem is not to use the x-y-axis, but an infinite set of axes, caring equally for all directions. Let the azimuth be denoted α (angle with the positive x-axis in direction of positive y-axis say). Then instead of having edge finders in the direction $\alpha = 0, \pi/2$, you have a continuous uniform distribution of them over the range $\alpha \in (0, 2\pi)$. Instead of two numbers you now have an activity function $f(\alpha)$ (say) as the result of running your edge finders locally. (Since the activity is actually present, we may forget about the specific definition of α!) It is easily shown that all the information is contained in the first order Fourier terms of $f(\alpha)$. Thus the activity severely over-represents the first order. However, since the Fourier coefficients are weighted averages of the activity you have obtained a very robust, stable and coordinate independent representation. If hardware is cheap, then the over-representation may well turn out to be an advantage.

The same trick works for any order. Use n^{th}-order directional derivatives in all directions. The information is contained only in the $(n, n-1, n-2, ...)^{th}$-order Fourier coefficients of the resulting activity. (Of course these coefficients depend on the coordinate system.) For the 2nd-order one has a set of "line finders" which closely resembles (part of the structure of) a hypercolumn of simple cells of the primate visual cortex.

The activities of oriented directional derivatives are more complicated than just numbers: there are $(n + 1)$ Fourier coefficients of the nth-order that transform as a single entity and depend on the choice of the coordinate system. For many purposes one would like a set of scalars that have a meaning independent of the coordinate systems. Such scalars exist, they are known as "differential invariants". We have already met a few examples, *e.g.* , $\| \nabla I \|$ that is the magnitude of the gradient, $\triangle I$ that is the Laplacian, and $(-I_y^2 I_{xx} + 2I_x I_y I_{xy} - I_x^2 I_{yy})(I_x^2 + I_y^2)^{-3/2}$ which has to do with isophote curvature. Although the coordinate axes (x, y) formally appear in these expressions, these numbers are actually independent of the specific choice of axes. It can be shown that complete sets of invariants of order n can be constructed, *i.e.*, such that any invariant of order n or less can be expressed in terms of members of the set. Several representations are possible. These invariants are nonlinear combinations of partial derivatives. It is possible to construct polynomial representations, although these need not be the most desirable in any case. Thus they cannot be implemented as simple weighted sums, but only as highly nonlinear combinations of the results of applying variously modulated weights.

5   Spatiotemporal Operators

The basic spacetime operators can easily be constructed via multiplication of spatial and temporal operators. Any operator is characterized by its spatiotemporal location (point and moment), spatial resolution, temporal resolution and delay, spatial and temporal orders and types (*e.g.* polar or Cartesian). Thus the space and time parts are always separable in these linear machines. Separability may be lost in the invariant combinations though. We have not come up with a complete set of spatiotemporal invariants so far. We have merely implemented local operators for motion, and the affine structure of image flow (divergence, vorticity and shear). In the latter case the operators work directly on local spatiotemporal derivatives of the image irradiance, *e.g.*, divergence is not computed via velocities (although this would also be a route worth exploring).

6   The Significance of an Observation

Until now we have mainly discussed the nature of the sampling and representation, *i.e.*, more or less the transformations to be found in the bottom up stage. Now we change gears and perspective and try to specify what can be asserted concerning the input, given a certain activity in the front end. This is more like a top down inference. This is important, because the *meaning* of the front end activity is ultimately contained in the constraints this activity puts on the assertions that can be made concerning what is really "out there". Since we *reverse* the direction of reasoning we have to switch to novel methods of description. In this section we provide a possible way of approaching these important issues. In the literature such problems are typically skipped altogether.

Suppose one has a set of operators up to (and including) order n. To be specific we take $n = 2$, but the reasoning works for any order. There are $(n + 1)(n + 2)/2 = 6$ independent operators of order $n = 2$. Running these operators locally on the image yields an ordered set of 6 numbers, $\{a_{00}, a_{10}, a_{01}, a_{20}, a_{11}, a_{02}\} = \mathcal{O}^2$ say. We refer to the set $\mathcal{O}^n$ as an "observation of order n". In the literature one would probably refer to such a set as a "feature vector', a term that had better be avoided in view of the fact that $\mathcal{O}^n$ doesn't transform as a vector. Notice that $\mathcal{O}^n$ is just a description of the activity in the front end hardware for a given input image.

The numbers depend upon the particular set of operators. If you had happened to pick a rotated set, a polar instead of a Cartesian representation, *etc.*, then you would have obtained different numbers, although the meaning of the observation would have been the same: the meaning is a function of the input image only and can't depend on the arbitrary choice of representation. The formal way to avoid this flaw is to use the mathematical notion of a *jet*. The "n-jet" of images at a location is the equivalence class of all images that would yield the same response for some given set of n-order operators. Such images necessarily agree among one another in all spatial derivations up to – and including – the order n. The particular set of operators is irrelevant. The images have the same initial terms in their local spatial Taylor series development. The "n-jet-space" at the location is the space of all n-jets, and the "n-jet-bundle" of the visual field is just the visual field with a n-jet space attached to each spatial location. This formal mathematical construct is very similar to the neurophysiological concept of the columnar structure of V-1: the hypercolumns are the hardware implementations of the local (4?-)jets.

Consider the following problem:

Given an observation $\mathcal{O}^n$, what can one assert about the image?

This question is rarely raised, which is remarkable in view of the fact that one most likely wants to come up with some description of the actual input. The reason is probably that very little indeed can be asserted about the image: typically *infinitely* many images can be constructed that would have yielded the same observation.

Some terminology: two images that yield the same observation are called "metameric". The equivalence class of all images that yield some given observation is referred to as the metamere specified by that observation. A metamere can be specified by giving any member, by specification of a canonical member, or by some algorithm that allows one to decide whether any given image belongs to the class. We use a canonical representative.

The question on the significance of an observation then is to solve the following problem (see figure 6):

- find all metameres and how they partition the space of all images into cells,

- find a canonical image (*e.g.*, the simplest one in some sense) for every metamere.

A solution of this problem completely characterizes the "observational power" of the n-jet space.

These problems are related to the question of whether the local operators can be said to be "feature detectors". *E.g.*, does an edge finder find edges? Yes, *if an edge is defined as that what an edge finder finds.* But what if an edge finder responds equally well to bars (as all edge finders do when the bar is broad enough)? Although the questions as posed here are silly, the problem matter is really serious. The present section bears on this.

One way to obtain an insight in the nature of metamery is to consider the construction of "invisible images". An invisible image yields an observation composed of all zeros, just like an image that were constantly zero (flat black) would yield. It is easy enough to construct examples. For example, the image $I(x, y) = x$ is invisible for the point operator at the origin. (The simplest case of the zeroth order jet.) Notice that this "image" has *negative* irradiances at $x < 0$ and thus is a physical impossibility. This is true for every invisible image. Thus the invisible images don't occur in isolation. However, you can *add them* to any given picture, as long as the total intensity remains positive. For instance, the image $I(x, y) = I_0 + \mu x + \nu y, I_0 > 0$ is a possible image near the origin. The point operator can't distinguish such images from the flat image $I(x, y) = I_0$, whatever the values of μ and ν are. Thus we have constructed an *infinity* of metameres for the zeroth order jet.

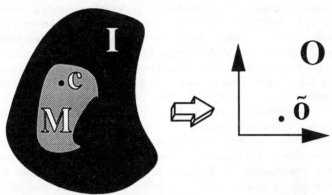

Figure 6: *The black blob symbolizes the space I of all possible input images. This is a very large (infinitely dimensional Hilbert) space. The space O is the space of all possible observations of a given order. These are finitely dimensional Cartesian spaces. (E.g., for order two in dimension two the space has dimension six.) An observation õ (say) could have been caused by many different input patterns. This collection of metameres is a (possibly very large) subset M of the space I. In this subspace you may pick any member c (say) as a canonical representative of the input for this observation. In practice you will characterize the canonical representative through very simple properties.*

The construction of "invisible images" may seem like a mere mathematical oddity. However, because of linearity, any two indistinguishable images must differ by exactly such an invisible image. Thus the images corresponding to given observations are determined only *modulo* the set of all invisible images. Consequently the set of invisible images neatly sums up the degeneracy and it makes a lot of sense to study this set in detail.

Consider the second order jet. There are six degrees of freedom. (*E.g.*, the Taylor expansion of the image, truncated after the second order, has six terms.) Suppose you are handed an *arbitrary* sextuple of numbers. Do they always comprise a possible observation? In order to answer this question we need a few preliminary insights.

First notice that the observations are obtained from the images via purely linear operations. Thus the value of an observation is simply multiplied by a factor when I multiply the image by that factor. If I add two different images, then I can find the observation by addition of the observations of the two original images. Not just any linear combination of images is a possible image though: *E.g.*, minus an image would be an image with negative irradiances, which is physically impossible. If A and B are possible images then the linear interpolate $\lambda A + (1-\lambda)B$ ($\lambda \in [0,1]$) is also a possible image though. But that means that if two observations belong to possible images (have good interpretations), then all observations on the line segment connecting these observations in observation space must also correspond to possible images. Thus the set of observations that corresponds to possible images is geometrically *convex*. (If two points belong to it, then the average of the two also belongs to it.) Thus the boundary of the set of possible observations will be the boundary of a convex body in observation space.

Observations that lie *outside* this convex volume in observation space do not correspond to possible images, whereas interior points admit (in general) of an *infinity* of possible interpretations. The boundary points are special: it can be shown that these are the *only* observations that admit of *unique interpretations*, i.e., there exists only a single possible image that leads to the boundary observation. This makes the boundary observations important in practice: if an observation is "near" a boundary observation, then it must admit of an interpretation that is "almost unique", *i.e.*, the degree of metamery must be small. That a boundary observation admits of only a single interpretation is shown by explicit construction of the singular image (see below).

If the images are subject to further constraints it is possible to be more specific. Consider the case that is most frequent in practice: you know a priori that the irradiance is strictly limited to the range $(0, I_{max})$. This further constraint shrinks the set of possible images in image space as well as the set of observations that admit of a possible interpretation to much smaller volumes.

For this constraint, and order two, it is possible to calculate exactly the nature of the boundary observations. It is especially interesting to consider the unique interpretations of the boundary observations. We find that these special images are blobs with a quadric outline, with vanishing irradiance on the exterior, I_{max} on the interior. (Thus they are *binary* images with elliptic, circular, parabolic or hyperbolic edges.)

The reasoning that leads to such conclusions is related to Schrödinger's and Ostwald's methods in the theory of object colors (Schrödinger 1920a, 1920b; Ostwald 1916, 1917). (Object colors involve the case of order two in dimension one.) Although the proof is not very difficult we do not give it *in extenso*, only the flavor of it.

The proof depends on the fact that the observation space for the second order is six-dimensional. This means – among other things – the following: given any sextuple of lightpoints on a black fond (in general position) you can achieve *any* observation by judicious adjustment of the intensities, although it will often be the case that one or more of the "intensities" will turn out to be negative. It is easy enough to find the required intensities: it boils down to the solution of six linear equations in six unknowns. This entails that if you can find six points in an image that you can perturb as you please, then you can move the corresponding observation into *all* possible directions in observation space. The only obstructions to this freedom occur if you cannot freely pick six points at which to perturb, or if the set of linear equations turn out to be dependent.

Here is a simple proof, based upon these preliminary observations, that the images belonging to boundary observations are *binary images*:

> Assume (for the sake of argument) that there is a region in which the illuminance *does* take on intermediate values. Then we can pick six points in general position in this region. As an immediate consequence we can perturb the observation in *any* direction in observation space by fiddling with the intensities. But then we can't be on the boundary. *Ergo*, there can't be a region of intermediate intensities; the image must be a binary one. **QED.**

The following reasoning leads to the shape of the circumference:

> Assume that the image is a binary one (either 0 or I_{max}, see above) and let the circumference of the light blob be Γ. We can pick 6 points on Γ as we please and slightly perturb the circumference at these points, moving in or out at pleasure. Again, we can move in *any* direction in observation space, thus we can't be on the boundary in observation space. An exception occurs when the six linear equations involved in this problem are dependent. This case defines the boundary points in observation space. Algebraic examination shows that it occurs for binary images for which Γ is a general quadric. **QED.**

As argued above, it is intuitively reasonable that if you meet with an observation that is close to a limiting one, you may make assertions concerning local image structure with

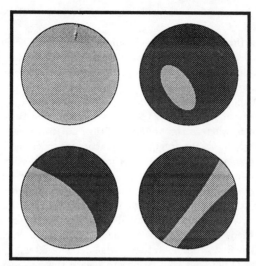

Figure 7: *The essentially different images that can be distinguished by an ensemble of order two in dimension two. These are: the uniform image, the blob, the edge and the bar. Notice that these images may be further characterized through various parameters, e.g., illuminance levels, position, orientation, and curvature or shape. Thus the value of an observation may be "a convex light edge of such and so a curvature passing through this or that point at such and so an orientation, separating the levels this and that", etc.*

some degree of certainty (you are clearly certain when the observation is a limiting one). If the observation is at all close to the "medium gray" one (that is $\{I_{\max}/2, 0, 0..., 0)\}$, and is as far from the boundary as you can get), then the degree of uncertainty will be exceedingly high. In that case you commit yourself least if you pick the medium gray image (irradiance $I_{\max}/2$ all over) as the canonical representative. (In general a useful canonical representative is the mixture of an image that belongs to an extreme observation and a gray image. For instance, you may consider mixtures with the image $I = 0$.) It is always possible to find such an image and it is unique.

We now have obtained a handle on the "feature detector" problem: since the collection of all operations up to order 2 can only distinguish between medium gray (by default, rather than evidence!) and quadric blobs (with various degrees of certainty), no single operator (*e.g.*, an "edge detector") can be expected to do better. Since quadric blobs are either ellipses or hyperbolas (Apollonius in Heath 1921; Loria 1910) (with various degenerated singular cases like circles, straight lines, etc.), it is the case that through the 2nd-order the front end encodes only curved edges and bars, or elongated blobs, of various sizes, orientations, and positions. (See figure 7.) Notice that although you may detect an edge, this cannot be said to be the result of running an edge detector on the image.

Because the local jet can only represent image structure *locally*, we only consider the image structure within a disc of total area $8\pi s$.

The term "feature" can be applied to the metameres if you want, or one may further coarsegrain and speak of an "edge", or an "edge at such and so an orientation and position", or even of an "edge at such and so an orientation and position, and such and so a curvature", *etc.* The important thing is that such "features" are canonical names of possibly very large classes of equivalent images. By just declaring something an "edge" one doesn't change the image, of course. The basic ambiguity in the observation remains pertinent.

If all one knows is that the illuminance is non-negative, then the situation is even more ambiguous. It is still true that possible observations are contained in a certain convex volume, but this volume is now infinitely large. A closer analysis reveals that much of the above analysis remains pertinent though.

It turns out that the general case of order n (say) is not essentially more complicated than the case of the second order ($n = 2$) considered above. The canonical representations are again binary blobs, whereas the boundaries turn out to be general n^{th}-order curves (quadrics for order two, cubics for order three, quartics for order four (Loria 1910; Newton 1706)). Because the general n^{th}-order curves include the mth-order ($m < n$) ones as a proper subset, we obtain a natural hierarchy of features. In the zeroth order you have only the uniformly gray pictures, in the first order you gain straight edges, in the second order bars and convex blobs, *etc.* Once again, it is *not* the case that features are being detected by "feature detectors". For example, suppose we detect an edge in the first order and a bar in the second order for the same input image. Such a case is perfectly possible of course. The first order operators ("edge detectors") will give exactly the same response, irrespective of whether you happen to regard them as members of a first or a second order assembly. Thus the feature "edge" can be "overruled" by the second order operators ("bar detectors"). The response of an edge detector does not by itself define the detected feature.

If we assume the primate visual system to be of at least the 4th-order (Young 1985), then the gamut of possible "features" already runs into the hundreds.

7   Conclusion

We have shown a principled method to obtain a complete characterization of neighborhood operators, their interrelations and transformation properties, as well as their role in inferences concerning the structure of the optical input on the basis of local measurements.

The method is founded upon a limited number of well understood symmetries of the visual front end, such as homogeneity, isotropy, and self similarity. Such symmetries express the preference for minimum commitment, such as preferred positions, directions, or sizes. Together with rather fundamental physical constraints (*e.g.*, the fact that illuminance is necessarily positive and in any given case strictly limited from above) these conditions constrain the possible front end structures to a great extent.

It is remarkable that very general symmetry considerations *constrain* the allowable operators to a great extent. This is essentially due to the "principle of non-commitment": you may well try to design operators with certain desirable qualities that will make their use preferable over the default operators derived from general symmetry principles. Indeed, there has been and still is quite an activity in machine vision in the design of optimum edge and corner detectors, higher order differential operators, and the like. Such operators may well have better localization properties (in the case of edge detectors) or better noise immunity. (Although in the cases studied by us the differences turn out to be hardly worth the effort.) They acquire such new properties by violating the principle of non-commitment though. Indeed, the operators are already constrained by the symmetries and you may only impose other constraints by lifting some of these natural ones. Thus one may gain noise immunity by focusing on a limited scale range, for instance. Then it will turn out that the design of the operators depends on the scale parameter and that the blurring implicitly involved in any operator will generate spurious detail. To summarize: it may indeed be useful to violate the front end symmetries, especially when you know a lot about the application (*e.g.*, its scale range, *etc.*). You do so at the cost of universality (*e.g.*, true size invariance, *etc.*).

The utility of these notions for image processing and machine vision should be obvious. The applicability as part of a theory of biological front ends is less well established. Clearly the present theory has to be regarded as an "ideal" limiting case. In any real system the symmetries cannot be expected to hold universally, moreover it may be an evolutionary *advantage* to violate these symmetries if the environment is biased towards specific features, sizes, and orientations. The present theory thus appears as a convenient reference from which to depart in all special cases.

References

Apollonius of Pergae, Conics. See: Th. Heath (1921) A history of Greek mathematics, Vol. 2. From Aristarchus to Diophantus. Oxford at the Clarendon Press

Bijl, P., Koenderink, J. J. (1989) Visibility of blobs with a gaussian luminance profile. Vision Res. 29, pp. 447-456

Horn, B. K. P. (1986) Robot Vision. The MIT Press, Cambridge Mass.

Jones, J. P., Palmer, L. A. (1987a) The two-dimensional spatial structure of simple receptive fields in cat striate cortex. J. Neurophysiol. 58, pp. 1187-1211

Jones, J. P., Palmer, L. A. (1987b) An evaluation of the two-dimensional Gabor filter model of simple receptive fields in cat striate cortex. J. Neurophysiol. 58, pp. 1233-1258

Loria, G. (1910) Spezielle algebraische und transzendente ebene Kurven, Theorie und Geschichte. 2nd ed., transl. F. Schütte, Vol. 1: Die algebraischen Kurven. Teubner, Leipzig and Berlin

Lotze, H. (1884) Mikrokosmos. Hirzel, Leipzig

Koenderink, J.J. (1984) The structure of images. Biol. Cybern. 50, pp. 363-370

Koenderink, J. J. (1990) The brain a geometry engine, Psychological Research 52, pp. 122-127

Koenderink, J. J., van Doorn, A. J. (1978) Visual detection of spatial contrast; Influence of location in the visual field, target extent and illuminance level. Biol. Cybern. 30, pp. 157-167

Koenderink, J. J., van Doorn, A. J. (1987) Representation of local geometry in the visual system. Biol. Cybern. 55, pp. 367-375

Koenderink, J. J., Richards, W. (1988) Two-dimensional curvature operators. J. Opt. Soc. Am. A5, pp. 1136-1141

Newton, I. (1706) Enumeratio linearum tertii ordinis, Londini

Ostwald, W. (1916, 1917) Das absolute System der Farben. Z. physik. Chemie 91, p. 132, 1916, and 92, p. 222, 1917

Powell, J. L., Chasemann, B. (1961) Quantum Mechanics. Addison-Wesley, Reading Mass.

Schrödinger, E. (1920a) Theorie der Pigmente von grösster Leuchtkraft. Ann. Physik 62, p. 603

Schrödinger, E. (1920b) Grundlinien einer Theorie der Farbenmetrik im Tagessehen. I, II. Ann. Physik 63, p. 397 and p. 427

Young, R.A. (1985) The gaussian derivative theory of spatial vision: Analysis of cortical cell receptive field line-weighting profiles. General Motors Res. Tech. Rep. GMR-4920, May 1985

The Analysis of Motion Signals and the Nature of Processing in the Primate Visual System[1]

Guy A. Orban

Laboratorium voor Neuro- and Psychofysiologie, Katholieke Universiteit te Leuven

Optic flow is generated by relative motion between the observer and objects in his surroundings (Gibson 1950). It thus represents the most general case of motion signals to be analysed by the primate brain. In the case of a subject moving through a scene with several moving objects, the flow can be quite complex, and its nature will vary in different parts of the retinal image. In our study of optic flow, we have assumed that the flow is homogeneous, i.e. of the same nature, throughout the image. The issue of segmentation – the fact that different parts can be distinguished by their difference in motion – has been addressed separately and only for differences in direction of translation.

Although optic flow is a rich source of information about the outside world – it contains information on the relative motion between the observer and the surrounding objects and also on the 3D structure of the surroundings – it is only one aspect of vision. It is quite clear that we and other primates can see without relative motion between ourselves and our surroundings. A natural consequence is that only a part of the visual system is devoted to the analysis of motion signals. I will therefore review the layout of the visual system and point out which of the areas are supposedly involved in the analysis of optic flow. This will also provide a framework for the experimental study the results of which are reviewed here. This review will be centred on two basic questions. First, what new properties do neurons in a higher-order area exhibit compared to lower-order ones, and second, what sort of analysis does the visual system perform upon the image.

1 Organisation of the Primate Visual System and the Interpretation of the Anatomical Blueprint of the Visual System

Here, I will concentrate on the cortical part of the primate visual system. The visual signals reach the cortex at the level of the primary visual cortex – termed V1 or striate cortex – via the retina and the lateral geniculate body, which is a thalamic relay nucleus

---

[1] Dedicated to Nicolas, born during the writing of this chapter, and to his mother Chantal

(figure 1). This retino-striate projection contains two separate, parallel streams, the magno- and parvocellular pathways (for review see Lennie *et al.* 1990; Schiller *et al.* 1990). These pathways operate over different ranges of a four-dimensional space with dimensions of space-time and wavelength. From the primary visual cortex, the visual signals are sent to the thirty or so extrastriate or higher-order visual cortical areas, among which about 300 connections have been reported (for review see Maunsell and Newsome 1987; Boussaoud *et al.* 1990; Baizer *et al.* 1991; Felleman and Van Essen 1991). There is considerable variation in the size of these areas, primary visual cortex (V1) is the largest, exceeding 10 cm², but most areas are less than a tenth this size (Van Essen *et al.* 1992). Some of these areas are also reached by retinal signals which bypass the retino-striate pathway, and travel from the retina via two other subcortical structures, the superior colliculus and the pulvinar (Rodman *et al.* 1989, 1990; Girard *et al.* 1991, 1992).

The organisation of the cortical visual system is derived from three types of observations. First, each cortical area, although connected with many other areas, is nonetheless restricted in its connectivity. For example, the primary visual cortex has connections with three areas: the second visual area (V2) and the third visual area (V3), which are both located in the immediate vicinity of V1, and the middle temporal (MT) area, located in the dorsal part of the superior temporal sulcus. These are the main connections of V1, on which all studies agree (Maunsell and Newsome 1987). Weaker and inconsistent connections have been reported between V1 and V4, MST (middle superior temporal area) and V3A (Zeki 1980; Desimone and Ungerleider 1986). Second, most if not all cortico-cortical connections are reciprocal and third, these connections have a laminar pattern (Rockland and Pandya 1979). Cortical areas have a laminar organisation, each area consisting of 6 layers, numbered from 1 to 6. Layer one is the most superficial of these, layer 6 is the deepest and adjoins the white matter. Forward cortico-cortical connections originate from the superficial laminae (2 and 3) of the lower-order area and terminate in the middle layer (layer 4) of the higher-order area. Backward connections – from higher-order cortices to lower-order ones – originate in the deepest layer (layer 6) and terminate in all layers outside layer 4. Connections which do not fit these patterns, even in the broadest sense, are labelled intermediate and are considered to link areas at the same level of processing. As a consequence of this laminar organisation of cortico-cortical connections, the laminar position of neurons reflect their role in cortical processing. In particular, any difference in properties between cells in the input layer (layer 4) and the output layers (layers 2 and 3), will provide indications of the nature of the transformation carried out by an area upon the incoming signals.

Using these constraints on cortico-cortical connections as guide-lines, one can distinguish at least four hierarchical levels of processing in the primate visual system (figure 1). Others (Felleman and Van Essen 1991) have described up to 9 levels, but here we describe the minimum number. The primary visual cortex (V1) occupies the lowest or first level in the hierarchy. At the second level, one finds V2, V3 and MT which all receive input directly from V1. This level is sometimes subdivided into V2 which receives only from V1, and V3 and MT which receive from both V1 and V2. At the third level, one finds MST and FST (fundus of superior temporal sulcus visual area) which both receive from MT, V4 which receives from V2 and V3, and V3A, which receives from V3, but also from V1 and V2. Finally, at the fourth level one finds on one hand the parietal areas such as area 7a, which receive from MST and FST, and on the other hand the infero-temporal areas, including the two architectonically defined areas TEO and TE (von Bonin and Bailey 1947). TE receives from V4, TEO and FST. Parietal and infero-temporal areas

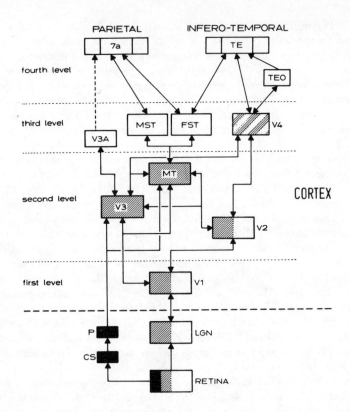

Figure 1: *Connectivity of the primate visual system. The subcortical structures as well as the four tentative levels of cortical processing are indicated. According to Ungerleider and Mishkin (1982) the dorsal stream leads from V1, the primary visual cortex located in occipital lobe, to parietal cortical areas, one of which is area 7a, while the ventral stream runs from V1 to infero-temporal cortices such as TEO and TE. The nature of the subcortical afferents are also indicated. At the retinal level three types of ganglion cells have been distinguished: parvocellular (light areas), magnocellular cells (striped hatching) and ganglion cells projecting to the brain stem (black areas). The structures receiving the different types of afferents are hatched as are the retinal ganglion cells. Those areas for which this information is not available are left blank. In the lateral geniculate nucleus (LGN), V1 and V2 magnocellular and parvocellular projections are anatomically segregated (for review see Lennie et al. 1990). V4 receives both magno- and parvocellular input (Nealey et al. 1991), but the segregation has yet to be demonstrated. MT, the middle temporal area receives almost exclusively magnocellular afferents (Maunsell et al. 1990), but also receives from colliculus (CS) and pulvinar (P) (Rodman et al. 1989, 1990; Girard et al. 1991). V3 also receives from P and CS (Girard et al. 1992), although its pure magnocellular input has yet to be demonstrated physiologically. Notice that while there is only a single pathway leading to infero-temporal cortex, there are several ones leading to parietal cortex: signals from V1 can run over MT, but also over V3A, which receives form V1 and V2 (not shown) in adittion to V3, or PO (V6, not shown) which also receives from V1. While the pathways over PO and V3A, which emphasize peripheral vision, belong exclusively to the dorsal stream, the pathways over MT which emphasizes central vision, distributes both to the dorsal and the ventral stream.*

are generally considered endstations of the visual system which then connect with limbic areas (Webster *et al.* 1991) and the basal ganglia (Saint-Cyr *et al.* 1990).

These structural data have received many different interpretations as more has become known not only of the connections but also of the physiology of the areas and their contribution to perceptual tasks. During the seventies, it was mainly the front end – I use this term in a somewhat broader sense than do Koenderink *et al.* in this volume – of the visual system – V1 and its targets – which were explored physiologically, leading to the emphasis upon the parallel processing taking place in the visual system (Zeki 1978). In this view, V1 acts as a clearing-house dispatching the same message to different areas for further processing of different aspects of the same image, such as colour in V4, motion in MT (or V5), form in V3, or V4 and stereo in V3. One of the major questions unresolved by this theory was that posed by the 'yellow Volkswagen': how are the different aspects, colour, shape, motion etc. pertaining to the same visual object again combined to yield a unitary percept? Neither did this theory account for the fact that several paths lead from one area to another, e.g. MT receives from V1 directly but also indirectly through V2 and V3. The question of perceptual unity was at least partially answered by the hypothesis put forward by Ungerleider and Mishkin (1982) on the basis of anatomical evidence and of behavioural studies. Lesions of parietal cortex produced deficits in a visual localisation task, while lesions in temporal cortex lead to deficits in visual identification tasks. Therefore Ungerleider and Mishkin proposed that two visual pathways originate from V1, a dorsal one, leading to parietal cortex, and a ventral one, leading to infero-temporal cortex. The function of the dorsal pathway is visual localisation or the construction of visual space – the 'where' pathway – and that of the ventral pathway is object identification – the 'what' pathway. Morel and Bullier (1990), as well as Baizer *et al.* (1991), have shown that in the case of areas sending projections into both pathways, the ventral pathway receives mainly from representations of the central visual field and the dorsal one from representations of the peripheral visual field. Furthermore it has been argued that the dorsal pathway is concerned more with 'how' – the close visual guidance of movements – than 'where' (Goodale *et al.* 1991). For the time being this hypothesis is subscribed to in one form or another by most researchers (e.g. Van Essen *et al.* 1992).

The gateway of the dorsal pathway is area MT, which is heavily implicated in motion analysis. There is both physiological evidence (Dubner and Zeki 1971; Maunsell and Van Essen 1983; Albright *et al.* 1984; Newsome *et al.* 1989; Lagae *et al.* 1992) and behavioural evidence for this claim (Newsome and Paré 1988; Siegel and Andersen 1986; Vandenbussche *et al.* 1991). Given the importance of optic flow in the recovery of 3D structure of the surroundings and the 3D trajectory of both moving objects and of the moving subject, the role of MT in motion analysis fits with its position in the dorsal pathway. However, motion also contributes considerably to identification of objects. First, motion can contribute to the definition of 2D shape (kinematic figures), 3D structure of objects (motion parallax) or the nature of visual objects (biological motion). This has led Zeki (1990) to consider different ways of integrating signals from the dorsal pathway into the ventral pathway. Zeki (1990) exhausted the logical possibilities by discussing integration by confluence i.e. convergence of motion and non-motion signals onto the same higher-order area, integration by lateral connections between areas at the same level, or integration by re-entry in which a higher-order area of one stream projects to a lower-order area of the other stream. Re-entry from MT into V1, V2 or V3 could underlie 2D shape from motion, i.e. the encoding of kinematic boundaries. On the other hand, integration by confluence, e.g. by FST projecting to infero-temporal cortex, could underlie 3D structure from motion and biological motion. Second, the motion pattern itself can be an identifying characteristic of a class – the flight of a species of bird – or of a member of a class – the style of running of a particular soccer player. This also might be accommodated by integration of signals from the dorsal pathway into the ventral pathway, in particular integration by confluence. Finally, motion and spatio-temporal changes in general are also characteristic of actions, and it is an open question how actions are represented in the visual system. Indeed, actions can be very typical of objects, almost to the point of defining the object, e.g. a knife and cutting, but in general the same visual object can perform or be used for many different actions. Depending on the degree of association between actions and objects, one could represent actions and objects in related parts of the same area, e.g. TE, or in different areas. In this respect, it is worth noting that the possibility of a third pathway leading to action identification and ending in the ventral part of the superior temporal sulcus has been suggested (Boussaoud *et al.* 1990). Many of these questions are presently unanswered because, amongst other reasons, so little physiology or behavioural work has been done with stimuli which engage this type of processing. Indeed, it should be stressed that the original Ungerleider and Mishkin (1982) hypothesis was based on perceptual tasks with static material.

The physiology of infero-temporal cortex has revealed that infero-temporal cells respond selectively to complex visual stimuli such as faces (Gross *et al.* 1972; Perrett *et al.* 1982; Desimone *et al.* 1984) or combinations of shape with texture or colour (Tanaka *et al.* 1991). This fits the more traditional idea that complex processing is achieved by higher order processing in the visual system. However, a substantial fraction of TE neurons respond to simple grating stimuli in monkeys trained to discriminate orientation of static gratings (Orban and Vogels 1990). The interpretation given to these latter results is that processing performed in TE is not purely visual, but that it incorporates the behavioural significance of the stimuli (Vogels and Orban 1990). Furthermore, our recent results show that the response latency of infero-temporal neurons is relatively short and, on average, only 40 msec longer than that of V1 neurons recorded under exactly the same circumstances in the same animal (Vogels and Orban unpublished). Thus the activity in the last visual station starts only 40 msec later than the activity in the primary visual cortex (V1). This corresponds to an average increase in latency per level of processing of about 10 msec, in agreement with other studies from our laboratory (Raiguel et al. 1989). These temporal data stress that, although the system is hierarchically organised, the hierarchy of the primate visual system is not very deep and that, given the spread of latencies in a single area, first and fourth level neurons can be active simultaneously. Thus, in addition to the notion of 'streams' which emphasizes the depth of the system, it is important to stress the width of the system as determined by parallel or even collateral processing. By collateral processing I mean that the number of processing steps can be increased by making an additional loop through another pathway, e.g. by going from V1 to V3 and from there to MT, to return to V1 to be processed in the ventral stream, or by including facultative, additional steps, e.g. by going from V4 to TEO before going to TE, rather than going directly from V4 to TE. One can capture this concept by depicting the visual system as a modular network with serial, parallel and collateral processing. The more complex the visual task in which the system is engaged, the more the parallel and collateral processing capacities will be used.

Within this framework, we have studied the responses of MT, MST, and FST neurons to optic flow stimuli and have also examined the selectivity of MT and TE neurons for orientation of kinematic boundaries.

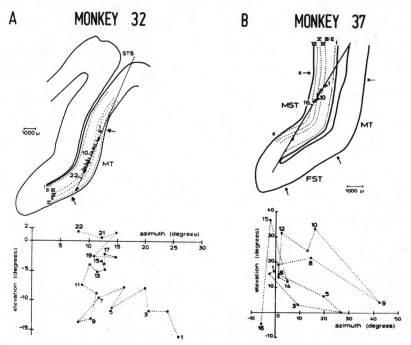

Figure 2: *Retinotopic organisation in MT (A) and MST (B) compared. The top graphs are reconstructions of penetrations on sagittal sections of the cerebra of two monkeys. The limits of the areas are indicated (arrows MT and FST, x's densely myelinated part of MST) as well as the layers. The ticks on the penetration indicate positions of different cells labelled by numbers. Dots on the penetration indicate lesions made to help reconstruction. The bottom graphs indicate RF centres of cells recorded in the two penetrations on a visual field graph, in which azimuth represents horizontal distance from the fixation point (origin) and elevation vertical distance from the fixation point. Retinotopy means that spatial relations present on the receptor sheet (retina) are maintained in the areas connected to the retina: neighbouring neurons thus have neighbouring RFs. In MT this is still true, although there is some local disturbance. The RF of neurons 4, 5 and 6 are located between those of neurons 7, 8, and 9 on one hand and those of neurons 1, 2, and 3 on the other hand; however, the RF of neuron 5 is not situated exactly between those of 4 and 6. In MST retinotopic order was not discernible even in long penetrations. Desimone and Ungerleider (1986) have reported a retinotopic organisation because, on average, central and peripheral parts of the visual field are represented in different parts.*

2 Properties Emerging at a Higher Level of Processing

If the hierarchy of the visual system is relatively shallow, each level of processing can be expected to introduce as many changes as possible. In fact, there have been relatively few systematic quantitative studies comparing response properties in two successive areas. Notable exceptions are the selectivity for orientation of illusory contours which has been observed in V2 but not in V1 (von der Heydt and Peterhans 1989; Peterhans and von der Heydt 1989), for pattern direction of MT cells (Movshon *et al.* 1985), and the selectivity

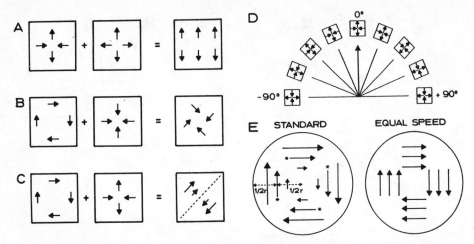

Figure 3: *Schematic representation of stimuli used in the optic flow experiments. A, B and C: the EFCs and some of their combinations. D: the 8 deformations tested in the cocktail tests. E: velocity vector configurations in standard and equal speed stimuli. In A, a deformation is added to an expansion to yield a unidimensional orthogonal deformation or stretch (Tanaka and Saito 1989). Notice that unless mentioned otherwise, deformation in the present study means a two-dimensional deformation. In B, a clockwise rotation is added to a contraction to yield a spiral. Notice that in a spiral the local vectors are turned by 45 deg with respect to the component flows. In C, a deformation, is added to a rotation yielding a unidimensional parallel shear. In D, the 8 deformations, corresponding to 4 axes of deformation, are defined with respect to the preferred direction of translation. The main axes of the deformations differed by 22.5 deg. To be able to represent the responses to the different deformations on a single polar plot, each deformation is plotted twice in mirror-image fashion. The deformations were labelled with respect to the optimal translation direction (indicated with an arrow). Positive deformations are deformations with axis of expansion clockwise to the optimal translation direction. Please note that opposite directions of deformation only differ by 90 deg e.g. deformation 0 and +90 deg are opposite on the same axis and form a pair. In E, the nominal speed of the standard EFC (rotation) is equal to the speed of the vectors (marked by asterisk) located halfway between the centre and the border of the EFC. In the equal speed stimuli the speed of all velocity vectors equal the nominal speed. r indicates the EFC radius. Similarity of velocity response curves obtained with these two types of stimuli indicate that only configurations of local directions are important.*

which V1 cells also lack. At the start of our study, these were the only two examples of properties emerging at a higher level of processing documented with quantitative electrophysiological techniques. The work of Tanaka and associates (Saito *et al.* 1986) had suggested that MST cells, but not MT cells, might be selective for rotation and expansion/contraction, but this was not supported by quantitative investigations at the two levels.

Of course it is well-established that at each level of processing, receptive fields (RFs) increase in size and that, in parallel, retinotopic organisation becomes less distinct. Retinotopy is very precise at the first level (V1), is still clearly present at the second level (V2, MT and V3), although less precise (figure 2). It is present only very coarsely at the third level (MST, FST – see figure 2 – and V4) and is absent altogether at the fourth level (infero-temporal and parietal cortex). With our work on analysis of optic flow, we have

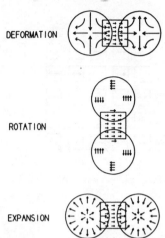

DEFORMATION

ROTATION

EXPANSION

Figure 4: *Schematic figure indicating the reversal of direction selectivity of a cell selective only to translation, when the EFCs are positioned on either side of the RF. The square refers to the RF of a hypothetical unit with a preferred translation direction to the right. For the 3 EFCs, direction selective responses of opposite sign can be obtained on either side of the RF. For instance, expansion responses will be obtained in positions on the left of the RF, while contraction responses will be obtained in positions on the right side. In this way, traversing the RF with an EFC will show a reversal of direction selectivity for this component. For clarity, the velocity gradient within the EFC stimuli has not been represented.*

been able to show that between MT and its target areas, MST and FST, the RFs increase in size and in irregularity, but more importantly, that two radically new properties emerged at the level of MST and FST: selectivity for elementary flow components (EFCs) and filling-in of the stimulus.

2.1 Selectivity of MST and FST Cells for EFCs

In our work on optic flow we wanted to test Koenderink and van Doorn's (1975) ideas on first order analysis of optic flow. Mathematically, optic flow can be decomposed locally into three differential invariants, depicted in figure 3, curl (or rotation), div (contraction/expansion) and def (deformation or shear), up to a translation. If the primate visual system were to perform a similar analysis, we should be able to show two things: first, that cells are selective for the three elementary flow components, rotation, expansion/contraction and deformation and second, that a neuron selective for an EFC can extract this component from a more complex flow i.e. that it responds to the EFC for which it is selective, independently of the presence of other EFCs. Here we will be concerned with the first question. Each of the EFCs can be produced in two fundamentally different ways: first by temporal changes in orientation and in size or spatial frequency (local, figural cues) or second, by spatial configurations of velocity vectors (global cues). By using random dot patterns we made certain that only the latter – pure motion – cues were present.

If EFCs are spatial configurations of local vectors (figure 3), any demonstration of selectivity for an EFC requires that the selectivity is not a trivial consequence of selectivity for a local velocity vector. Indeed, selectivity for local velocity vectors is present at the

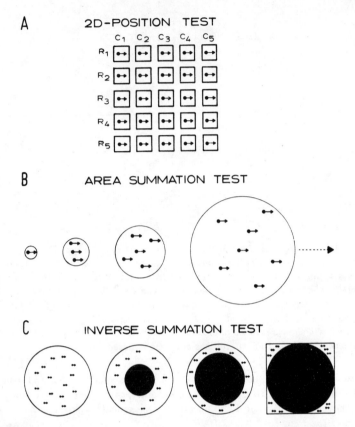

Figure 5: *Schematic representation of the three tests with which the RFs of MT, MST and FST cells were investigated, illustrated for translation of random dots to the right. A: 2D-position test, B: summation test and C: inverse summation test. In A, squares 3, 5, 10 or 20 deg in side were presented at 25 locations on a 5 by 5 grid. The stepsize was usually equal to the size of the squares, although in MST a number of cells were tested with 20 deg squares spaced 10 deg because of the summation requirements of MST cells. In B and C, diameters ranged from 6 to 51 deg. Notice that in these tests the size of the dots remained constant, generally 0.3 deg in MT and 0.6 deg in MST or FST.*

level of MT, since MT neurons are direction selective and many of them are velocity tuned (Maunsell and Van Essen 1983; Lagae *et al.* unpublished). Tanaka and co-workers (Saito *et al.* 1986) were well aware of this problem and required the cells they labelled selective for rotation or expansion/contraction to be non-responsive to translation. This stringent requirement is not necessary if one finds another way of distinguishing between selectivity for the configuration of vectors and selectivity for a vector. The position invariance test, in which one requires the cell to maintain its preference for the same direction of EFC over its entire RF, particularly on either side of the translation RF centre, is such a test, since any cell selective for a local vector will fail the test (figure 4).

Therefore our experimental strategy was to record from MT, MST, and FST cells and to compare selectivity for EFCs and translation, and then once such an EFC selectivity was found, to test its position invariance. Selectivity for an EFC means either that the

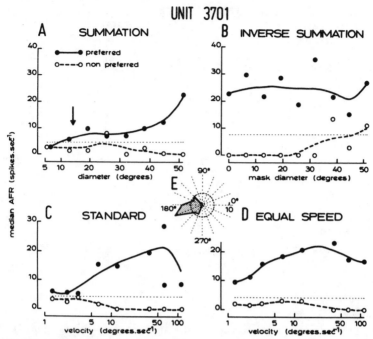

Figure 6: *Summation (A) and inverse summation (B) curves, velocity response curves for standard (C) and equal speed (D) stimuli, and directional tuning (E) of an expansion selective MST cell (3701). The stimuli in A to D were expansion/contraction and translation in E. The speed of stimuli in A, B and E was 10 deg/sec, the size of stimuli in C and D 51.2 deg diameter. Once the directional tuning was measured, further testing was done with the display turned to align its vertical or horizontal axis – as done for this cell – with the preferred direction of the neuron. Thus in subsequent figures, left corresponds to the preferred direction for this neuron. The similarity of the velocity response curves obtained with standard and equal speed EFCs, compare C and D, indicates that only the pattern of local directions determines the response of the EFC selective cells.*

cell responds only to one EFC and not to the other EFCs or to translation or, in a broader sense, that the cell is direction selective for an EFC, responding, for example, to clockwise rotation but not to anti-clockwise rotation. In order to challenge our hypothesis that MST and FST cells are selective for an EFC, we optimised the translation as much as possible, by determining quantitatively its optimum direction and speed, as well as the optimum location and size of the random dot pattern. Optimal direction and speed were obtained by preparing directional tuning curves at different speeds. Optimal location was obtained by the so-called 2D-position test (figure 5) in which small square patches of translating random dots were presented at 25 different positions on a 5×5 deg grid. Optimal size was determined in the area summation test in which circular patches of different diameters were presented (figure 5). An example of this translation optimisation, for cell 3701, is shown in figure 6 for the directional tuning (figure 6E) and summation curve (figure 6A), and in figure 8 for the 2D-position test. The optimal translation was then compared to rotation, expansion/contraction and four deformations with axes spaced 22.5 deg apart (figure 3D), in a so-called cocktail test. Finally, if a direction selective response to an EFC

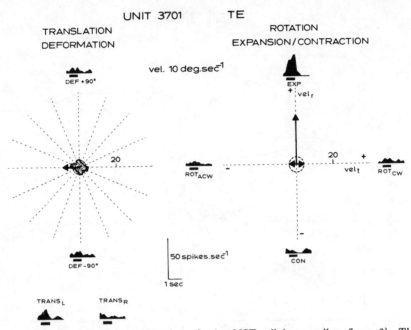

Figure 7: *Cocktail test of an expansion selective MST cell (same cell as figure 6). The net responses to translation and the different optic flow components are shown in two polar plots and illustrated by representative peristimulus time histograms (PSTHs). Stimulus diameter was 51 deg and speed for all stimuli was 10 deg/sec – for EFCs this is the speed of the dots in the middle of the stimulus. The circles in the polar plots indicate the significance level. On the left, responses to translation and the four deformations are shown. The deformations elicited no significant response. In contrast translation evoked a small response with clear direction selectivity. Calibration bars for the PSTHs are indicated in the middle. Trans$_l$: translation to the left, trans$_r$: translation to the right. On the right side responses to expansion/contraction and rotation are shown. Significant and direction selective responses were obtained only for expansion. Rot$_{acw}$: rotation anti-clockwise, rot$_{cw}$: clockwise rotation. Since this cell responded in a direction selective manner to expansion and translation it was labelled TE.*

was present, its position invariance was tested by performing a 2D-position test for that EFC. This test was similar to the 2D-position test for translation except that the patches now contained dots moving according to the EFC, rather than undergoing translation. The few MST or FST cells which did not respond to translation were directly tested with a cocktail test which was repeated after optimisation of position and size for the EFC indicated by the first cocktail test.

This regimen of testing, which required more than three hours per cell, was performed on 66 MT cells, 66 MST cells and 19 FST cells. In all cases, the assignment of neurons to an area was confirmed histologically (figure 2). The results were unequivocal (Orban *et al.* 1992). None of the MT cells were selective for an EFC, 14 of the MST cells (figures 7 and 8) were selective for an EFC, only one of which was deformation, and 2 of the FST cells were selective for an EFC, which was deformation in both cases (figures 9 and 10). It should be noted that the position invariance test could not be performed on all cells, because the recording contact was lost before reaching this point, so that the percentage of cells selective for EFCs in MST and FST is underestimated. However, it is quite clear

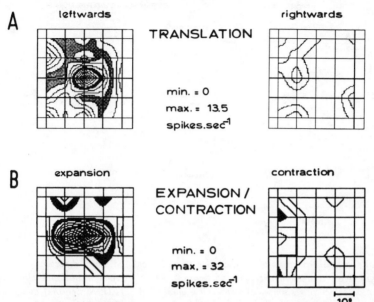

Figure 8: *Position invariance of an expansion selective MST cell (same cell as figures 6 and 7). Two-dimensional position tests were performed for translation (A) and expansion/contraction (B) with 20×20 deg size stimuli spaced 10 deg. The response strength for different positions is indicated by contour lines equally spaced between the maximum and minimum response. The actual strengths corresponding to these extremes are indicated next to the contour plots. For translation the RF was about 10 by 10 deg at the 50% of maximum response level, indicated by the thick black contour. The stippled contour indicates 25% of maximum level. For expansion a somewhat larger RF was observed which was basically overlapping that for translation. No responses for contraction were observed.*

from the results that MST and FST cells can be selective for EFC, something MT cells cannot be. Thus, selectivity for EFCs emerges at the third level of visual processing.

It is worth mentioning that these results have been modeled by Vincent Torre and his group, and that position invariant selectivity for EFCs can be modeled by 'anding' inputs from a configuration of MT cells matching the configuration of vectors in the EFC (Orban *et al.* 1992; see also Verri *et al.* 1992). This is noteworthy, in that other selectivities emerging at a given cortical level, such as selectivity for orientation of luminance contours in V1 (Ferster 1986) and for orientation of illusory contours in V2 (Peterhans *et al.* 1986), also seem to be modelable by 'anding' configurations of inputs matching the configuration in the stimuli for which selectivity arises. This could then be an important principle of operation in the cortex, restricting the role of intracortical inhibition to not so much creating new selectivities as to sharpening and controlling them.

In their chapter, Verri *et al.* (1992) suggest that the brain could encode the type of EFC present in part of the visual field as well as the position of the singularity, and in particular the focus of expansion. Therefore we designed a test in which only the focus of the expansion was moved rather than the whole EFC as we did in the 2D-position test. Notice that in this test (figure 11), the speeds at the edge of the stimulus remained

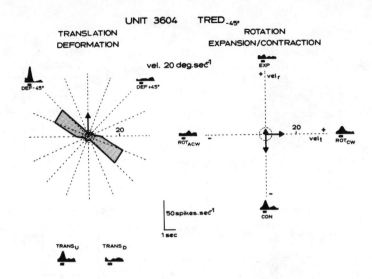

Figure 9: *Cocktail test of a deformation selective FST cell. Stimulus diameter was 51 deg and speed 20 deg/sec. Same conventions as in figure 6. All four types of motion elicited significant and direction selective responses and the cell was labelled TRED. Notice that deformation was the EFC eliciting the strongest response and that the tuning for deformation was narrow.*

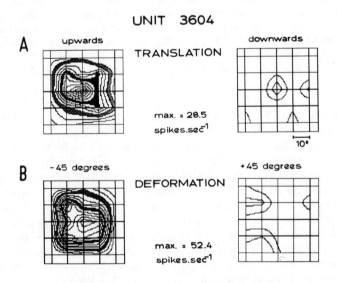

Figure 10: *Position invariance of a deformation selective FST cell. Same cell as figure 9, same conventions as figure 8. Speed was 20 deg/sec, stepsize 10 deg and stimulus size 20 deg. For deformation no reversal of direction selectivity was seen.*

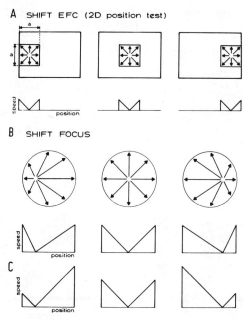

Figure 11: *Schematical representation of expansion stimuli in 2D-position test (A) and in shift-of-focus test (B) and after adding a horizontal translation to the expansion (C). In A and B, both the flow patterns (trajectories of dots) and the distribution of speeds along the horizontal diameter are shown. In C, only the speed gradients are shown since the flow pattern is identical to that in B. Notice that in A only three positions are shown, while in the actual test 25 positions were tested (see figure 5). In A, a indicates the size of the stimulus patch, b the step of displacement.*

unaffected, contrary to what happens when a translation is added to the expansion. Given the low incidence of expansion selective cells, and the many different tests we had to run on them, only two cells were tested in this way. The results are shown in figure 12, in which the outcome of the 2D-position test for the preferred direction (expansion) is shown for comparison. Notice that in this figure the central position of the shift-of-focus test corresponds to the central position in the 2D-position test. In cell 3701, a cell selective for translation and expansion (figure 7), shifting the focus influences the response level considerably and shifts of the focus to the right are optimal (figure 12B). Contrarily, cell 3922, a pure expansion cell, showed a preference for a centred focus (figure 12B). It thus seems that at least some cells might be signal the focus of expansion position in addition to EFC position.

2.2 Filling-in of the Stimulus by MST Cells

In many mathematical studies of optic flow and the recovery of information from the optic flow, special attention has been devoted to singularities (Verri *et al.* 1989, see also this volume). In EFCs, the singularity is the point where speed is zero, i.e. the centre of the stimuli. Another way to find out how important these points are for MST cells selective for EFCs was to remove them. We did this by blanking out increasingly large central discs in the EFC (figure 5C). Given the surprising result we obtained, in fact,

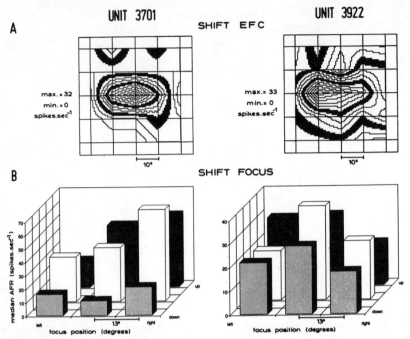

Figure 12: *Outcome of the 2D-position test for expansion (A) and the shift of focus test (B) in two expansion selective MST cells 3701 and 3922. Cell 3701 same cell as figures 6, 7 and 8. Only responses in preferred direction are shown. In A, the response strengths corresponding to maximum response and minimum response in the contour plots are indicated (same conventions as figure 8). In B, the responses in median average firing rate obtained at different focus positions are plotted.*

no effect, we extended this investigation to other MST cells, not selective for EFCs, by blanking out central discs of a translation stimulus. We performed this latter test on MT as well as MST cells, and for these cells we also had the results of the 2D-position test and the area summation test. The importance of the summation test is that it allows one to distinguish between neurons with and without antagonistic surrounds which occur in all three areas. The area summation curve of neurons with surrounds is maximal for small diameters and decreases to less than 50% of the maximum response at larger diameters. Responses of cells without surround increase with increasing diameter and then level off. For these cells the stimulus of maximum diameter (51 deg) was optimal. It should be mentioned that most of the area summation and inverse summation tests were performed at two or more speeds to ensure the velocity invariance of the results, and to ensure, in the case of optic flow stimuli, that the effects were truly spatial effects.

The outcome of the inverse summation test was clear-cut. For simplicity we will restrict the discussion of the results to cells without antagonistic surrounds. In MT, as expected, the response decreased as increasing areas were masked off (figure 13). When the diameter of the blanked out area exceeded the diameter of the RF corresponding to 50% of the maximum response in the 2D-position test, the response decreased significantly. Further increase in the mask diameter reduced the response further, and at 25 deg diameter the

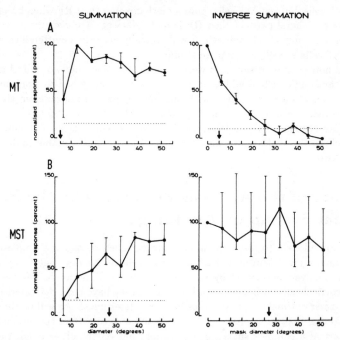

Figure 13: *Average response diameter curves of 5 MT cells (top) and 15 MST cells (bottom) obtained in the summation (left) and inverse summation (right) test. Responses of each cell were normalised and medians (dots) and quartiles (vertical bars) were calculated for each diameter. Horizontal stippled line equals median significance level. Vertical arrows indicate median RF diameter measured at 50% of maximum response in the 2D-position test.*

response was negligible. This is what one expects when the stimuli cover only the fringes of the RF, which are least sensitive.

The result was very different in MST: here the inverse summation test yielded an almost flat line: the response hardly decreased, even with most of the stimulus blanked out. At the diameter corresponding, on average, to 50% of the optimum response in the 2D-position test, the response in the inverse summation test was still equal to the response with no blanking out (figure 13). The previous results were obtained with translation. Figure 6B shows that the same applies to EFCs and particularly to expansion, for which 3701 was selective. The 2D-position test of this cell was shown for expansion in figure 8: the largest dimension of the 50% contour was 25 deg. Yet when the central 30 deg were masked off, the response remained unchanged. Another indication that the inverse summation result in MST is not a simple consequence of an increase in RF size between MST and MT is the following consideration. With the diameter of the blanking central disc at 51 deg, only the four corners of the stimulus remained. The surface area of the remaining stimulus corresponds to a full stimulus of 14 deg diameter. This latter stimulus yields only weak responses in the MST cells (less than 50% of maximum), while the response in the inverse summation test at 51 deg blanking is still 75% of the maximum response (figure 13). Again we can verify this, in particular for cell 3701. Its response to a 14 deg stimulus is rather weak (figure 6A, arrow) and clearly less than that for the maximum blanking in the inverse summation (figure 6B). These results indicate that somehow MST cells, but not MT cells, have a mechanism in their RF whereby summation over the more peripheral parts of their RF is more efficient than over the central part of their RF. Hence incomplete stimuli, which will typically occur in a cluttered environment, will drive MST cells as well as complete stimuli.

As a final note of this section it is worth mentioning that the two radical differences between MT and MST cells would have escaped us completely had we continued to use the classical translation stimuli, even with random dot patterns of different sizes. Hence in order to reveal the processing capacity of cortical cells it is absolutely essential to use appropriate stimuli. This underscores the progress which is possible now that new technology allows the synthesis of almost any spatio-temporal stimulus configuration at a reasonable cost.

3 Nature of the Analysis Performed by the Primate Visual System on the Image

Usually the stimuli used to test cortical cells are designed to find the simplest stimulus which is optimal for the neuron under study. We believe this does not really reveal the nature of the analysis performed by the visual system. This can be revealed only by stimulating cortical cells with complex stimuli, after they have been characterised physiologically, i.e. their optimal simple stimulus has been determined. As will be shown below, such studies reveal that on one hand cortical cells are able to extract the optimal stimulus when it is part of a complex spatial configuration, but on the other hand, the neurons are unable to decompose within their RF a more complex stimulus into its constituents. Thus cortical neurons are active when the part of the stimulus configuration over their RF is close enough to their optimal stimulus, in effect performing a template matching over their RF.

3.1 Spatial Extraction of the Optimal Stimulus

MT cells are not selective for EFCs, even if they give direction selective responses to an EFC which is optimally centred. Indeed, these responses to the EFC are not position invariant. In the 2D-position test with the EFC, opposite directions of the EFC evoke strong responses on either side of the RF (figure 14), when the local velocity vectors present over the cell's RF are well-matched to the optimal velocity vector specified by the speed and direction tuning of the neuron. This clearly shows that MT cells extract the optimal stimulus from the EFC configuration of velocity vectors. This extraction functions not only when the EFC is located optimally i.e. on either side of the RF, but also when the EFC is centred on the RF. The responses in this case are explained by the asymmetry of the RF whereby there is a better – i.e. over a larger area – match with the optimal stimulus in one direction of the EFC than in the opposite direction.

Responses of MT cells to EFCs depended very little on the nature of the luminance pattern used to test them. In particular there was virtually no difference between responses to the random dot pattern and a checkerboard undergoing similar spatio-temporal changes (figure 15). Of course the checkerboard was aligned so that one set of contours was orthogonal to the preferred direction of translation. Furthermore, in both test patterns (random dot pattern and checkerboard) higher-order spatial derivatives of the luminance distribution differ from zero (see Nagel in this volume).

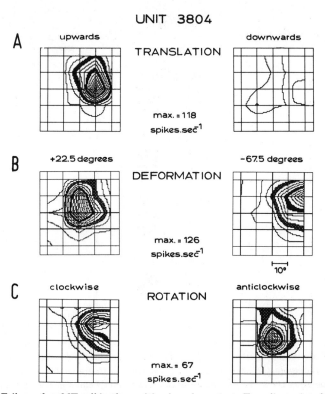

Figure 14: *Failure of an MT cell in the position invariance test. Two-dimensional position tests were performed for translation (A), deformation (B) and rotation (C) with 10 deg size stimuli and 10 deg stepsize. Stimulus speed was 10 deg/sec. Response at the maximum sensitivity point is indicated for the three types of motion. Same conventions as figure 7. Notice that the response areas for the two opposite directions of deformation and rotation sandwich the translation RF.*

We made a related observation on MT cells in our other series of experiments, in which we looked for orientation selectivity for kinematic boundaries in MT cells, specified by opposite directions of translation. In these experiments we optimised direction, speed, localisation and stimulus size for translating random dot patterns, just as we had done for the optic flow stimuli, the only difference being that we used smaller dots. Using stimuli with optimal translation, we then tested for orientation selectivity of kinematic edges or gratings in two ways: with stimuli in which the local vectors were orthogonal to the kinematic boundary and with stimuli in which the local vectors were parallel to the boundary. The prediction was that if the cells responded only to the local velocity vectors, the tunings would shift 90 deg with a change in stimulus type. The result was the same for all MT cells: those which responded to the kinematic boundary and showed a tuning for kinematic boundary orientation shifted their optimum orientation by 90 deg when changing the type of kinematic boundary, exactly as predicted (figure 16). This shows that the responses were due to the local vectors and not to the configuration of vectors – the kinematic boundaries. Thus, this shows in turn that again the MT cells extract the local vector matching their tuning from a spatial configuration. That MT

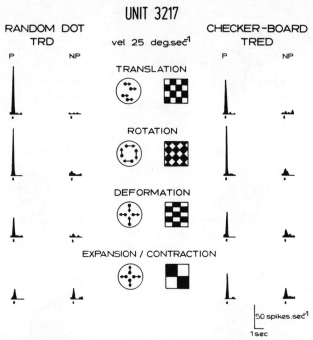

Figure 15: *Comparison of responses of an MT cell to the standard random dot pattern and a checkerboard undergoing translation and EFC changes. Histograms are peristimulus time histograms representing the average response to the preferred (P) and non-preferred (NP) direction of change. The transformations for the two patterns are depicted graphically in the middle. The calibration bars for the histograms are indicated in the lower right. This cell was the MT cell in which the discrepancy between responses to the two luminance patterns was the largest.*

cells do this, even after centring, can again be attributed to the asymmetries in the RF of MT cells.

One could of course argue that the test used for MT cells was too complex, that cortical cells do not extract the orientation of kinematic boundaries. However, TE neurons tested in monkeys discriminating the orientation of two successively presented kinematic gratings behave differently from the MT cells. Generally they are tuned to the orientation of one type of kinematic boundary and show little selectivity for the other type of kinematic boundaries. Exceptionally, one can find cells with a similar optimum for the two types of kinematic gratings, although one type will be more effective than the other. These very recent results (Vogels *et al.* 1992) show that motion signals indeed reach infero-temporal cortex. In fact, not just signals corresponding to 2D shape from motion do so, but simple translation signals reach infero-temporal cortex as well. Indeed, a number of TE neurons respond to uniformly translating random dots. We are presently recording in V1 and V2 to find out how these different signals enter the ventral stream, leading to infero-temporal cortex.

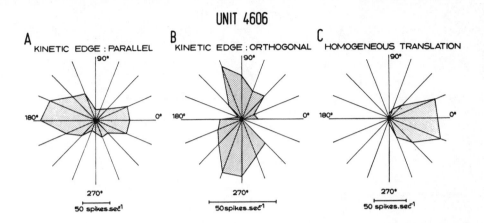

Figure 16: *Polar plots for responses to kinetic edges (A: velocity vectors parallel to boundary and B: orthogonal to boundary) and to homogeneous translation (C). In A and B, each orientation of the boundary is plotted twice, once for each direction of local vectors. In polar plots distance from the centre indicates response strength. Calibration of responses is indicated. In A, 0 deg corresponds to a horizontal edge with vectors above pointing to the right and below to the left, while 180 deg corresponds to the same orientation with vectors reversed. In B, 0 deg corresponds again to a horizontal edge but with vectors above corresponding to upward motion and below to downward motion; 180 deg is still a horizontal edge but with vectors opposite. In C, 0 deg corresponds to translation to the right.*

3.2   Cortical Neurons do not Decompose Complex Stimuli into Mathematically Simpler Constituents

The initial evidence for this statement also originated from the optic flow experiments. Indeed the second question we pursued, once we had answered the first one positively, was the following: can cortical neurons selective for an EFC extract this EFC from a more complex stimulus, as required by the mathematical scheme proposed by Koenderink (1986).

To answer this question we tested MST cells, once it was established that they gave position invariant and direction selective responses to an EFC, with mixtures of the optimal EFC and either another EFC or translation. In either case, the second stimulus was one to which the cell was not responsive when presented in isolation. The relative strength of the two stimuli was altered by keeping the speed of one constant and increasing the speed of the other one. Either the optimal or the other stimulus could be the one remaining constant. The prediction was that if the neuron extracts the EFC from the mixture, it will respond equally well to all mixtures containing a constant optimal EFC. This was never observed (Orban *et al.* 1992); on the contrary, the responses of the MST cells decreased as more of the non-optimal stimulus was added. This was true when adding rotation to expansion for an expansion selective cell (figure 17), as well as when adding translation to rotation for a rotation selective cell (figure 18). In fact the responses of the neuron were determined by the ratios of the speeds of the two stimuli in both tests – whether the optimal EFC or the other stimulus was held constant – indicating that the similarity to the optimal EFC was the key. Indeed, as the speed of the non-optimal stimulus increases relatively, the directions of the local vectors in the complex stimulus depart from their original value. The fact that the response of a rotation selective cell decreased for both

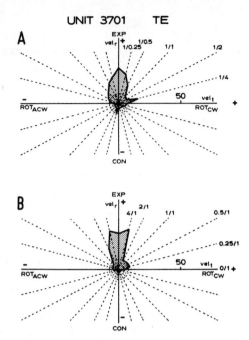

Figure 17: *Responses of an expansion selective MST neuron (figures 6, 7, 8 and 12) to mixtures of expansion/contraction and of rotation in which either expansion was kept constant (A), or rotation was kept constant (B). The relative strength of the two EFCs is determined by the ratio of their speeds, the first number indicating that of expansion, and corresponds to a rotation of the local vectors in the stimulus. The stippled circle indicates the significance level. Rot_{cw}: rotation clockwise, Rot_{acw}: rotation anti-clockwise, exp: expansion, con: contraction. Notice that as soon as the rotation was as strong as the expansion, the cell completely failed to respond.*

the addition of expansion/contraction and the addition of translation makes it unlikely that the interaction between expansion/contraction and rotation in MST cells is due to an antagonistic organisation between these two EFCs, similar to what is known for cone inputs in retinal ganglion cells (H. Barlow, personal communication). In fact, two rotation selective cells were also tested with unidimensional deformations, which are mixtures of expansion/contraction or rotation on one hand and deformation on the other (figure 3A and 3C). Both cells barely responded to parallel unidimensional shear which is an equal speed (ratio 1/1 in the figure 17 terminology) mixture of rotation and deformation. This is illustrated in figure 19 for cell 4207, for which responses to mixtures of rotation and translation are shown in figure 18 and to mixtures of rotation and expansion/contraction in figure 2 of Orban *et al.* (1992). Thus adding any other motion type to rotation has a deleterious effect upon the response of rotation selective cells, underscoring the importance of local vector direction for MST cells. We and others have established that only the direction of local vectors, and not their magnitude, is important in the selectivity of MST cells for EFCs (see figure 6).

Up to now we have considered only mixtures in which non-optimal motion types were added to an EFC for which the cell was selective. A number of MST cells were selective for an EFC and for translation, and so the question arises what happens to the response

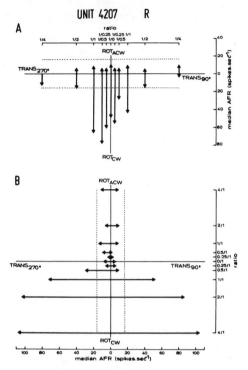

Figure 18: *Responses of a rotation selective MST cell to mixtures of rotation and translation in which rotation was kept constant (A) or translation was kept constant (B). In A, the length of the arrows indicate the strength of the response to clockwise and anti-clockwise rotation as more and more vertical translation was added. In B, the length of the arrows indicate the strength of the response to vertical translation when more and more rotation was added. Again the relative strength of the components is indicated by the speed ratio, the first number referring to the rotation speed. Notice that as soon as rotation was as strong, i.e. as fast, as translation, the cell responded.*

to a mixture in this case. Unfortunately we never tested a combination of translation and an EFC in a cell selective for both. However, cell 3701 was selective for translation and expansion and was tested with shifts in focus. As shown in figure 11, shifts in focus were similar to addition of translation, except that speed gradients differ. As demonstrated in figure 6, for cell 3701 speed gradients are not important; only the pattern of local directions determines the EFC response. Thus the shift in focus could be considered a test of mixture of translation and expansion. This would explain why the central focus was optimal in the pure expansion cell (figure 12), since addition of translation decreases the response, just as noted for other EFC selective cells above. However cell 3701, in addition to expansion, preferred translation to the left which corresponds to shift in focus to the right, which was in fact the optimal condition (figure 12).

In MST cells for which EFC responses were not position invariant, the mixture of translation with and EFC yielded totally different results than in case of EFC selective cells. This is shown in figure 20. For either cell the addition of rotation does not decrease the response to translation but decreases the direction selectivity since the translation

UNIT 4207

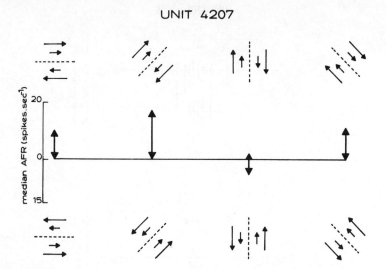

Figure 19: *Responses of a rotation selective MST cell (same as figure 18) to unidimensional parallel shear for different orientations of the deformation. Arrows indicate response strength as in the previous figure. The deformations are indicated schematically. This test was performed immediately after the one shown in figure 18 and the responses to unidimensional parallel shear can be compared to the response to pure rotation shown in figure 18.*

component becomes smaller. Thus for these cells there is a gradual transition from a response dominated by translation to that dominated by the EFC. However, the amount of the other component which must be added is relatively large (more than equal speed) before any effect is seen. Again, however the local directions of motion determine the response.

The second example showing that cortical cells, this time MT cells, do not decompose stimuli within their RF, comes from experiments in which we measured directional tuning for random dots translating at different speeds. Indeed, vectorial decomposition shows that a larger – i.e. faster motion – vector in a non-optimal direction can have a component of optimal size – i.e. of optimal speed – in the preferred direction of a cell (figure 21). Furthermore, the faster the translation, the larger the angular difference which is allowed between the actual stimulus and the optimal direction. Thus one would expect that if neurons could decompose a vector of any direction into an optimal component and second component, that the directional tuning would widen considerably as speed of translation increases. This was observed in some of the V2 cells (figure 21). The results obtained in MT were exactly the opposite. In MT the width of tuning is remarkably constant with speed: it changes hardly at all, even for speeds double or triple the optimal speed (figure 22). Thus the decomposition of direction into a few cardinal directions is not the sort of operation MT cells perform. Neurons in MT indeed have optima in all directions (figure 23). There is not a set of privileged directions which would suffice to encode direction of translation in a case where the neurons decompose any translating stimulus along these privileged directions. Here again the response of the MT neurons decreases as the stimuli deviate from the optimal stimulus. Thus, these two experiments lead to a similar conclusion, namely that there is no local decomposition of stimuli into simpler

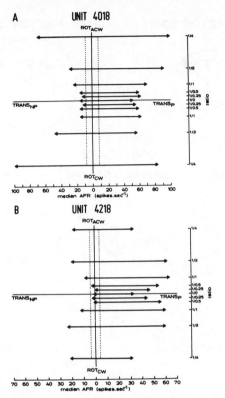

Figure 20: *Responses of two non-selective MST cells to mixtures of translation and rotation. Same conventions as in figure 18B. Both cells responded to rotation in addition to translation but the rotation responses failed to meet the position invariance test.*

components by cortical cells. However, there is an important difference between the two experiments. In the case of directional tuning for translation we and many others have observed the alternative to decomposition, which is that neurons encode all intermediate directions so that decomposition is not required. For the EFCs, expansion/contraction and rotation in particular, we failed to observe cells with demonstrated EFC selectivity and tuned to stimuli intermediate between rotation and expansion/contraction, i.e. spiral cells, as others have claimed (Andersen *et al.* 1991). One logical conclusion would be that for the primate brain there is no dimension spanning tangential and radial velocity as Andersen *et al.*'s (1991) graphical formalism could suggest, and that only the two extreme values of this arbitrary dimension are represented in the brain.

4 Conclusions

1. We now have a much better understanding of what MT cells do: they analyse the local velocity vectors as a consequence of their tuning for direction and speed of translation. We have shown that at least in one direction, the two dimensions were separable since directional tuning does not depend on speed, in agreement with others' observations

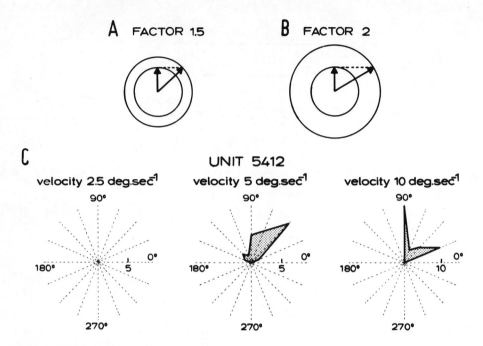

Figure 21: *Schematic representation of the effect of increasing speed beyond the optimum on directional tuning. The same vector can be produced on the optimal direction by faster speeds angled away from the optimum: 45 deg for a speed 1.5 times faster (A) and 60 deg for a speed twice as fast (B). In C, the directional tunings obtained at three speeds are shown for a V2 cell in which direction tuning widens with speed, in accordance with the scheme shown in B.*

(Rodman and Albright 1987). Furthermore, MT neurons are able to extract these local vectors from spatial configurations.

MT cells are selective neither for EFCs nor for kinematic boundaries. Apparently, selective responses are mere consequences of the asymmetries in their RF profile. As such, MT neurons seem rather unsophisticated. It is worthwhile to recall here that they can have two additional properties. A number of them are selective for the pattern direction rather than the component direction (Movshon *et al.* 1985). Secondly, a number of them have an antagonistic surround, and since these cells occur mainly outside layer 4 (Lagae *et al.* 1989, 1990) this indicates that this property is a result of processing within MT itself.

2. One thing we now know about MST cells is that a fraction, about one third, are selective for EFCs. Since most EFCs for which MST cells are selective are either rotation or expansion/contraction and not deformation, our results suggest that these neurons are involved in encoding relative motion between the observer and his surroundings, which includes both object and self-motion. Indeed, deformation depends only on 3D structure of the environment and not on relative motion between observer and his surroundings (Koenderink 1986). Thus if MST cells were encoding 3D structure, one would expect more MST cells to encode deformation. This was not the case, but our restricted observations in FST suggest that FST cells could play that role.

In addition, we have shown that all MST cells, those selective for EFCs as well as others, have a special mechanism in their RF to boost summation over peripheral parts of

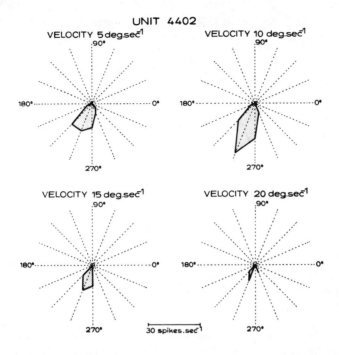

Figure 22: *Directional tuning of an MT cell obtained at four different speeds: 5 deg/sec (A), 10 deg/sec (B), 15 deg/sec (C) and 20 deg/sec (D). The calibration bar applies to all four speeds of which testing was interleaved. The optimum velocity is 10 deg/sec, yet tuning at 15 or 20 deg/sec is not broader than at 10 deg/sec.*

the RF in such a way that stimulation of the RF centre – usually the most sensitive region – is not required. These cells can therefore operate well on cluttered images in which several motions, even those of a different nature are present simultaneously. Even the relatively simple situation in which a moving object is moving behind a stationary object still presents a problem in computer vision. This is illustrated in figure 24, showing the result of an analysis performed in H.H. Nagel's group by D. Koller on image sequences taken at a road intersection. The program successfully tracked the different cars and described their motion, but when a car moved behind a traffic light it came to be considered as two separate objects. The biological system might face similar problems. Indeed, assuming that objects are continuous, the visual system might consider activity in neighbouring neurons in topographically organised areas as representing the same object or the same part of an object unless the contrary is indicated by a motion discontinuity or kinetic edge (figure 25). This will yield a number of image fragments within which coherent motion occurs, and some of these pieces will belong together, others not. MST could provide the signals to bind those fragments which move coherently, either according to a translation or to an expansion or rotation. It is important to remember that the motion providing the binding signal need not to be a translation. Indeed, this is an instance in which all schemes supposing that oscillations of local velocity vector detectors support segmentation (Gray *et al.* 1989) will fail dramatically. Lower-order neurons, such as V1 or even MT

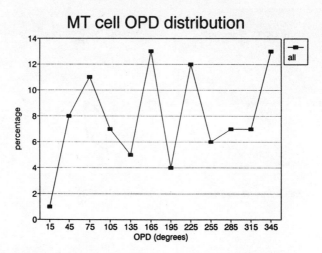

Figure 23: *Distribution of optimal directions (OPD) of 94 MT cells. The optimum direction was determined from directional tunings at optimal speed such as shown in figure 22. The stimulus were random dots. Similar findings have been made with moving bars (Albright 1989; Lagae et al. 1992).*

cells, cannot decide whether to associate with and hence to oscillate in synchrony with a neuron having the same preferred direction or one displaying an orthogonal preference: the first case corresponds to binding by homogeneous translation, the other to rotation. MST cells experience no such difficulty and can make the decision, even with incomplete stimuli.

3. In our search for neurons encoding orientation of kinematic boundaries we have been only partially successful. We have established that MT cells, even those with antagonistic surrounds, are not capable of encoding the orientation of these boundaries, although they might provide the local velocity vector signals which are assembled somewhere else. This would fit the results of recent lesion experiments (Marcar and Cowey 1992). Infero-temporal neurons are selective for orientation of kinematic boundaries which cannot be explained by a simple directional tuning for translation, but they seem not to fit our initial prediction for cells detecting kinematic boundaries. They are usually selective for only one type of kinetic boundary, either parallel or orthogonal (Vogels *et al.* 1992). We are presently investigating early levels of the system to find out whether or not, at these earlier levels, cells encode kinematic boundaries independently of the direction of translation.

4. We have shown that between MT, a second order area, and MST, a third order area, important new properties emerge. This is exactly what one would expect in a system with little depth – the primate visual system seems to have only four levels. In such a system each change in level should correspond to as large a change in representation as possible. Results from other groups show that compared to MT, an additional elaboration occurs in MST, in which extraretinal pursuit signals (Newsome *et al.* 1988) and head position signals (Thier and Erickson 1990) have been recorded. In fact, it seems that the transition from MT to MST corresponds to a change from a representation which closely fits what is present on the retina, to a more interpretative representation. The inverse summation data reveal that, given the presence of many local velocity vectors signaling translation or

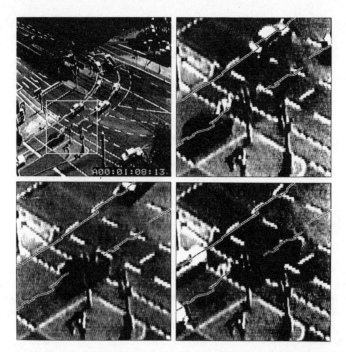

Figure 24: *Results of a computer vision program tracking cars at a street intersection. The program was developed by D. Koller at the Institut für Algorithmen und Kognitive Systeme of the Universität Karlsruhe (Dir. Prof. H.H. Nagel). Left up: overall view, other views: details of car moving behind traffic light, in three frames, before the car reaches the light (upper right), when it is behind the light (lower left) and when it has passed the light (lower right). The zigzag lines drawn by the computer on the image is the calculated trajectory of the cars.*

an EFC, MST cells continue to respond, even if a large number of vectors are missing in the centre: in this sense one could say that the absence of signals in the centre is interpreted in view of the signals present around it. Notice that this strategy is reminiscent of the use of peripheral vision for recovering ego-motion in computer vision (Eklundh *et al.* 1992).

It might well be that this difference between MT and MST reflects a general change between second level and third order areas. Also, in V4, the major third order area in the ventral stream, several lines of evidence suggest that cells 'interpret' visual signals. V4 cells seem to be able to encode the reference orientation in delayed matching to sample experiments (Maunsell *et al.* 1988) as well as cross modality signals (Haenny *et al.* 1988). The true colour selectivity which has been claimed by Zeki (1983) to exist in V4 can also be seen as an interpretation of retinal signals: only wavelengths, and not colour, exist on the retina. Thus neurons in second order and, *a fortiori*, in first order areas – even those with surrounds – represent more or less literally what is present on the retina, in agreement with the retinotopic organisation of these areas. In contrast, third order neurons behave less literally and interpret signals at one position as a function of what is present elsewhere: this is not just a modulation of responses from the RF centre, as antagonistic surrounds do, but true integration and in the extreme, spreading-in. This fits the absence of a clear retinotopy at the third level but requires the presence of retinotopy at the level just below.

SOLUTION BINDING PROBLEM

1   Detection of discontinuities

2   Binding together separate pieces

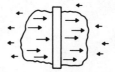

Figure 25: *Schematic representation of the solution of the binding by motion problem. The first step (top) is to segment regions in which motion is uniform. This can be done by assuming that motion is uniform until evidence to the contrary is provided by cells signalling kinematic boundaries. The second step (bottom) is to link together those regions which move according to the same rule. This would be achieved by the MST cells. Since cells in MST can be selective for rotation or expansion/contraction in addition to translation, the law of motion need not to be restricted to translation as is assumed in schemes basing segmentation on cell synchronization between cells with similar translation preferences.*

5. The primate visual system seems not to follow mathematical decomposition rules and does not decompose a stimulus locally, i.e. within the RF, into simpler constituents, although it can break down a stimulus into spatially distinct parts. The likely reason is that the wiring required for local decomposition is simply too complex. The wiring, and thus the development, of template matching mechanisms is much simpler and thus feasible. The template corresponds in this context to the canonical metamere of Koenderink *et al.* (1992). The main difference between lower and higher-order mechanisms is then that the template is more involved in higher-order mechanisms, thus that the set of metameres is more restricted. Other recent evidence suggesting template matching comes from the experiments of Tanaka et al. (1991) showing that response of infero-temporal cells to real objects can be reduced to stimuli associating a number of simple features of outline, colour, or texture. Also the reports of 'face' cells (Perrett *et al.* 1982; Gross *et al.* 1972) fit completely with the notion of templates. The identity of a single individual face will still be represented by the distributed activity of a number of neurons. However, only face cells or in general a restricted set of neurons with a certain type of template, will be contributing to the representation, perhaps in a way similar to the vector addition used to represent simple parameters such as orientation in the visual system (Vogels 1990). This implies a sort of developmental labelling of activities which has received support from the electric stimulation experiments of Salzman and Newsome (1991), in which stimulation of a site, probably a single direction column, in MT favours the choice by the monkey of a single direction out of eight possibilities. Thus template matching based on natural components, which was the initial suggestion after early discoveries of visual processing, and which has been abandoned for more mathematical schemes, such as Fourier analysis

or other linear component analyses, might after all be the way by which the cortex achieves its formidable feats.

Acknowledgements

The author is much indebted to L. Lagae, S. Raiguel, D. Xiao, H. Maes, V. Marcar, G. Sáry and R. Vogels who collaborated with him on the research described here and carried out most of the actual experiments. Thanks are also due to G. Vanparrijs, P.Kayenbergh, G. Meulemans and Y. Celis for technical assistance. In addition to Esprit BR Project Insight, this research was supported by grants from the Medical Research Council of Belgium (FGWO), the Belgian Ministry of Science and the Regional Ministry of Education to G.A.O.

References

Albright, T.D. (1989) Centrifugal directional bias in the middle temporal visual area (MT) of the macaque. Visual Neuroscience 2, 177-188

Albright, T.D., Desimone, R., Gross, C.G. (1984) Columnar organization of directionally selective cells in visual area MT of the macaque. Journal of Neurophysiology 51, 16-31

Andersen, R., Graziano, M., Snowden, R. (1991) Selectivity of area MST neurons for expansion/contraction and rotation motions. Investigative Ophthalmology and Visual Science 15, 823

Baizer, J.S., Ungerleider, L.G., Desimone, R. (1991) Organization of visual inputs to the inferior temporal and posterior parietal cortex in macaques. The Journal of Neuroscience 11, 168-190

Boussaoud, D., Ungerleider, L.G., Desimone, R. (1990) Pathways for motion analysis: Cortical connections of the medial superior temporal and fundus of the superior temporal visual areas in the macaque. The Journal of comparative Neurology 296, 462-495

Desimone, R., Ungerleider, L.G. (1986) Multiple visual areas in the caudal superior temporal sulcus of the macaque. The Journal of comparative Neurology 248, 164-189

Desimone, R., Albright, T.D., Gross, C.G., Bruce, C. (1984) Stimulus-selective properties of inferior temporal neurons in the macaque. The Journal of Neuroscience 4, 2051-2062

Dubner, R., Zeki, S.M. (1971) Response properties and receptive fields of cells in an anatomically defined region of the superior temporal sulcus in the monkey. Brain Research 35, 528-532

Eklundh, J.O., Gårding, J., Lindeberg, T., Bergholm, F. (1992) Extraction of Shape and Experiments on Cue Integration. This volume, 350-379

Felleman, D.J., Van Essen, D.C. (1991) Distributed hierarchical processing in the primate cerebral cortex. Cerebral Cortex 1, 1-47

Ferster, D. (1986) Orientation selectivity of synaptic potentials in neurons of cat primary visual cortex. The Journal of Neuroscience 6, 1284-1301

Gibson, J.J. (1950) The perception of the visual world. Houghton Mifflin Co., Boston

Girard, P., Salin, P.A., Bullier, J. (1991) Visual activity in areas V3a and V3 during reversible inactivation of area V1 in the macaque monkey. Journal of Neurophysiology 66, 1493-1503

Girard, P., Salin, P.A., Bullier, J. (1992) Response selectivity of neurons in area MT of the macaque monkey during reversible inactivation of area V1. Journal of Neurophysiology (in press)

Goodale, M.A., Milner, A.D., Jacobson, L.S., Carey, D.P. (1991) A neurological dissociation between perceiving objects and grasping them. Nature 349, 154-156

Gray, C.M., König, P., Engel, A.K., Singer, W. (1989) Oscillatory responses in cat visual cortex exhibit inter-columnar synchronization which reflects global stimulus properties. Nature 338, 334-337

Gross, C.G., Rocha-Miranda, C.E., Bender, D.B. (1972) Visual properties of neurons in infero-temporal cortex of the macaque. Journal of Neurophysiology 35, 96-111

Haenny, P.E., Maunsell, J.H.R., Schiller, P.H. (1988) State dependent activity in monkey visual cortex. II. Retinal and extraretinal factors in V4. Experimental Brain Research 69, 245-259

Koenderink, J.J. (1986) Optic flow. Vision Research 26, 161-180

Koenderink, J.J., van Doorn, A.J. (1975) Invariant properties of the motion parallax field due to the movement of rigid bodies relative to an observer. Optica Acta 22, 773-791

Koenderink, J.J., Kappers, A., van Doorn, A.J. (1992) Local Operations: The Embodiment of Geometry. This volume, 1-23

Lagae, L., Gulyás, B., Raiguel, S., Orban, G.A. (1989) Laminar analysis of motion information processing in macaque V5. Brain Research, 496, 361-367

Lagae, L., Raiguel, S., Xiao, D., Orban, G.A. (1990) Surround properties of MT neurons show laminar organization. Society for Neuroscience Abstracts 16, 6

Lagae, L., Raiguel, S., Orban, G.A. (1992) Velocity and direction selectivity of macaque middle temporal (MT) neurons. Journal of Neurophysiology (submitted)

Lennie, P., Trevarthen, C., Van Essen, D., Wässle, H. (1990) Parallel processing of visual infor-mation. In: L. Spillman and J.S. Werner (eds.) Visual perception. The neurophysiological foundations. Harcourt Brace Jovanovich, San Diego, pp. 103-128

Marcar, V.L., Cowey, A. (1992) The effect on motion perception of removing cortical visual area MT in the macaque monkey: II. Motion discrimination using random dot displays. European Journal of Neuroscience (submitted)

Maunsell, J.H.R., Newsome, W.T. (1987) Visual processing in monkey extrastriate cortex. An-nual Review of Neuroscience 10, 363-401

Maunsell, J.H.R., Van Essen, D.C. (1983) Functional properties of neurons in middle tempo-ral visual area of the macaque monkey. I. Selectivity for stimulus direction, speed and orientation. Journal of Neurophysiology 49, 1127-1147

Maunsell, J.H.R., Sclar, G., Nealey, T.A. (1988) Task-specific signals in area V4 of monkey visual cortex. Society for Neuroscience Abstracts 14, 10

Maunsell, J.H.R., Nealey, T.A., DePriest, D.D. (1990) Magnocellular and parvocellular contri-butions to responses in the middle temporal visual area (MT) of the macaque monkey. The Journal of Neuroscience 10, 3323-3334

Morel, A. and Bullier, J. (1990) Anatomical segregation of two cortical visual pathways in the macaque monkey. Visual Neuroscience 4, 555-578

Movshon, J.A., Adelson, E.H., Gizzi, M.S., Newsome, W.T. (1985) The analysis of moving visual patterns. In: C. Chagas, R. Gattass and C.G. Gross (eds.) Pattern Recognition Mechanisms. Pontificia Academia Scientiarium, Vatican City, pp. 117-151

Nagel, H.H. (1992) Direct Estimation of Optical Flow and of Its Derivatives. This volume, 193-224

Nealey, T.A., Ferrera, V.P., Maunsell, J.H.R. (1991) Magnocellular and parvocellular contri-butions to the ventral extrastriate cortical processing stream. Society for Neuroscience Abstracts 17, 525

Newsome, W.T., Paré, E.B. (1988) A selective impairment of motion perception following lesions of the middle temporal visual area (MT). The Journal of Neuroscience 8, 2201-2211

Newsome, W.T., Wurtz, R.H., Komatsu, H. (1988) Relation of cortical areas MT and MST to pursuit eye movements. II. Differentiation of retinal from extraretinal inputs. Journal of Neurophysiology 60, 604-620

Newsome, W.T., Britten, K.H., Movshon, J.A. (1989) Neuronal correlates of a perceptual deci-sion. Nature 341, 52-54

Orban, G.A., Vogels, R. (1990) Coding of orientation by infero-temporal neurons studied in the discriminating monkey. Society for Neuroscience Abstracts 16, 621

Orban, G.A., Lagae, L., Verri, A., Raiguel, S., Xiao, D., Maes, H., Torre, V. (1992) First order analysis of optical flow in monkey brain. Proceedings of the National Academy of Sciences of the United States of America 89, 2595-2599

Perrett, D.I., Rolls, E.T., Caan, W. (1982) Visual neurones responsive to faces in the monkey temporal cortex. Experimental Brain Research 47, 329-342

Peterhans, E., von der Heydt, R. (1989) Mechanisms of contour perception in monkey visual cortex. II. Contours bridging gaps. The Journal of Neuroscience 9, 1749-1763

Peterhans, E., von der Heydt, R., Baumgartner, G. (1986) Neuronal responses to illusory contour stimuli reveal stages of visual cortical processing. In: J.D. Pettigrew, K.J. Sanderson and W.R. Levick (eds.) Visual Neuroscience. Cambridge University Press, Cambridge, UK., pp. 343-351

Raiguel, S.E., Lagae, L., Gulyás, B., Orban, G.A. (1989) Response latencies of visual cells in macaque areas V1, V2 and V5. Brain Research 493, 155-159

Rockland, K.S., Pandya, D.N. (1979) Laminar origins and terminations of cortical connections of the occipital lobe in the rhesus monkey. Brain Research 179, 3-20

Rodman, H.R., Albright, T.D. (1987) Coding of visual stimulus velocity in area MT of the macaque. Vision Research 27, 2035-2048

Rodman, H.R., Gross, C.G., Albright, T.D. (1989) Afferent basis of visual response properties in area MT of the macaque. I. Effects of striate cortex removal. The Journal of Neuroscience 9, 2033-2050

Rodman, H.R., Gross, C.G., Albright, T.D. (1990) Afferent basis of visual response properties in area MT of the macaque. II. Effects of superior colliculus removal. The Journal of Neuroscience 10, 1154-1164

Saint-Cyr, J.A., Ungerleider, L.G., Desimone, R. (1990) Organization of visual cortical inputs to the striatum and subsequent outputs to the pallido-nigral complex in the monkey. The Journal of Comparative Neurology 298, 129-156.

Saito, H., Yukie, M., Tanaka, K., Hikosaka, K., Fukada, Y., Iwai, E. (1986) Integration of direction signals of image motion in the superior temporal sulcus of the macaque monkey. The Journal of Neuroscience 6, 145-157

Salzman, C.D., Newsome, W.T. (1991) Microstimulation of MT during an eight-alternative motion discrimination: directional tuning of the behavioral effect. Society for Neuroscience Abstracts 17, 525

Schiller, P.H., Logothetis, N.K., Charles, E.R. (1990) Role of the color-opponent and broad-band channels in vision. Visual Neuroscience 5, 321-346

Siegel, R.M., Andersen, R.A. (1986) Motion perceptual deficits following ibotenic acid lesions of the middle temporal area (MT) in the behaving rhesus monkey. Society for Neuroscience Abstracts 12, 1183

Tanaka, K., Saito, H. (1989) Analysis of motion of the visual field by direction, expansion/contraction, and rotation cells clustered in the dorsal part of the medial superior temporal area of the macaque monkey. Journal of Neurophysiology 62, 626-641

Tanaka, K., Saito, H., Fukada, Y., Moriya, M. (1991) Coding visual images of objects in the inferotemporal cortex of the macaque monkey. Journal of Neurophysiology 66, 170-189

Thier, P., Erickson, R.G. (1990) Vestibular input to visual-tracking neurons in area FST of awake rhesus monkeys. Society for Neuroscience Abstracts 16, 7

Ungerleider, L.G., Mishkin, M. (1982) Two cortical visual systems. In: D.J. Ingle, M.A. Goodale and R.J.W. Mansfield (eds.) Analysis of visual behavior. MIT Press, Cambridge, pp. 549-586

Vandenbussche, E., Saunders, R.C., Orban, G.A. (1991) Lesions of MT impair speed discrimination performance in the Japanese monkeys (Macaca Fuscata). Society for Neuroscience Abstracts 17, 8

Van Essen, D.C., Anderson, C.H., Felleman, D.J. (1992) Information processing in the primate visual system: an integrated systems perspective. Science 255, 419-423

Verri, A., Girosi, F., Torre, V. (1989) Mathematical properties of the two-dimensional motion field: from singular points to motion parameters. Journal of the Optical Society of America A/6, 698-712

Verri, A., Straforini, M., Torri, V. (1992) Computational Aspects of Motion Perception in Natural and Artificial Vision Systems. This volume, 71-92

Vogels, R. (1990) Population coding of stimulus orientation by striate cortical cells. Biological Cybernetics 64, 25-31

Vogels, R., Orban, G.A. (1990) Effects of task related stimulus attributes on infero-temporal neurons studied in the discriminating monkey. Society for Neuroscience Abstracts 16, 621

Vogels, R., Sáry, G., Orban, G.A. (1992) Responses of infero-temporal units to luminance, kinetic and texture boundaries. Investigative Ophthalmology and Visual Science 33(4), 1131.

von Bonin, G., Bailey, P. (1947) The Neocortex of Macaca Mulatta. University of Illinois Press, Urbana, Illinois.

von der Heydt, R., Peterhans, E. (1989) Mechanisms of contour perception in monkey visual cortex. I. Lines of pattern discontinuity. The Journal of Neuroscience 9, 1731-1748

Webster, M.J., Ungerleider, L.G., Bachevalier, J. (1991) Connections of inferior temporal areas TE and TEO with medial temporal-lobe structures in infant and adult monkeys. The Journal of Neuroscience 11, 1095-1116.

Zeki, S.M. (1978) Functional specialisation in the visual cortex of the rhesus monkey. Nature 274, 423-428

Zeki, S.M. (1980) A direct projection from area V1 to area V3A of rhesus monkey visual cortex. Proceedings of the Royal society of London. B: Biological sciences 207, 499-506

Zeki, S.M. (1983) Colour coding in the cerebral cortex: the reaction of cells in monkey visual cortex to wavelengths and colours. Neuroscience 9, 741-765

Zeki, S.M. (1990) Functional specialization in the visual cortex: The generation of separate constructs and their multistage integration. In: G.M. Edelman, W.E. Gall and W.M. Cowan (eds.) Signal and sense: Local and global order in perceptual maps. Wiley-Liss, New York, pp. 85-130

Motion Perception and Eye Movement Control

Klaus-Peter Hoffmann

Allgemeine Zoologie und Neurobiologie, Ruhr-Universität Bochum

1   Introduction

Vision of artificial systems can be improved by using movable cameras. The biological substrate for this task is a close link between perception of visual motion and smooth pursuit eye movements. Without seeing a moving target we are unable to move our eyes smoothly in a non-saccadic fashion. Without keeping the moving target in the foveae of our eyes, perception is greatly hampered. To understand the underlying neuronal mechanisms for the generation of smooth eye movements, knowledge about visual motion perception has to be improved and expanded.

But what can we learn from these biological facts for the design of artificial systems? As an example we want to describe one stream in a parallel information processing system. In this stream a powerful general purpose movement analysing structure (the middle temporal area (area MT) in the superior temporal sulcus of the primate's cortex) provides the specific signals for a dedicated velocity error measuring system (nucleus of the optic tract and dorsal terminal nucleus (NOT-DTN) in the subcortical midbrain), which in turn provide the input to a velocity integrator which tries to keep the eye velocity as close to target velocity as possible. To do so NOT-DTN neurons relay visual information from the retina and the cortex to the vestibular nuclei (VN) via the nucleus prepositus hypoglossi (NPH) (VN and NPH form one type of integrator) and to the flocculus in the cerebellum via the dorsal cap of the inferior olive. The latter path calibrates the compensatory eye movements during turning of the head (see Simpson et al. 1988 for review). This visual motion analysing subsystem is completely directionally asymmetric, to a large extent position invariant, and centered with respect to the fovea. It gives us the opportunity to study the different levels of one of the essential cortico-subcortical interfaces for visuo-motor control.

It is well documented that neurons in area MT are movement sensitive and to a high degree, direction selective. Their properties vary depending on laminar position. Lagae *et al.* from the Leuven group, reported in 1989 that 3 different types can be distinguished depending on their differential response to moving bars and texture in anesthetized and paralysed monkeys. The authors suggested that these types may play different roles in movement perception and eye movement control. Experimentation in the awake and trained animal is required to further explore these different functional roles. Of particular interest to us are cells in lamina V because only they send their axons to the NOT-DTN and from there to the oculomotor system.

In this report we describe the cortical analysis of moving visual stimuli in the middle temporal area (MT) and demonstrate presence and functional characteristics of strong projections from specific cortical areas to the NOT-DTN in macaque monkeys. In addition to neurons in the striate cortex (V1) only neurons located in the lower bank and fundus of the superior temporal sulcus (STS), presumably area MT, could be shown to project to the NOT-DTN. This subpopulation of cortical cells exclusively codes ipsiversive horizontal stimulus movement and may thus constitute the specific pathway whose destruction leads to the well-known velocity deficit during ipsiversive tracking eye movements after posterior cortical lesions. MT and the other motion analysing structures in the superior temporal sulcus of the cortex contain, of course, many more processing streams. To understand their functional relevance we have to find the other output structures making use of the specific information available in this region of the brain.

2   Methods

Experiments were performed in awake and trained, or anaesthetized and paralyzed macaque monkeys. The procedures for surgery, electrical and visual stimulation have been described elsewhere (Hoffmann *et al.* 1988, 1991; Ilg and Hoffmann 1991).

2.1   Definitions

Orthodromic activation: refers to an electrical transsynaptic activation of a neuron via its input fibers. Shows from where the neuron under study gets its input.
Antidromic activation: refers to the electrical activation of a neuron via its own axon. This involves no synapse and shows to where the neuron under study sends its output.

3   Results

3.1   Responses of Neurons in Extrastriate Area MT During Smooth Pursuit and Optokinetic Nystagmus in Awake Monkeys

First we want to present data concerning the dynamic role of MT neurons during slow eye movements. We recorded from 239 single MT units during smooth pursuit and optokinetic nystagmus in awake and trained monkeys. These neurons respond strongly and always direction specifically to the retinal slip of small targets within their restricted receptive fields. Movement of the entire visual scene as used to elicit optokinetic nystagmus leads to direction specific responses and neuronal modulations similar to single targets. In presenting both stimuli simultaneously, there is however only a weak influence from passive background movement during pursuit of a single target. In all MT neurons tested the modulation of the neuronal activity was entirely dependent on the direction of target motion if the monkey's eyes followed the target. The presence of the background movement across the retina due to the eye movements never lead to a change of the preferred direction of the neurons (see Figure 1).

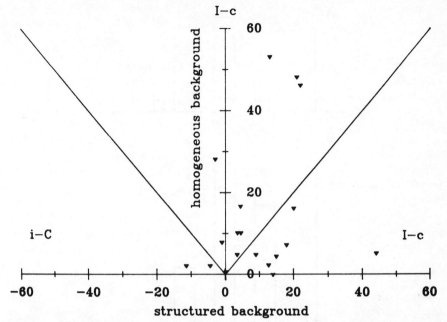

Figure 1: *Neuronal modulation of 19 MT neurons during smooth pursuit over homogeneous (ordinate) and structured background (abscissa). i-C, I-c: neuronal modulation as difference between the neuronal response to ipsi- and contraversive target movement. Positive values indicate preference for ipsiversive, negative values preference for contraversive target movement. Background structure has no significant influence on the neuronal modulation of MT neurons.*

In 148 neurons complete directional tuning curves were accumulated while the animal was fixating. The distribution of preferred directions did not show any bias. All directions are evenly represented by MT-neurons.

During optokinetic after nystagmus (OKAN) the neuronal activity goes back to spontaneous level. MT neurons thus seem to code the velocity error between eye and the visual target to pursue, and not the eye velocity per se.

What happens when the two stimuli moving in opposite directions are of equal quality, e.g. two random dot stimuli. Our preliminary results on MT-neurons agree completely with those recently published by Snowden et al. (1991). Simultaneous presentation of two random dot patterns, one moving in the preferred and the other in the opposite direction (transparency paradigm), invariably leads to a decrease in response strength when compared to the preferred stimulus only condition.

3.2 The Transparency Paradigm in Human Psychophysics

In addition, psychophysical experiments have been carried out using this transparency paradigm. In humans, horizontal and vertical eye positions were measured using a corneal reflection technique (OBER2). The non-human primates were implanted chronically to record the eye position by the magnetic search coil method (SKALAR).

Each human observer decoded the transparency stimulus immediately into two planes of depth. Eye velocity of human subjects was reduced if they had to track the dots of

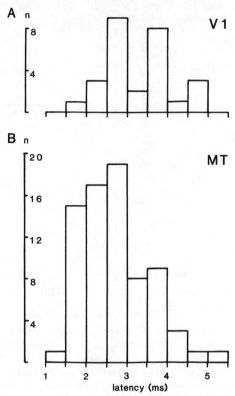

Figure 2: *Frequency distributions of orthodromic latencies measured at NOT-DTN neurons after electrical stimulation in area V1 (A) and in area MT (B). Abscissa: latency in ms; Ordinate: number of cells.*

one of the stimulus planes in the transparency paradigm as compared to unidirectional stimulation with one plane of dots. The degree of reduction depended on the attention of the subject. If subjects are asked to track attentively one group of moving dots, a statistically significant difference between eye speed during transparent and during unidirectional stimulation can be observed only for high stimulus velocities (24°/s). These results agree with our observation that monkeys displayed a decreased smooth pursuit eye speed and an increased amount of catch-up saccades during pursuit across a structured background.

In conclusion, the reduction in neuronal activity seen in MT-neurons correlates with a reduced eye velocity during pursuit in the transparency versus one-stimulus paradigm. Which are, however, the signals selected from the wealth of the cortical display in MT to control slow eye movements? We tried to answer this question by analysing the cortical output to one of the subcortical visuo-motor interfaces for slow eye movements, the NOT-DTN complex.

3.3 Cortical Areas Projecting to the Direction Specific Neurons of the NOT-DTN

In a first approach in anaesthetized and paralyzed monkeys NOT-DTN neurons were identified physiologically and tested for cortical input by orthodromic electrical stimulation in various cortical areas. Successful stimulation sites to activate NOT-DTN neurons lie in the primary visual cortex (V1) and in the motion processing areas in the superior temporal sulcus (STS). In contrast, electrical stimulation in area V4 and in parietal areas never yielded any spike discharges in the NOT-DTN.

3.3.1 Striate Cortex (V1)

Both central and peripheral field representations in area V1 were electrically stimulated in 2 experiments (Figure 3B). All but one NOT-DTN unit tested could be orthodromically activated from V1. The latency distribution is presented in figure 2A.

3.3.2 Superior Temporal Sulcus (STS)

For the stimulation of extrastriate cortex we used two different approaches. In the first two animals we recorded from cortical visual areas in STS to identify the location of direction selective cells and the representation of the fovea, presumably in area MT. The physiological response properties and receptive field locations of cortical neurons recorded at certain sites which later on were electrically stimulated, were also used for identifying the location of these cortical stimulation sites. The recording electrode was then replaced by stimulating electrodes which were cemented into place. This procedure allowed us to sample orthodromic latencies from a large number of subcortical cells in response to electrical stimulation of one particular part of the extrastriate visual cortex (Figure 2B). According to the physiological data, the two stimulation sites in the first animal correspond to the region spanning the border between foveal MT and FST. Shocks delivered between these electrodes were indeed successful in eliciting action potentials in NOT and DTN cells.

The second monkey had a total of 7 stimulation sites. Three effective ones were in the left STS corresponding to 2 foveal MT locations and 1 FST location. Of the 4 stimulating electrodes placed in the right hemisphere, three were in STS. Only one was effective and corresponded to foveal MT. The other two were ineffective and were in peripheral MT and the white matter underlying MST. The fourth site was also ineffective and lay within area 7a on the inferioparietal gyrus.

In the remaining 3 monkeys (cases 3, 4, and 5), we used another procedure. In these animals, 3, 10, and 15 penetrations, respectively, were made in the temporal and parietal cortex to map cortical areas from where single NOT-DTN neurons could be orthodromically activated. Furthermore, this approach enabled us to compare the orthodromic latencies and thresholds of a limited number of individual NOT-DTN cells in response to electrical stimulation of various cortical areas.

Case 3 had 3 parallel cortical penetrations. It is interesting to note that the stimulating electrode went through posterior parietal areas, areas within the lateral sulcus and MST, or area V4 and V4t without activating NOT-DTN neurons. Only with stimulation in areas MT and FST were clear orthodromic responses obtained in NOT-DTN. Their latencies are presented in figure 2B.

In case 4 (Figure 3A and C) cortical penetrations 1-6 went parallel to each other and to the banks of the superior temporal sulcus, whereas penetrations 7-10 went almost

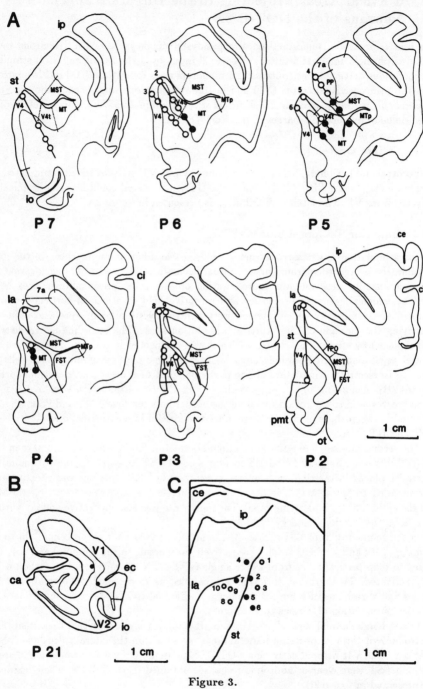

Figure 3.

◀**Figure 3:** *Case 4: A: Reconstruction of frontal sections through the left hemisphere of monkey 4. The sections are presented from posterior (top left) to anterior (bottom right), the stereotaxic levels are indicated. Numbers 1-10: penetration tracks; filled circles: effective stimulation sites; open circles: ineffective stimulation sites; dotted lines and arrows: borders between visual areas. B: Section through the right hemisphere of monkey 4 showing the location of two stimulation electrodes in area V1 between which electrical stimuli were delivered effectively. Conventions as in A. C: Reconstruction of the lateral view of the left hemisphere of monkey 4 showing the location of penetration tracks (numbered as in A) along which electrical stimulation was effective (filled circles) or non-effective (open circles) in orthodromically activating NOT-DTN cells. Note that effective stimulation sites never occurred in superficial cortex. For abbreviations see list. The bars correspond to 1cm.*

in the dorso-ventral direction and reached only superficial cortical areas. Again, only stimulation sites in MT, and this time additionally in MST were effective. Stimulation in areas V4, V4t, posterior parietal area, and at the lower lip of the lateral sulcus were ineffective even at the highest stimulus strengths.

In case 5, a widespread area of the parietal and superior temporal cortex was stimulated along 15 penetration tracks. Generally, the results of cases 3 and 4 were confirmed. Surprisingly, we could activate NOT-DTN cells also from the extreme posterior parietal area, from the depth of the lateral sulcus, and from the anterior part of STS. This could be due to spread of electrical current to fibers originating for example in the intraparietal sulcus. Stimulation of area V4, V4t, area 7a, and the lower lip of the posterior lateral sulcus were not effective.

3.3.3   Differences in Cortical Projections

The efficacy of electrical stimulation was very similar for the positive sites in different cortical areas. Single shocks always elicited at least two action potentials and often a burst of up to 5-8 spikes with intervals of 2-4ms. Applying two electrical stimuli at V1 or the optic chiasm and the STS sites with variable intershock delays (5-100ms) never demonstrated inhibitory interactions. Therefore, the cortical projection to NOT-DTN seems to be exclusively excitatory in action. The latencies of the first spikes elicited from either V1 or STS stimulation sites were not identical, however. Latencies from V1 were on average 0.5ms longer than latencies from STS (3.51 ± 0.81ms and 2.99 ± 0.85ms, respectively). This latency difference is statistically significant at the $p<0.01$ level (t-test). In one animal (case 1) we had placed stimulating electrodes in both V1 and in MT. For 22 NOT-DTN neurons we obtained latencies from both sites. The latency differences between V1 and MT varied between 0 and 2.5ms, with V1 having always the longer latency. Longer latencies of 1ms or less make a relay of V1 activity in V2 and/or in MT en route to the NOT-DTN very unlikely. This was the case in 16 of the 22 cells. We take this as evidence for a direct projection of V1 to the NOT-DTN (see also anatomy).

3.4   Anatomical Demonstration of Cortical Projections to the NOT-DTN

Orthodromic electrical stimulation does not unequivocally prove a direct connection between stimulation and recording site. We therefore made rather large horseradish peroxidase (HRP) injections into area V1 as well as into STS to investigate if any terminals in the NOT-DTN were labelled by these injections. Since HRP as used as in our protocol is

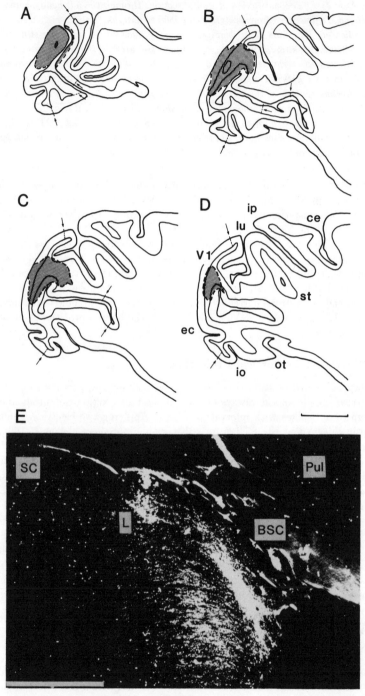

Figure 4.

◄**Figure 4:** *A-D: Serial reconstruction of parasagittal sections through the right hemisphere of case 6 with an HRP injection largely confined to area V1. The sections are arranged from medial (A, top left) to lateral (D, bottom right), the occipital pole of the brain is left. Shaded areas represent the extent of the injection site. Arrows indicate the borders of area V1. The bar represents 1cm. E: Darkfield microphotograph of a frontal section through the BSC and pretectum of case 6 demonstrating massive anterograde labelling in the region of the NOT-DTN marked by microlesion (L). The bar represents 1mm.*

not transported transneuronally, such label within the NOT-DTN would indeed indicate a cortical axon projecting directly to this area.

Two injections were made into area V1. Both resulted in massive fiber labelling in the brachium of the superior colliculus (BSC) and labelled terminals around microlesions at recording sites of direction specific NOT-DTN neurons (Figure 4). The projections demonstrated to the NOT-DTN add further weight to our notion, from the physiological experiments, that area V1 projects directly to the NOT-DTN.

One injection into STS was centered in the lower bank and fundus of STS with diffusion into the white matter and small parts of the inferio-occipital sulcus. This injection resulted in labelled fibers running through the BSC and the NOT-DTN region, and reaching also more medial parts of the pretectum. Some labelled terminals were found in close vicinity to a microlesion marking a NOT-DTN recording site where direction specific neurons could be activated orthodromically by STS stimulation. Interestingly, another lesion a little deeper in the pretectum lay clearly outside the region of labelled fibers and terminals. This lesion marked the recording site of a non direction-selective cell which could also not be activated orthodromically from extrastriate cortex.

The other injection was centered in the lower bank of STS including MT, V4t, parts of FST and V4 as well as the white matter underlying these areas. This injection resulted in rather limited labelling of fibers and terminals along inferior parts of the BSC only. Nonetheless, the label surrounded penetration tracks through, and a microlesion within, the NOT-DTN.

Thus, our anatomical results support the electrophysiological data by disclosing direct projections from both area V1 as well as areas in STS to the NOT-DTN.

3.5 Receptive Fields of Cortical Neurons Projecting to the NOT-DTN

In a second approach we tried to identify the cortical neurons giving rise to the projection to the NOT-DTN. Electrical stimulation was applied to the NOT-DTN and antidromically activated single units in extrastriate cortex were analyzed. Here we only want to present the description of characteristics relevant for extent and position of receptive fields as well as coding of ipsiversive stimulus movement in the NOT-DTN.

The data presented in figure 5A and B stem from such antidromically identified cortical neurons recorded in the floor of the left STS of two normal monkeys. The receptive fields include, in addition to the contralateral hemifield and the fovea, up to about 15 ° of the ipsilateral hemifield also. These data from extrastriate cortical cells projecting to the NOT-DTN are very similar to receptive field characteristics and visual field covering of neuronal populations in the NOT-DTN of normal monkeys.

In addition, these cortical cells resemble those in the NOT-DTN very closely with respect to their direction selectivity and their direction tuning. All 41 neurons antidromically identified as cortex-to-NOT-DTN projection neurons preferred ipsiversive stimulus

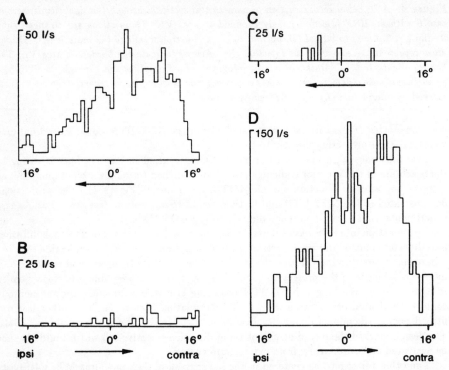

Figure 5: *Population responses of cortical neurons from the left hemisphere projecting (A, B) and not projecting (C, D) to the NOT-DTN during horizontal movement of a random dot pattern. A and C: ipsiversive stimulus movement, B and D: contraversive stimulus movement, ordinate: neuronal activity in impulses per second, abscissa: position in the visual field, 0°: vertical 0-meridian, ipsi: ipsilateral visual field, contra: contralateral visual field.*

movement (Figure 5A) whereas 43 not antidromically-driven-neurons recorded in the same penetration, and interleaved with the antidromic ones, preferred other directions of stimulus movements (Figure 5D). Therefore, it seems very likely that this subpopulation of cortical cells constructs its large central receptive fields via callosal connections and then provides the NOT-DTN with the very specific and unmistakable visual information about ipsiversive stimulus movements in the central visual field on both sides of the vertical 0-meridian.

This type of neuron could be recorded exclusively in the floor of STS, presumably in area MT. In the two-dimensional reconstructions of the left STS of the two monkeys enclosed in this part of the study (Figure 6) lesions marking cortical recording sites are indicated by asterisks. Area MT and the densely myelinated zone (DMZ) within area MST are marked by broken outlines. Lesions at recording sites of antidromically activated neurons (solid asterisks) were always in infragranular layers, whereas at least some of the lesions at other cortical sites (open asterisks) were located in supragranular layers.

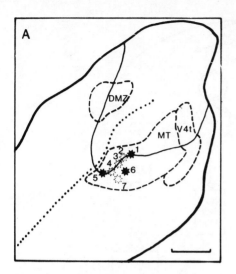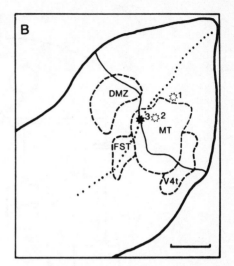

Figure 6: *A: Two-dimensional map of the superior temporal sulcus (STS) derived from sagittal sections. Myeloarchitectonic borders are given by broken lines, the fundus of the sulcus by the dotted line. Solid asterisks (1, 5, 6) indicate locations of microlesions at recording sites of cells antidromically activated from the NOT-DTN. Open asterisks (2, 3, 4, 7) indicate sites where cells could not be antidromically activated. The scale indicates 500μm. DMZ: densely myelinated zone of MST; MT: middle temporal area; V4t: transitional zone of area V4. B: Two-dimensional map of STS derived from frontal sections. The scale bar represents 500μm. DMZ: densely myelinated zone of MST; FST: visual area in the fundus of STS; MT: middle temporal area; V4t: transitional zone of area V4. All other conventions as in A.*

4 Discussion

Taken together, our results indicate a strong cortical input from striate as well as extrastriate visual areas to the NOT-DTN in the macaque. The input from extrastriate areas, mainly MT, was found to provide the NOT-DTN with visual information about ipsiversively moving visual stimuli from the contra- as well as ipsilateral hemifield and thus promote NOT-DTN's role for gaze stabilizing eye movements.

With orthodromic electrical stimulation we could demonstrate activation of the NOT-DTN cells from a number of cortical areas (Hoffmann *et al.* 1991). The broad latency range of orthodromic potentials in NOT-DTN cells would allow a number of cortical relays before an excitatory corticofugal fiber projects to the NOT-DTN. So far we have found antidromically activated cells only in the subregion of MT, but negative results are not very convincing. Further anatomical work has to show whether the cortical extrastriate motion-analyzing pathway converges onto the corticopretectal cells shown in this study before the information is transmitted to the NOT-DTN. These cells would be then the only source from STS which provides information to the NOT-DTN about ipsiversive stimulus movement across large central receptive fields reaching far into the ipsilateral visual field.

In conclusion, the described projection of a specific group of cells in the fundus of STS to the NOT-DTN would be the solution to two so far unexplained findings concerning eye movements after lesions of posterior cortical structures:

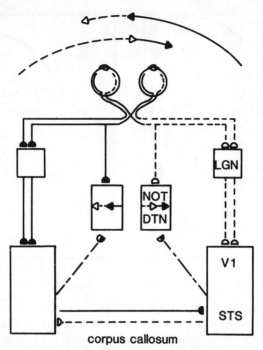

corpus callosum

Figure 7: *Wiring diagram of the subcortical and cortical components of the pathway subserving the optokinetic reflex in normal monkeys. The flow of visual information from the left hemifield is indicated by broken lines, that from the right hemifield by solid lines. With intact corpus callosum the corticopretectal projection carries information from the entire visual field and NOT-DTN receptive fields extend far into the ipsilateral hemifield. LGN: lateral geniculate nucleus; V1: primary visual area; STS: visual areas in the superior temporal sulcus.*

1. Ipsiversive directional deficits in smooth pursuit eye movements after lesions of the posterior cortex despite the fact that the lesioned areas contained neurons selective for all stimulus directions (Dürsteler and Wurtz 1988; Komatsu and Wurtz 1988a, 1988b, 1989; Newsome *et al.* 1988).

2. Asymmetry of OKN with hemifield stimulation after corpus callosum transection (Pasik and Pasik 1964; Mehdorn 1984).

Figure 7 displays the wiring diagram for the cortical influence on the subcortical pathway for the generation of slow eye movements. Neurons in each NOT-DTN sharply increase their firing rate during ipsiversive pursuit if eye velocity is lower than stimulus velocity. They do so for stimuli presented to either eye and in both hemifields because of their cortically generated binocular and bilateral receptive fields. Let us construct, for example, the cortical input to the left NOT-DTN. An ipsiversive movement (to the left) across the visual field is split at the vertical 0-meridian. As long as the stimulus moves in the right (contralateral) hemifield, it is transmitted through the left geniculo-cortical system (solid arrows and lines). When the stimulus moves in the left (ipsilateral) hemifield it is transmitted through the right geniculo-cortical system (open arrows and dotted lines). However, to get to the left NOT-DTN all axons have to leave from the left cortex. The information from the right cortex (left hemifield) has to cross via the corpus callosum. Then in the left STS, the whole field information is reconstituted and sent to the NOT-DTN (solid line with dashes). The defect in ipsiversive tracking eye movements to stimuli moving on both sides of the fovea after MT and MST lesions is fully explained by the destruction of these neurons or the input to these neurons in STS projecting to the NOT-DTN. The remaining direct retinal input to the NOT-DTN is too weak to maintain high gain OKN. Why smooth pursuit also suffers the same direction-specific deficit has to be further investigated. One possibility is that directionally biased cortical output through the NOT-DTN has an influence on the velocity of smooth pursuit as well.

The general implication of our findings is that grouping of cortical activity for certain visuomotor functions could be achieved by subcortical selection of a specific cortical output. In our example only cortical neurons with specificity for ipsiversive stimulus movement are connected with the NOT-DTN.

What we have described makes sense to a neurobiologist as well as to an engineer. But one of the most important questions has to remain unanswered. How was the system designed and how are the connections in such a system established. A self-organising mechanism involving genetically determined, direction-specific retinal input and experience dependent "learning" cortical input to the NOT-DTN has been proposed (Hoffmann 1987).

References

Dürsteler, M.R., Wurtz, R.H. (1988) Pursuit and optokinetic deficits following chemical lesions of cortical areas MT and MST. J. Neurophysiol. 60, 940-965

Hoffmann, K.-P. (1987) Influence of visual experience on ontogeny of the optokinetic reflex in mammals. In: J. Rauschecker, P. Marler (eds.) Imprinting and Cortical Plasticity. New York: John Wiley & Sons. Inc., pp. 267-286

Hoffmann, K.-P., Distler, C., Erickson, R. (1991) Functional projections from striate cortex and superior temporal sulcus to the nucleus of the optic tract (NOT) and dorsal terminal nucleus of the accessory optic tract (DTN) of macaque monkeys. J. Comp. Neurol. 313, 707-724

Hoffmann, K.-P., Distler, C., Erickson, R.G., Mader, W. (1988) Physiological and anatomical identification of the nucleus of the optic tract and dorsal terminal nucleus of the accessory optic tract in monkeys. Exp. Brain Res. 69, 635-644

Ilg, U.J., Hoffmann, K.-P. (1991) Responses of monkey nucleus of the optic tract neurons during pursuit and fixation. Neurosci. Res. 12, 101-110

Komatsu, H., Wurtz, R.H. (1988a) Relation of cortical areas MT and MST to pursuit eye movements. I. Localization and visual properties of neurons. J. Neurophysiol. 60, 580-603

Komatsu, H., Wurtz, R.H. (1988b) Relation of cortical areas MT and MST to pursuit eye movements. III. Interaction with full-field visual stimulation. J. Neurophysiol. 60, 621-644

Lagae, L., Gulyas, B., Raiguel, S., Orban, G.A. (1989) Laminar analysis of motion information processing in macaque V5. Brain Res. 496, 361-367

Mehdorn, E. (1984) The importance of the corpus callosum for the optokinetic nystagmus in man. Fortschr. Ophthalmol. 81, 157-160

Newsome, W.T., Wurtz, R.H., Komatsu, H. (1988) Relation of cortical areas MT and MST to pursuit eye movements. II. Differentiation of retinal from extraretinal inputs. J. Neurophysiol. 60, 604-620

Pasik, T., Pasik, P. (1964) Optokinetic nystagmus: an unlearned response altered by section of chiasm and corpus callosum in monkeys. Nature 203, 609-611

Simpson, J.I., Giolli, R.A., Blanks, R.H.I. (1988) The pretectal nuclear complex and the ac-
    cessory optic system. In: J.A. Buettner-Ennever, (ed.) Neuroanatomy of the oculomotor
    system. Elsevier-Science Pub., pp. 333-362
Snowden, R.J., Treue, S., Erickson, R.G., Andersen, R.A. (1991) The response of area MT and
    V1 neurons to transparent motion. J. Neurosci. 11, 2768-2785

Computational Aspects Motion Perception in Natural and Artificial Vision Systems

Alessandro Verri, Marco Straforini and Vincent Torre

Dipartimento di Fisica dell' Università di Genova

Abstract

This paper is divided into two parts. In the first part a computational scheme for motion perception in artificial and natural vision systems is described. The scheme is motivated by a mathematical analysis in which first order spatial properties of optical flow, such as singular points and elementary components of optical flow, are shown to be salient features for the perception of visual motion. Singular points and elementary flow components are used to compute motion parameters, such as time-to-collision and angular velocity, and to segment the visual field into areas which correspond to different motions. In the second part a number of biological implications are discussed. Electrophysiological findings suggest that the brain perceives visual motion by detecting and analysing optical flow components. However, the cortical neurons which seem to detect elementary flow components are not able to extract these components from more complex flows. A simple model for the organisation of the receptive field of these cells, which is consistent with anatomical and electrophysiological data, is finally described.

1 Introduction

The understanding of changing images is a remarkable feature of natural visual systems (Gibson 1950; Hassenstein and Reichardt 1956; Barlow and Levick 1965; Torre and Poggio 1978; Marr and Ullman 1981), and a major goal of artificial intelligence is to build machines with similar capabilities. The classical notion of optical flow (Gibson 1950), or the motion of the image brightness pattern on the retina of the visual sensor, appears to be the bridge between research in natural and artificial vision systems. In recent years many aspects of motion understanding in artificial vision systems have been clarified (see (Koenderink and van Doorn 1975; Horn and Schunck 1981; Nagel 1983), for example) and it is now possible to compute optical flow almost in real time. On the other hand, recent electrophysiological recordings from cortical neurons of the monkey (Tanaka and Saito 1989; Andersen *et al.* 1990; Andersen *et al.* 1991; Lagae *et al.* 1991; Duffy and Wurtz 1991a, 1991b; Lagae 1991) have revealed the existence of units tuned to rotating or expanding stimuli, thus shedding new light on how the brain perceives motion. This paper outlines a theory of motion perception which has inspired a number of papers on the processing of image sequences in machine vision (Uras *et al.* 1988; Verri *et al.* 1989; Verri *et al.* 1990; Campani and Verri 1990; Campani and Verri 1992; Rognone et al. 1992), and, at the same time, is also relevant for the understanding of motion mechanisms in the brain (Orban *et al.* 1992) (see also the chapters of Orban, Nagel and Koenderink of the same volume).

A general analysis of the many stages of motion perception is beyond the scope of the present research. This paper deals with the segment of motion perception which, in the monkey, occurs along the visual pathway that goes from the visual area V1 (and V2) to the Medial Superior Temporal area, or MST, through the Middle Temporal area, or MT (Livingstone and Hubel 1988; Zeki and Shipp 1988; De Yoe and Van Essen 1988). From the computational standpoint, this is taken as being equivalent to the computation and analysis of optical flow.

The paper is divided into two parts. In the first part, basic mathematical facts about optical flow and methods for the computation and use of optical flow for motion analysis are discussed. It is shown that the spatial structure of the optical flow produced by rigid, opaque objects is usually simple. Consequently, optical flow, with the exception of surface boundaries, can often be described in terms of its first order spatial properties, such as singular points and elementary components of optical flow. These properties can be used to compute motion parameters, such as time-to-collision and angular velocity, and segment the visual field into the different moving objects. Furthermore, the computation of optical flow is neither difficult nor critical. Due to the simple spatial structure of the apparent motion of rigid (or piecewise rigid) objects, different methods for the computation of optical flow produce similar results which can equivalently be used in later processing. The implications of this analysis on the design of artificial systems and on the understanding of how the brain analyses visual motion are discussed in the second part of the paper. It is firstly observed that both natural and computer vision systems overcome the intrinsic noise of visual motion estimation by using the regularity and density of motion estimates, and integrating visual information over large areas of the visual field. Then, it is argued that the brain does not have to necessarily compute optical flow. Instead, it may immediately extract the first order spatial properties of optical flow from the changing image. Finally, a model for the organisation of the receptive field of cells in area MST, which are likely candidates for the proposed analysis, is described. The model cells respond selectively to the stimuli presented and do so almost independently of the position of the stimuli on the receptive field. In addition, in agreement with the experimental evidence, the model cells do not extract elementary flow components from more complex flows.

The paper is organised as follows. Section 2 reviews the mathematical properties of optical flow. Section 3 is dedicated to the first order spatial properties of optical flow, that is, the singular points and the elementary components. Since the spatial structure of the optical flow is usually very simple, different methods for the recovery of optical flow produce very similar results. This fact is discussed in section 4 where four different methods are briefly analysed. Scope and goals of motion perception are then clarified in section 5. In section 6, the scheme, in the framework of computer vision, is tested on real images. Section 7 is devoted to the analysis of the biological implications of the proposed analysis. A simple model of the organisation of the receptive field of neurons sensitive to elementary flow components or their combinations is proposed in section 8. Finally, section 9 summarises the results obtained and discusses similarities and differences of motion perception in natural and computer visual systems.

2 Mathematical Properties of the Optical Flow of Rigid Objects

In this section it is shown that the spatial structure of the optical flow of rigid objects is often rather simple. Then, the notion of structural stability is briefly commented. The mathematical properties of optical flow described in this section have several implications for the understanding of motion perception in artificial and biological visual systems, implications which will be discussed in the next sections.

2.1 Preliminaries

Let us define the *motion field* $\vec{v}$ as the perspective projection onto the image plane of the *velocity field* $\vec{V}$, the velocity of the moving surfaces in space. Since the apparent motion of the image brightness pattern does not necessarily coincide with the perspective projection of the velocity field, optical flow can only be considered as an estimate of the motion field. Let $\vec{X}$ be a point in space on one of these surfaces and $\vec{x}$ its perspective projection onto the image plane. In what follows if $\vec{y}$ is a vector in space, let $y_i = \vec{y} \cdot \vec{e_i}$, $i = 1, 2, 3$, where the $\vec{e_i}$ are unit vectors parallel to the three mutually orthogonal axes of a system of coordinates. The image plane is orthogonal to $\vec{e_3}$ (that is, the optical axis is parallel to $\vec{e_3}$) and the focus O and the origin of the system of coordinates coincide. Thus, we have

$$\vec{x} = f\frac{\vec{X}}{X_3} \tag{1}$$

where f is the focal length. Note that $x_3 = f$, for every $\vec{x}$, and that $X_3 > f$, for every $\vec{X}$, since otherwise $\vec{X}$ would not be visible. If $\vec{X}$ is moving with velocity $\vec{V}$, then the motion field $\vec{v}$ can also be obtained by differentiating Eq. 1 with respect to time; that is,

$$\vec{v} = f\frac{\vec{e_3} \times (\vec{V} \times \vec{X})}{X_3^2} \tag{2}$$

As v_3 is always zero, we write $\vec{v} = (v_1, v_2)$. Thus, the motion field $\vec{v}$ of a smooth surface, with the exception of the occluding boundaries, is a smooth two-dimensional (2D) vector field. In the presence of more than one moving surface the motion field is only piecewise smooth, and to obtain a globally smooth motion field a filtering step may be required.

2.2 The Spatial Structure of the Motion Field

The motion field $\vec{v} = (v_1, v_2)$ of a moving planar surface at the point $\vec{x} = (x_1, x_2, f)$ of the image plane can always be written as a quadratic polynomial of the image coordinates

$$
\begin{aligned}
v_1(x_1, x_2) &= a_1 x_1^2 + a_2 x_1 x_2 + a_3 x_1 f + a_4 x_2 f + a_5 f^2 \\
v_2(x_1, x_2) &= a_1 x_1 x_2 + a_2 x_2^2 + a_6 x_1 f + a_7 x_2 f + a_8 f^2
\end{aligned} \tag{3}
$$

where the a_i, $i = 1, ..., 8$, depend on the motion and structure parameters (Longuet-Higgins 1984). From the qualitative point of view the spatial structure of the vector field

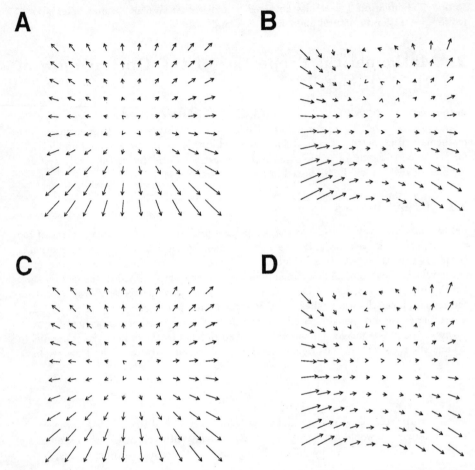

Figure 1: *The motion fields of a translating (A) and rotating (B) plane. The linear approximation of the vector fields in A (C) and B (D).*

of Eq. 3 is fairly simple as there are three singular points at most (Verri *et al.* 1989) and there cannot be limit cycles (Verri and Aicardi 1990). Most importantly, as f, the focal length, is usually much larger than x_1 and x_2, the motion field $\vec{v}$, over relatively large patches of the image plane, and apart from special values of the a_i, $i = 1, ..., 8$, is well approximated by a linear vector field (Campani and Verri 1992). The motion fields of a translating and a rotating planar surface are shown in figure 1A and 1B respectively. In figure 1C and 1D the piecewise linear vector fields which best approximate the vector fields of figure 1A and 1B are reproduced. The vector fields of figure 1C and 1D are obtained by computing the best linear approximation of the vector fields of figure 1A and 1B over 8×8 non-overlapping patches of the image plane. It is evident that the piecewise linear approximations are almost indistinguishable from the actual vector fields.

Theoretical and experimental evidence shows that, with the exception of occluding boundaries, the same argument holds true for smooth non-planar surfaces (Campani and Verri 1992). In this case, the size of the patches of the image plane on which the motion field can be approximated by a linear vector field changes depending on the motion parameters and the apparent curvature of the viewed surface.

2.3   Motion Field, Optical Flow, and Structural Stability

Motion field and optical flow are usually different (Verri and Poggio 1989), but the notion of structural stability provides a natural framework within which they can be considered equivalent (Verri *et al.* 1989). Intuitively, a sufficiently smooth vector field is structurally stable if its qualitative properties (like the number and the topological structure of its singular points and limit cycles) remain unchanged for small (and smooth) perturbations. The emphasis on structural stability is due to a fundamental theorem of the theory of dynamical systems (Hirsh and Smale 1974), which says that, among planar dynamical systems, the property of structural stability is generic. This result suggests that certain qualitative and, to some extent, quantitative properties of motion field and optical flow, are likely to be the same.

3   First Order Properties of the Motion Field

This section analyses the relationship between the linear part, or the first order spatial properties of the motion field, and some relevant features of the viewed motion. From these properties it is possible to obtain a qualitative understanding of the 3D motion, to distinguish between translation, rotation, and general motion, and to recognise different moving objects. Moreover, relevant quantitative parameters, such as time-to-collision and angular velocity, can be robustly estimated.

It may be useful to give an intuitive notion of first order properties before going into the mathematical details. A vector field, which is not constant, has a different orientation and amplitude at each position. An important property of a vector field is the way in which the orientation and amplitude change at each position. When changes of the amplitude and direction of vectors are the same for similar displacements, first order properties of the field are constant all over the field. In this case the essence of the vector field is captured by the rate of change of the amplitude and direction, or in other words by first order properties. When changes of the amplitude and direction vary significantly at each position, it is also necessary to consider higher order properties. The main reason why first order properties of the motion field are usually adequate to describe the motion field is that the motion field of solid opaque objects is usually coherent and its rate of change is almost constant. In the presence of a relative motion between two different objects, the motion field can be discontinuous at some locations and therefore it is also necessary to detect discontinuities.

3.1   Singular Points

The linear part of the motion field in the neighbourhood of a point $\vec{x}$ can be described by the 2×2 matrix

$$\mathbf{M}_{ij} = \frac{\partial v_i}{\partial x_j} \quad i = 1, 2 \tag{4}$$

computed at $\vec{x}$. If $\vec{x}$ is a singular point, that is, a point in which the motion field vanishes, the matrix $\mathbf{M}$ can be useful to distinguish between different kinds of motion and estimate motion parameters. If $\vec{x}$ is not a singular point, the matrix $\mathbf{M}$ can still be used to understand qualitative properties of the viewed motion and segment the image in the different moving surfaces. Let us consider the case in which $\vec{x}$ is a singular point first.

The qualitative nature of 3D motion (like translation or rotation) can be described by looking at the temporal evolution of the spatial structure of the motion field in the neighbourhood of the singular points. Intuitively, translation is often associated with expanding (or contracting) motion fields, and rotation with circulating motion fields around a singular point (which may or may not be visible). Let us study these assumptions in detail.

3.1.1 Translation

If a surface is translating in space with velocity $\vec{T}$, the resulting motion field $\vec{v}$ has at most one singular point $\vec{p} = (p_1, p_2, f)$ (which can be obtained by substituting $\vec{V} = \vec{T}$ in Eq. 2 and setting $\vec{v} = 0$) such that (Verri *et al.* 1989):

1. The spatial location of $\vec{p}$ on the image plane does not change over time;

2. The heading direction is related to $\vec{p}$ as for $i = 1, 2$

$$p_i = f \frac{T_i}{T_3}$$

3. The matrix $\mathbf{M}$ at $\vec{p}$ is a multiple of the identity matrix, that is, $\mathbf{M} = \lambda \mathbf{I}$. In the language of the theory of dynamical systems, $\vec{p}$ is always a non-degenerate focus (Hirsh and Smale 1974) with the only eigenvalue λ of $\mathbf{M}$ given by

$$\lambda = \frac{T_3}{X_3} = \frac{1}{\tau} \tag{5}$$

where τ is the time-to-collision.

From these properties it is evident that useful information on translational motion can easily be obtained from the singular point of the motion field.

3.1.2 Rotation

In the case of a rotating surface it is useful to distinguish between two different kinds of singular points. Let $\vec{p}$ be the singular point perspective projection of a point $\vec{P}$ in the 3D space and $\vec{V}(\vec{P})$ the velocity of $\vec{P}$. A point $\vec{p}$ can be a singular point because either $\vec{P}$ lies on the rotation axis (that is, $\vec{v}(\vec{p}) = 0$ and $\vec{V}(\vec{P}) = 0$), or $\vec{V}(\vec{P})$ lies on the straight line which goes through $\vec{P}$ and the center of projection (that is, $\vec{v}(\vec{p}) = 0$ but $\vec{V}(\vec{P}) \neq 0$). A singular point of the first kind is named *immobile* point.

Let us briefly discuss the case in which $\vec{p}$ is an immobile point (a more general analysis of the singular points of rotation can be found in (Verri *et al.* 1989)). If the rotation axis is orthogonal to the plane tangential to the surface at $\vec{P}$, $\vec{p}$ is always a *center* and the motion field in the neighbourhood of $\vec{p}$ is tangential to closed orbits. In this case, that can be named *orthogonal rotation*, the angular velocity ω can be written in terms of the complex conjugate eigenvalues λ and $\bar{\lambda}$ of the matrix $\mathbf{M}$ computed at $\vec{p}$ as

$$\omega^2 = \lambda \bar{\lambda}$$

In all the other cases, the spatial structure of the motion field in the neighbourhood of $\vec{p}$ changes over time depending on the angle θ between the rotation axis and the unit normal to the surface at $\vec{p}$ and on the relative position of the rotating surface with respect to the viewing point.

3.1.3 General Motion

The case of general motion is more complicated. Whilst it is well known that any rigid motion can be instantaneously decomposed into a translational and a rotational term, it is not evident whether this classical result of kinematics is relevant to motion perception. Let us restrict the present analysis of general motion to the simple but interesting case in which the viewed object is moving on a flat surface S. In this case, which can be termed *passive navigation*, the rotation axis is orthogonal to both the surface S and the translational component of motion and useful information can again be obtained from the singular points of the motion field.

Firstly, the 3D motion is instantaneously indistinguishable from a rotation around a fixed axis. Therefore, the singular points of the motion field must lie on the straight line of the image plane which is parallel to S and contains the projection of the optical axis. At a singular point $\vec{p}$ we can write

$$DetM = \frac{Tr^2M}{4} - \frac{1}{4}\Big(\frac{\vec{\alpha} \times \vec{V} \cdot \vec{\omega}}{\vec{\alpha} \cdot \vec{V}}\Big)^2 \qquad (6)$$

where $DetM$ and TrM are the determinant and the trace of the matrix M respectively, $\vec{\alpha}$ is the unit vector normal to the moving surface at $\vec{P}$, $\vec{\omega}$ is the angular velocity, and $\vec{V} = \vec{V}(\vec{P})$. From Eq. 6 it follows that the eigenvalues of the matrix M are always real and thus that the singular points cannot be *spirals*.

Concluding, the understanding of motions such as rotations and constrained general motion requires a rather deep knowledge of mathematics which is probably beyond the relatively simple stages of *motion perception*. It may thus be argued that at these preliminary stages biological visual systems are mainly tuned for the purpose of recognising translational motion and orthogonal rotation. More general kinds of motion probably require more sophisticated analysis and processing.

3.2 Elementary Components

Let us now consider the case in which the matrix M of Eq. 4 is computed at $\vec{x}$, where $\vec{x}$ is not a singular point. According to a classical theorem of the theory of deformable bodies (Helmholtz 1858; Koenderink and van Doorn 1975) the spatial structure of a vector field over a sufficiently small patch can be described as the sum of a rigid translation, a uniform expansion, a pure rotation, and two components of shear. Therefore, the motion field $\vec{v}'$, at a point $\vec{x}'$ in a sufficiently small neighbourhood of $\vec{x}$, can be written as

$$\vec{v}' = \vec{v} + M(\vec{x}' - \vec{x}) \qquad (7)$$

where $\vec{v}$ is the motion field at $\vec{x}$ and the matrix M is meant to be computed at $\vec{x}$. It is clear that in Eq. 7 the rigid translation is given by the term $\vec{v}$. Since $\vec{x}$ is not a singular point, $\vec{v} \neq 0$. The matrix M can be written as

$$M = \alpha I_1 + \beta I_2 + \gamma_1 I_3 + \gamma_2 I_4$$

where

$$\mathbf{I_1} = \begin{pmatrix} 1 & 0 \\ 0 & 1 \end{pmatrix}; \quad \mathbf{I_2} = \begin{pmatrix} 0 & 1 \\ -1 & 0 \end{pmatrix}; \quad \mathbf{I_3} = \begin{pmatrix} 1 & 0 \\ 0 & -1 \end{pmatrix}; \quad \mathbf{I_4} = \begin{pmatrix} 0 & 1 \\ 1 & 0 \end{pmatrix}.$$

and $\alpha = (M_{11} + M_{22})/2$, $\beta = (M_{12} - M_{21})/2$, $\gamma_1 = (M_{11} - M_{22})/2$, and $\gamma_2 = (M_{12} + M_{21})/2$. The matrices $\mathbf{I}_i$, $i = 1, ..., 4$ correspond to the terms of uniform expansion, pure rotation, and components of shear respectively. Since $\mathbf{I}_1$ and $\mathbf{I}_2$ are left unchanged for arbitrary orthogonal transformation, the quantities α and β are independent of the system of coordinates. Consequently, the amount of uniform expansion and pure rotation are intrinsic properties of the motion field, that is they are independent of the choice of the coordinate system (Koenderink 1986). On the contrary, since $\mathbf{I}_3$ and $\mathbf{I}_4$ are not invariant, γ_1 and γ_2 do not describe intrinsic properties of the motion field. An invariant measure of the amount of shear can be given by $\gamma = \sqrt{\gamma_1^2 + \gamma_2^2}$ (since $\alpha^2 + \beta^2 + \gamma_1^2 + \gamma_2^2$ is invariant).

The relevance of elementary components to the analysis of visual motion stems from to the fact that the motion field is usually well approximated by a linear vector field over rather large patches of the image plane (see section 2). In other words, the amount of expansion, rotation, and shear is a nearly piecewise constant function over the image plane. This makes it possible to segment the motion field into regions where the relative amount of expansion, rotation, or shear is larger than a fixed value. These regions may correspond to different moving objects, and can be used to qualitatively describe the observed motion and identify motion discontinuities (see section 6).

4   The Computation of Optical Flow

Let us briefly review four different methods for the computation of optical flow in computer vision. These methods produce optical flows with very similar qualitative features and which differ for minor quantitative aspects. This is a consequence of the simple spatial structure of the motion field discussed in section 2. (See also the chapter of Nagel for an extended review on the computation of optical flow)

4.1   The Spatial Gradient Constancy Method

Let us assume that the spatial gradient of the image brightness, $\vec{\nabla}E$, is stationary over time, *i.e.*,

$$\frac{d}{dt}\vec{\nabla}E = 0 \tag{8}$$

Eq. 8 can be rewritten as a pair of linear algebraic equations for the optical flow components (u_1, u_2), or

$$E_{xx}u_1 + E_{xy}u_2 + E_{xt} = 0$$
$$E_{xy}u_1 + E_{yy}u_2 + E_{yt} = 0$$

where the subscripts x, y, and t denote partial derivatives with respect to the spatial and temporal coordinates respectively. The solution to Eq. 8 can be written as

$$\vec{u} = -\mathbf{H}^{-1}\frac{\partial}{\partial t}\vec{\nabla}E$$

where

$$\mathbf{H} = \begin{pmatrix} E_{xx} & E_{xy} \\ E_{yx} & E_{yy} \end{pmatrix}$$

is the Hessian matrix. Eq. 8, which can be derived through an analogy with the theory of deformable objects (Uras *et al.* 1988; Verri *et al.* 1990), has been proposed by a number of authors. The optical flow $\vec{u}$ is uniquely determined when $\mathbf{H}$ is invertible and can be written in terms of the true motion field $\vec{v}$ as

$$\vec{u} = \vec{v} + \mathbf{H}^{-1} \left(\mathbf{M}^T \vec{\nabla} E - \vec{v} \frac{dE}{dt} \right) \tag{9}$$

where $\mathbf{M}^T$ is the transpose of $\mathbf{M}$ and $dE/dt = \vec{\nabla} E \cdot \vec{v} + E_t$. From Eq. 9 the true motion field and the optical flow of Eq. 8 coincide if the image brightness pattern is stationary ($dE/dt = 0$) and the motion field is spatially constant ($\mathbf{M} = 0$). Strictly speaking, both of the conditions above do not usually hold, but it can been shown (Verri et al. 1989) that provided dE/dt and the entries of $\mathbf{M}$ are small, then the larger the eigenvalues of the Hessian matrix $\mathbf{H}$, and the smaller the relative difference between $\vec{u}$ and $\vec{v}$.

Figure 2 shows four frames of a sequence in which the viewing camera is translating toward a picture posted on the wall. Figure 3A shows the optical flow obtained from Eq. 8 associated with the third frame of figure 2. It is evident that the computed optical flow is qualitatively consistent with the observed motion.

4.2 The Image Brightness Constancy Method

Many approaches to the computation of optical flow are based on the view that the image brightness E is stationary over time (Horn and Schunck 1981),

$$\frac{d}{dt} E = 0 \tag{10}$$

Eq. 10 neither implies, nor is implied by, Eq. 8 and can be rewritten as

$$\vec{\nabla} E \cdot \vec{u} + E_t = 0$$

from which it is clear that the vector field $\vec{u}$ cannot be determined uniquely from Eq. 10. In the search for additional constraint the most common approach (Horn and Schunck 1981) looks for the vector field $\vec{u}$ which minimises the functional

$$\Phi(\vec{u}) = \int \int \left(\frac{dE}{dt} \right)^2 dx dy + \alpha \int \int \left(|\vec{\nabla} u_1|^2 + |\vec{\nabla} u_2|^2 \right) dx dy$$

where α is a positive parameter. Intuitively, the vector field which minimises Φ is the smoothest vector field, which nearly satisfies Eq. 10. Quite interestingly, the conditions for which Φ is minimum are similar to those which ensure that the optical flow computed as the solution to Eq. 8 is close to the true motion field (as $|dE/dt|$ and the $|\partial u_i/\partial x_j|$ must be small). This suggests that the two different schemes provide similar results due to the relatively simple spatial structure of the motion field of rigid surfaces. Figure 3B shows the optical flow obtained through the technique described in Horn and Schunck (1981) and associated with the third frame of the sequence of figure 2. It is evident that the vector field of figure 3B is qualitatively correct and quantitatively similar to the optical flow of figure 3A.

Figure 2: *Four frames of an image sequence taken while the viewing camera is translating toward a picture posted on the wall.*

4.3   A Matching Method

A third approach to the computation of optical flow is based on the matching of features from two images. A possible method (Poggio *et al.* 1986) assumes that optical flow is locally constant, that is, for each point, the displacement of nearby points under the optical flow is the same. At each point in the image, under each integer displacement, the images in the two frames are compared and a measure of the matching between points is computed, and summed over a small region. This can be interpreted as matching small patches from the first image with small patches in the second. The resulting flow field is spatially coherent as a result of the fact that the support regions, the patches, have large overlap. The displacement is chosen to maximise the matching measure over all displacements. Accuracy can be improved by interpolating the matching measure. Once again the conditions under which this method produces an optical flow similar to the motion field are those where the image brightness is nearly stationary ($dE/dt \simeq 0$) and where spatial variation of the motion field is locally negligible ($\partial u_i/\partial x_j \simeq 0$). The optical

A

B

C

D

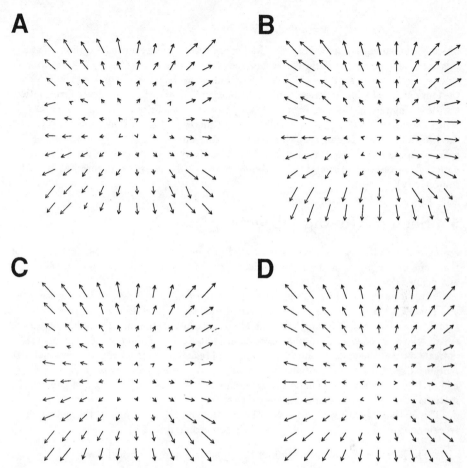

Figure 3: *The optical flows associated with the third frame of the sequence of figure 2 computed through the spatial gradient constancy method (A), the image brightness constancy method (B), the matching method (C), and the shape constraint method (D).*

flow of figure 3C has been obtained by means of this method. A computational advantage of matching approaches is that they perform equally well even if the dynamic range and the intensity distribution of the image brightness pattern are far from being optimal.

4.4 Shape Constraint Method

In the attempt to combine the simplicity of both the constraint (Eq. 10) and the spatial structure of the motion field of rigid surfaces, a method has been proposed in which optical flow is estimated as the solution to a highly overconstrained system of linear algebraic equations (Campani and Verri 1990). The image plane is divided into overlapping regions in which it is assumed that the image brightness is stationary and that the motion field is at most linear. Let $\vec{x}_0$ be the central point of a region r which contains N points. Then at each point $\vec{x}$ of r we have

$$\vec{\nabla} E \cdot \vec{v}(\vec{x}_0) + \vec{\nabla} E \cdot \mathbf{M}(\vec{x} - \vec{x}_0) = -E_t \qquad (11)$$

which is a linear equation for $\vec{v}(\vec{x}_0)$ and $\mathbf{M}$ (whose entries are meant to be computed at $\vec{x}_0$). It is possible to employ standard least mean square techniques to solve the overconstrained system of N equations, such as Eq. 11, for six unknowns (the two components of $\vec{v}(\vec{x}_0)$ and the four entries of $\mathbf{M}$). As a result, both the motion field and the elementary deformations can be estimated at the same time. Again it should be noticed that the conditions for which the method is likely to provide good estimates of the motion field coincide with the previous cases. It can easily be seen that the optical flow of figure 3D, which has been computed through this method, is qualitatively correct and very similar to the optical flows obtained through the techniques previously described (figure 3A, B, and C respectively).

5 Seeing Moving Objects

Seeing moving objects is a very sophisticated property of vision systems. In this section the scope of the paper in the investigation of the processing of visual motion is clarified and defined.

Firstly, the present research is restricted to the analysis of rigid (or piecewise rigid) and opaque objects. In the brain, the visual pathway we are looking at goes from the visual areas V1 (and V2), to MT and MST. In computational terms, this processing stage, which we may call *motion perception*, includes the computation and analysis of optical flow and the understanding of a number of relatively simple three-dimensional (3D) motions, like translational motions (which may also cause expanding or contracting flows), and orthogonal rotations. Therefore, deformable or transparent objects and complex 3D motions are beyond the scope of this research.

The mathematical results presented in the previous sections suggest that an adequate representation of motion perception is likely to be constituted by:

1. The location and identification of the singular points of optical flow;

2. The computation of the linear terms of optical flow at every location of the image plane. In other words the estimation of the average component of translation, rotation, expansion, and shear;

3. The detection of optical flow discontinuities.

In section 3 it has been shown that the information obtained in the first and second point makes it possible to distinguish between translation and rotation, and to evaluate 3D motion parameters like the time-to-collision and angular velocity. The information obtained in the third point, instead, can be used to understand the complexity of the viewed scene and identify the different moving objects (Francois and Bouthemy 1990). Let us now test the proposed scheme in computer vision. The biological implications are considered in section 7.

6 Motion Perception in Machines

In this section an experiment is reported in which the proposed scheme is tested on a sequence of natural images. The experiment indicates that optical flow can be computed

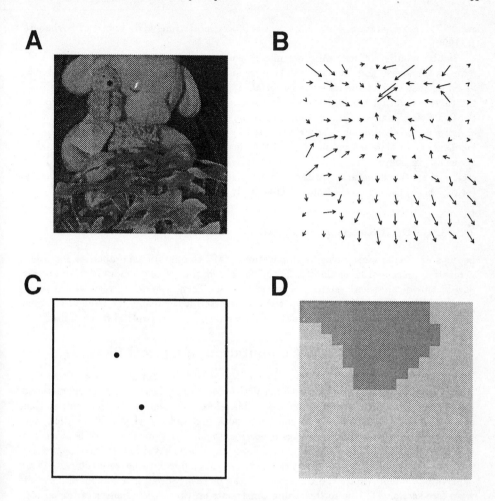

Figure 4: *A) A frame of a sequence in which the puppets are moving away from the camera, while the camera was translating toward the plant. B) The optical flow associated with the frame of A) and computed through the spatial gradient constancy method. C) The two singular points which have been detected in the optical flow of B). The motion segmentation which has been obtained through a technique described in (Orban et al. 1992). The darker area indicates a region of contraction and corresponds to the puppets motion, the lighter area a region of expansion and correspond to the rest of the viewed scene.*

from real images and that the proposed representation of motion perception can be effectively obtained.

Figure 4A illustrates a frame of a sequence of images in which the viewing camera was moving toward the scene (constituted by a plant in the foreground and two puppets in the background) while the puppets were moving away. Figure 4B shows the optical flow computed according to the technique described in subsection 4.1. The singular points are shown in figure 4C, while the optical flow segmentation is shown in figure 4D. The two homogeneous regions correspond to the apparent motion of the viewing camera (expansion, grey area), and to the motion of the puppets (contraction, darker area). The segmentation was obtained by a relaxation technique described in (Orban *et al.* 1992).

Similar results obtained on other image sequences indicate that the relative motion between the viewer and the scene can be recognised from the segmentation of optical flow based on first order properties (Orban *et al.* 1992). Quantitative motion estimates, like the time-to-collision between the viewing system and the point in the scene imaged in the focus of expansion, can be computed through equations like Eq. 5. In the example of figure 4, the eigenvalue λ can be estimated by integrating over quite a large number of independent motion measurements, and the obtained time-to-collision is thus reasonably accurate. In some cases, such as passive navigation, it may be possible to extend the region of spatial integration to the entire image, and a rather high degree of accuracy (nearly 99 %) can be reached almost effortlessly.

7   Biological Implications

Let us now discuss some biological implications of the computational properties of motion perception presented in sections 2, 3, and 4. The nature of the computation of optical flow in biological vision systems is firstly analysed. Then, a basic strategy which seems to be used by the brain in motion perception is described. Finally, the issue of whether or not, the brain actually needs and, therefore, really computes optical flow is raised.

7.1   The Nature of the Computation of Optical Flow

The analysis of the mathematical properties of the motion field of rigid objects (see section 2 and 3) has shown that the first order spatial properties of the motion field characterise the entire vector field almost completely. Therefore, the problem of the computation of image motion does not require going beyond a first order analysis. In addition, the property of structural stability makes it possible to tolerate some difference between the true motion field and the estimated optical flow. These facts have important consequences.

Firstly, the computation of optical flow does not appear to be very difficult. The "aperture problem" (Marr and Ullman 1981) can be easily solved in a number of different ways (see section 4). Due to the strong constraints on the spatial changes of the optical flow of rigid bodies, many of the proposed methods for the computation of optical flow do not require the use of sophisticated techniques such as those provided by regularisation theory (Tichonov and Arsenin 1977; Bertero *et al.* 1989). Therefore, it is difficult to find a single solution which the brain must adopt. In addition, a very accurate estimation of the true motion field is not necessary. Due to the property of structural stability, the 3D motion parameters can be recovered robustly from the singular points of an optical flow which is not exactly equal to the true motion field.

7.2   Spatial Integration

A striking feature, which indicates a possibly basic strategy adopted by the brain in motion perception, is the progressive integration of visual information over larger portion of the visual field in the visual pathway from V1 to MT and MST (Livingstone and Hubel 1988; Zeki and Shipp 1988; De Yoe and Van Essen 1988). The receptive field of neurons in V1 is usually not larger than 1° or 2°, while the average receptive field is about 15° in MT and 50° in MST (Maunsell and Van Essen 1983; Saito *et al.* 1986; Tanaka *et al.* 1986; Ungerleider and Desimone 1986; Boussaud *et al.* 1990). In the previous section it has been shown that the role of spatial integration in machine vision is to improve accuracy in the estimation (and understanding) of the viewed motion. A similar strategy may be also present in the brain. Because of the different circuitry, however, the actual extent of spatial integration in artificial and natural visual systems could be different.

7.3 Does the Brain Compute Optical Flow?

According to the computational model described in the previous sections, motion perception consists of two steps. In the first step optical flow is computed, while in the second step its first order spatial properties, like singular points and elementary flow components, are extracted and used to accomplish a number of relatively simple visual tasks. Although there is little doubt that the brain uses some kind of optical flow information when looking at moving images (Movshon *et al.* 1985), it is not obvious whether the brain computes and represents optical flow in a specific cortical area, or not.

Neurons in the visual area V1, for example, are orientation and directional selective and have been described as having the "component directional selectivity" (Movshon *et al.* 1985). Many neurons in the visual area MT, instead, have being reported as "pattern direction selective". Therefore, MT seems to be the area in which the aperture problem is solved. However, the receptive fields of MT and MST neurons are very large and the representation of optical flow information in MT and MST must be fundamentally different from the mathematical notion of planar vector field. The analysis of sections 2 and 3 suggests that the first order spatial properties of optical flow, instead of the optical flow itself, are likely candidates for the representation of motion information at the level of the areas MT and MST. This is because the first order spatial properties of the motion field of rigid objects are almost piecewise constant over rather large regions of the field of view and can thus be conveniently described by cells with large receptive fields (see section 2 and figure 1).

8   The Analysis of Optical Flow in the Brain

Recent electrophysiological experiments (Tanaka and Saito 1989; Andersen *et al.* 1990; Andersen *et al.* 1991; Lagae *et al.* 1991; Duffy and Wurtz 1991a, 1991b; Lagae 1991) have shown that units in area MST are tuned to expanding and rotating stimuli. In this section the main properties of these units are summarised. Thus, a simple model of MST cells, which is consistent with anatomy and electrophysiology, is discussed in some detail.

8.1   Properties of MST Neurons

Many neurons of area MST, which seems to be devoted to motion analysis, are very good candidates for the processing of optical flow information in the brain. In what follows we list the main properties of these neurons.

1. *Receptive field size.* The size of the receptive field of MST units is rather large, ranging from 30° to as much as 100° (Tanaka and Saito 1989), and does not increase with eccentricity (Tanaka and Saito 1989).

2. *Selectivity.* MST units appear to be selectively tuned to rotation, expansion and combinations of these stimuli (Andersen *et al.* 1990). Furthermore, many of these neurons are also sensitive to translation in a given direction, indicating a *continuum* of selectivity (Duffy and Wurtz 1991a, 1991b).

3. *Position invariance.* The response and selectivity of MST units to a given stimulus does not change appreciably when the stimulus is displaced with respect to the center of the receptive field (Lagae *et al.* 1991; Duffy and Wurtz 1991b; Lagae 1991). The response amplitude of a unit tuned to expansion, for example, decreases by about 30÷50% for a relative displacement of 30°.

4. *Non-linearity.* MST cells do not extract an optical flow component from a complex stimulus (Lagae 1991; Orban *et al.* 1992). A unit tuned to rotation, for example, decreases its response when an expansion of increasing strength is added to a rotating stimulus.

It is evident that these units appear to detect first order properties of the optical flow, almost in the mathematical sense of section 3. Depending on the observed degree of selectivity a neuron can be thought of as detecting either a singular point (e.g. a focus of expansion) or an elementary flow component (e.g. expansion). Interestingly, it seems that the brain analyses optical flow in a way which does not use the decomposition of a complex flow into the linear superposition of simple components. This is in agreement with the existence of MST units which are selectively tuned to spirals (Andersen *et al.* 1990). Let us now discuss a simple model of the organisation of the receptive field of cells in MST which takes into account the properties above.

8.2   Modelling MST Neurons

Let us illustrate a simple model of MST cells. It is assumed that the input to these cells is constituted by units with properties similar to those of cells in area MT. These units, or MT-like units, have thus a rectangular receptive field of about $10° \times 20°$ and are directionally selective (Maunsell and Van Essen 1983).

If $\phi(\vec{x})$ is the optical flow direction at the location $\vec{x}$ on the receptive field of the unit u, the response R_u is assumed to be

$$R_u = \sum_{\text{subunits}} f(\phi - \phi_p) \tag{12}$$

where the summation extends to all the subunits of u, ϕ_p is the preferred direction, and $f(\cdot)$ is a tuning function. Experimental recordings suggest a tuning function of the type

$$f(y) = \exp(y^2/\sigma^2) \tag{13}$$

where σ is about 30°. It is also assumed that each unit u is composed of 10×20 equally spaced subunits (probably V1 cells). Finally, the cell response, R_{cell}, is simply

$$R_{cell} = \Theta(\sum_{\text{units}} R_u) \tag{14}$$

with $\Theta(\cdot)$ a suitable threshold function. In the computer experiments the function Θ was a logistic function.

The behaviour of the model cell depends primarily on two factors:

1. the spatial organisation of the units within the receptive field of the cell;

2. the "shape" of the tuning function.

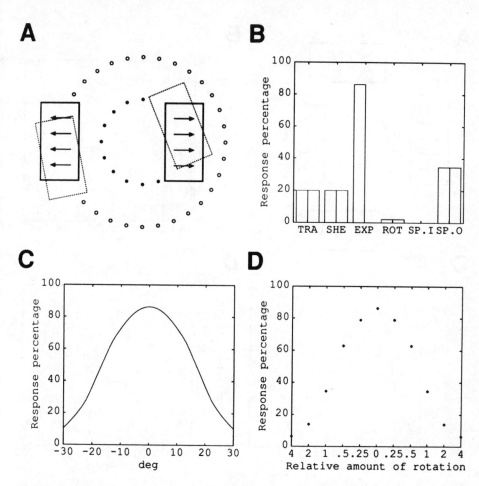

Figure 5: *A) A model of the organisation of the receptive field of an MST neuron tuned to expansion. The neuron is receiving inputs from 16 MT-like units located on an inner ring (10° of radius) and 32 MT-like units on an outer ring (20° of radius). The arrows indicate the unit (and subunit) preferred direction. B) Normalised responses to pure translation (TRA), shear (SHE), expansion (EXP), rotation (ROT), contraction and rotation (SP.I), and expansion and rotation (SP.O), in the forward and reversed direction respectively. C) Normalised response to expansion as a function of the displacement between the receptive field center and the focus of expansion. D) Normalised responses to a stimulus composed by expansion and clockwise rotation ("−" sign) or anticlockwise rotation ("+" sign).*

The type of selectivity of the cell is essentially due to the first factor, that is, the arrangement of the MT-like units within the receptive field, whereas the extent to which the properties of selectivity, position invariance, and non-linearity hold depends primarily on the second factor. A tuning function of the type of Eq. 13, in agreement with experimental evidence, produces a response which is moderately position invariant and shows a clear non-linear behaviour (Orban *et al.* 1992). A tuning function with negative side lobes (ideally a "cosine" function), however, produces an almost perfectly position invariant response and a linear behaviour (Poggio et al. 1991).

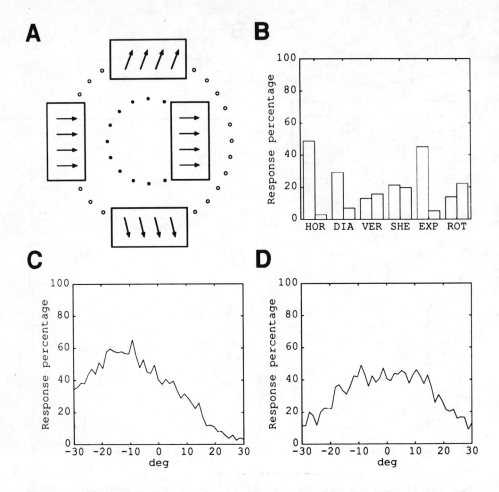

Figure 6: *A) A model of the organisation of the receptive field of an MST neuron tuned to expansion and horizontal translation. B), C), and D) as in figure 5.*

Figure 5A illustrates the organisation of the MT-like units in the receptive field of a model cell tuned to expansion. The MT-like units are located in two concentric rings (16 units in the inner and 32 in the outer ring) and the radius of the two rings is 10° and 20° respectively. The preferred direction of each unit is shown by the solid arrows. Figure 5B shows the stimulus selectivity for $\sigma = 30°$. The responses are given in percentage of maximum response, normalised to unity, and both the preferred and reversed direction of motion are shown (i.e., preferred/null direction of translation, expansion/contraction; clockwise/anticlockwise rotation, etc.). Figure 5C illustrates the property of position invariance when the focus of expansion is displaced over the receptive field. Finally, figure 5D reproduces the change of the response when a rotation of increasing intensity is added to a fixed amount of expansion. The parameters of the threshold function were chosen to closely match the behaviour of MST cells, as described in (Orban *et al.* 1992), particularly with respect to position invariance and non-linearity. Unlike a previous model (Tanaka *et*

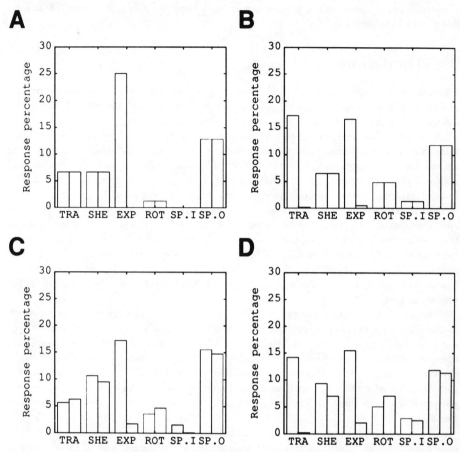

Figure 7: *The stimulus selectivity of the cell of figure 5 to a stimulus of 30° of radius centered on the receptive field (A) and displaced by 30° degree (B). C) and D) The stimulus selectivity of the cell of figure 6 as in A) and B) respectively.*

al. 1989), the property of position invariance is obtained without the repetition of similar arrangements of neurons.

The regular arrangement of units within the receptive field of figure 5A is not critical for the model. The addition of random noise in the unit preferred direction does not affect any of the properties of selectivity, position invariance, and non-linearity.

A cell tuned to expansion and translation can be modelled by adding a bias along the preferred direction of translation. A model cell with an arrangement of units as in figure 6A, for example, is tuned to expansion and horizontal translation (figure 6B) and still has a remarkable degree of position invariance (figure 6C) and non-linear behaviour (figure 6D).

The selectivity and position invariance illustrated in figures 5 and 6 were obtained with large stimuli completely covering the receptive field of the cell. If the cells of figures 5 and 6 are tested with a relatively small stimulus, the cell selectivities do not change significantly when moving the stimuli across the receptive fields (see figures 7A and B, and C and D respectively).

9 Discussion

The aim of this paper was to analyse some computational aspects of motion perception in biological and computer vision systems. Let us summarise the results obtained and discuss their relevance to the understanding of visual motion.

A mathematical analysis has shown that the spatial structure of optical flow is usually rather simple since, in most cases, a piecewise linear vector field is sufficient to approximate optical flow over large areas of the visual field. This observation has a few computational and biological implications. Firstly, the estimation of optical flow from the time-varying image brightness is not a difficult problem and many different (and simple) methods for the computation of optical flow produce similar results. Secondly, since the first order spatial properties of optical flow are almost piecewise constant, a coarse representation of these properties is usually adequate. Therefore, the increasingly large size of the receptive fields of the cells of the visual areas MT and MST, which seem dedicated to motion analysis, suggests the hypothesis that first order properties of optical flow (like elementary flow components), and not the optical flow itself, are computed and represented in the processing of visual motion.

The segment of motion perception which is discussed in this paper can be described as the perception of coherent motion, which can best be analysed with units with a large receptive field. The perception of the motion of a small spot of light, which avoids the aperture problem, can be obtained with units with small receptive fields such as those in the visual area V1. As a consequence, it is likely that motion perception may require the combined use of units with small receptive fields (as those in V1) and units with large receptive fields (as those in V5) as suggested by the mutual connections between these two areas.

The characteristic properties of stimulus selectivity and position invariance of MST cells have been explained by means of a very simple arrangement of MT-like units in the receptive field of these cells. According to the model, the non-linear behaviour of MST cells, which are tuned to specific elementary flow components but do not extract these components from complex flows, originates primarily from the shape of the tuning function (which controls the response in the preferred and null direction of the MT-like units). Consequently, it is possible to explain the behaviour of MST cells in terms of the direction mosaic hypothesis (Tanaka *et al.* 1989), which is in agreement with the properties of MT cells, the main input to MST cells.

Concluding, it is evident that, through the analysis of first order properties of optical flow, natural and artificial vision systems seem to rely on very similar tools for the processing of visual motion. This is hardly surprising since the physical properties of image formation and the data available to both of the systems are almost identical. An interesting difference, however, can be found when looking more closely at the analysis of the elementary components of optical flow. Artificial vision systems can be designed to exploit linear algebra and the Helmholtz theorem (an arbitrary motion can be obtained by means of a suitable linear combination of a translation with four elementary components). In the brain, instead, MST cells do not behave as linear detectors, but as non-linear detectors of the AND type. This computationally less efficient strategy adopted by the brain can probably be explained in terms of the evolutionary need to disambiguate quickly between different kinds of stimuli in order to detect and avoid danger.

Acknowledgements

We would like to thank H. Hildreth, for very useful comments on the manuscript. This work was partially supported by grants from the EEC (ESPRIT II VOILA), ESPRIT B.R.A. Insight Project 3001, EEC BRAIN Project No. 88300446/JU1, Progetto Finaliz-zato Trasporti PROMETHEUS and Progetto Finalizzato Robotica. Clive Prestt checked the English.

References

Andersen, R.A., Snowden, R.J., Treue, S., Graziano, M. (1990) Hierarchical processing of motion in the visual cortex of monkey. In: The Brain, Cold Spring Harbor Laboratory Press

Andersen, R.A., Graziano. M., Snowden, R.J. (1991) Selectivity of area MST neurons for expansion/contraction and rotation motions. Inv. Ophthalmol. Vis. Sc. 32, 823

Barlow, H.B., Levick, R.W. (1965) The mechanism of directional selectivity in the rabbit's retina. J. Physiol. 173, 477-504

Bertero, M., Poggio, T., Torre, V. (1989) Ill-posed problems in early vision. Proc. IEEE 76, 869-889

Boussaud, D., Ungerleider, L.G., Desimone, R. (1990) Pathways for Motion Analysis: Cortical Connections of the Medical Superior Temporal and Fundus of the Superior Temporal Visual Areas in the Macaque Monkey. J. of Comp. Neurology 296, 462-495

Campani, M., Verri, A. (1990) Computing optical flow from an overconstrained system of linear algebraic equations. Proceedings of the Third International Conference on Computer Vision, Osaka, Japan

Campani, M., Verri, A. (1992) Motion analysis from first order properties of optical flow. CVGIP: Image Understanding (in press)

De Yoe, E.A., Van Essen, D.C. (1988) Concurrent processing streams in monkey visual cortex. TINS 11, 219-226

Duffy, C.J., Wurtz, R.H. (1991a) Sensitivity of MST neurons to optic flow stimuli I: A continuum of response selectivity to large field stimuli. J. Neurophysiol. 65, 1329-1345

Duffy, C.J., Wurtz, R.H. (1991b) Sensitivity of MST neurons to optic flow stimuli II: Mechanisms of response selectivity revealed by small field stimuli. J. Neurophysiol. 65, 1346-1359

Francois, E., Bouthemy, P. (1990) The derivation of qualitative information in motion analysis. Image and Vision Computing Journal 8, 279-287

Gibson, J.J. (1950) The perception of the visual world. Houghton Mifflin, Boston

Hassenstein, B., Reichardt, W. (1956) Systemtheoretische Analyse der Zeit-, Reihenfolgen und Vorzeichenauswertung bei der Bewegungs-perzeption des Russelkäfers. Chlorophanus. Z. Naturforsch. IIb, 513-524

Helmholtz, H. (1858) Uber Integrale der hydrodynamischen Gleichungen welche den Wirbelwe-gungen entsprechen. Crelles J. 55, 25

Hirsh, M.W., Smale, S. (1974) Differential equations, dynamical systems, and linear algebra. (Academic Press, New York).

Horn, B.K.P., Schunck, B.G. (1981) Determining optical flow. Artif. Intell. 17, 185-203

Koenderink, J.J., van Doorn, A.J. (1975) Invariant properties of the motion parallax field due to the movement of rigid bodies relative to an observer. Optica Acta 22, 773-791

Lagae, L. (1991) A neurophysiological study of optic flow analysis in the monkey brain. Ph. D. thesis Leuven

Lagae, L., Xiao, D., Raiguel, S., Maes, H., Orban, G.A. (1991) Invest. Ophthalmol. Vis. Sci. 32, 823

Livingstone, M., Hubel, D. (1988) Segregation of.Form, Color, Movement, and Depth. Anatomy, Physiology and Perception. Science 240, 740-749

Longuet-Higgins, H.C. (1984) The visual ambiguity of a moving plane. Proc. Roy. Soc. London B/223, 165-175

Marr, D., Ullman, S. (1981) Directional selectivity and its use in early vision processing. Proc. R. Soc. London B/211, 151-180

Maunsell, J.H.R., Van Essen, D.C. (1983) Functional properties of neurons in middle temporal visual area of the macaque monkey I: selectivity for stimulus direction, speed, and orientation. J. Neurophysiol. 49, 1127-1147

Movshon, J.A., Adelson, E.H., Gizzi, M.S., Newsome, W.T. (1985) The analysis of moving visual patterns. Pontificia Academiae Scientianum Scipta Varia 54, 117-151

Nagel, H.H. (1983) Displacement vectors derived from 2nd order intensity variations in image sequences. Comput. Vision Graph. Image Process. 21, 85-117

Orban, G.A., Lagae, L., Verri, A., Raiguel, S., Xiao, D., Maes, H., Torre, V. (1992) First order analysis of optical flow in monkey brain. PNAS 89, 2595-2599

Poggio, T., Little, J., Gamble, E. (1986) Parallel optical flow. Nature 301, 375-378

Poggio, T., Verri, A., Torre, V. (1991) Green theorems and qualitative properties of the optical flow. MIT AI Lab. Memo 1289

Rognone, A., Campani, M., Verri, A. (1992) Identifying multiple motions from optical flow. Proc. 2nd European Conference on Computer Vision, S. Margherita (Italy)

Saito, H., Yukio, M., Tanaka, K., Hikosaka, K., Fukada, Y., Iwai, E. (1986) Integration of direction signals of image motion in superior temporal sulcus of the macaque monkey. J. Neurosci. 6, 145-157

Tanaka, K., Saito, H.A. (1989) Analysis of motion of the visual field by direction, expansion/contraction, and rotation cells illustrated in the dorsal part of the Medial Superior Temporal Area of the Macaque Monkey. J. Neurophysiol. 62, 626-641

Tanaka, K., Hikosaka, K., Saito, H., Yukie, M., Fukada, Y., Iwai, E. (1986) Analysis of local and wide-field movements in the superior temporal visual areas of the macaque monkey. J. Neuroscience 6, 134-144

Tanaka, K., Fukada, Y., Saito, H. (1989) Underlying mechanisms of the response specificity of expansion/contraction and rotation cells in the dorsal part of the medial superior temporal area of the macaque monkey. J. Neurophysiol. 62, 642-656

Torre, V., Poggio, T. (1978) A synaptic mechanism possibly underlying directional selectivity to motion. Proc. R. Soc. London B/202, 409-416

Tichonov, A.N., Arsenin, V. (1977) Solution of ill-posed problems. (John Wiley, New York)

Ungerleider, L. G., Desimone, R. (1986) Cortical connections of visual area MT in the macaque. J. Comp. Neurology 248, 190-222

Uras, S., Girosi, F., Verri, A., Torre, V. (1988) A computational approach to motion perception. Biol. Cybern. 60, 79-87

Verri, A., Aicardi, F. (1990) Limit cycles of the two-dimensional motion field. Biol. Cybern. 64, 141-144

Verri, A., Poggio, T. (1989) Motion field and optical flow: qualitative properties. IEEE Trans. Pattern Anal. Mach. Intell. 11, 490-498

Verri, A., Girosi, F., Torre, V. (1989) Mathematical properties of the two-dimensional motion field: from singular points to motion parameters. J. Opt. Soc. Am. A/6, 698-712

Verri, A., Girosi, F., Torre, V. (1990) Differential techniques for optical flow. J. Opt. Soc. Am. A/7, 912-922

Zeki, S., Shipp, S. (1988) The functional logic of cortical connections. Nature 335, 311-317

Four Applications of Differential Geometry to Computer Vision

Rachid Deriche, Olivier Faugeras, Gérard Giraudon,
Théo Papadopoulo, Régis Vaillant and Thierry Viéville

Institut National de Recherche en Informatique et en Automatique, Sophia Antipolis

1   Introduction

This chapter summarizes part of the work done at INRIA during the duration of the Insight project. We have focused on the work done during the last year of this project and selected four topics because we think that each of them represents a significant improvement with respect to previous methods either conceptually, or experimentally, or sometimes both. The attentive reader will note that the common methodological thread between the four topics is the use of differential geometry as a solid basis for modeling different aspects of machine perception and as a means for designing robust algorithms.

Edges are most important in images. In section 2, we describe an edge model which accounts for a much broader class of edges than just step-edges. We show that the edge can be accurately *localized* and *geometrically characterized* using derivatives of the intensity function up to the third order. In order to compute these derivatives in an unbiased and robust fashion we have developed optimal discrete derivative operators that generalize and justify results by previous authors.

Corners and junctions are also features of great interest in both machine and biological vision. In section 3, we describe an approach for modeling and detecting corners and trihedral junctions. It starts with an analytical representation which allows us to make *predictions* about the behaviour of differential operators when applied to the image intensity function in the vicinity of such features and to *localize* and *identify* them with a much greater accuracy than previously published methods.

We then set out to study in section 4 a special kind of contour in images called occluding edges: these are contours for which the optical rays are tangent to the object surface. It has been known for some time that if these contours can be identified in the image, a complete three-dimensional description of the surface can be obtained in their vicinity, including location, orientation, and curvature. We have implemented a robust technique that allows us, given a minimal number of slightly different views of the same scene, to identify occluding edges in the scene and compute the surface description of the objects in their vicinity.

In section 5, we describe our work on the motion of rigid and non-rigid three-dimensional curves. This is particularly important for understanding the relationship between the optical flow and the 3-D motion that gave rise to it. We study in detail the differential properties of the spatio-temporal surface generated by an image curve. This has turned out to be a very useful tool for studying motion problems in general. In the case where the motion is rigid, we establish the fundamental equations that relate the kinematic screw of the moving three-dimensional curve to quantities which can be measured in the image.

2   Robust Edge Detection

2.1   Introduction

Edges are important features in an image. Detecting such edges in static images is now a well understood problem (Hildreth 1980; Tsai *et al.* 1982; Deriche 1987; Francois and Bouthemy 1990). In particular, an optimal edge-detector using Canny's criterion has been designed (Deriche 1987) and implemented as a fast algorithm (Deriche 1990) in real time on the Depth from Motion Analysis European machine (Faugeras *et al.* 1988). In subsequent studies this method has been generalized to the computation of 3D-edges (Monga *et al.* 1991). This edge-detector, however, has not been designed to compute edge geometric and dynamic characteristics, such as curvature and velocity.

It is also well known that robust estimates of the image geometric and dynamic characteristics should be computed at points in the image with a high contrast, that is, edges. Several authors have attempted to combine an edge-detector with other operators, in order to obtain a relevant estimate of some components of the image features, or the motion field (Horn and Schunk 1981; Hildreth 1984; Nagel 1985; Deriche and Giraudon 1990), but they use the same derivative operators for both problems.

However, it is not likely that the computation of edge characteristics has to be done in the same way as edge detection, and we would like to analyse this fact in this paper.

Since edge geometric and dynamic characteristics are related to the spatial and temporal derivatives of the picture intensity (Hildreth 1980; Horn and Schunk 1981; Tsai *et al.* 1982; Nagel 1985; Deriche and Giraudon 1990), we have to study how to obtain dedicated derivative operators.

This work is related a new theory of the motion of 3D curves (Faugeras 1990b). According to this theory, operators for computing the different parameters about edge location and motion are now available (Faugeras 1991b). This theory is elaborated without making any assumption about the photogrammetric characteristics of the edge, which is known to be very difficult to obtain, while actual models are only approximate (Verri and Poggio 1986), since they are all based on the intensity constancy assumption. This new theory is valid in the continuous case, that is, when information about the image is available at any time and location. Although this is the best way to derive theoretical properties about moving edges, it is not obvious how to implement these operators on a sampled image sequence. The present work was dedicated to the implementation of such operators.

One specificity of the present approach is to consider the case of actual available sequences of time-varying images sampled at 5 to 60 Hz. In such cases the rigid motion between two consecutive views is small, but not small enough to consider this discrete sequence as a good approximation to the continuous case, as in differential approaches. Thus, while spatial derivatives will be computed directly from the intensity function,

$$I(r) = c0 + c1 \qquad \diagup \qquad +c2 \qquad \underline{}\overline{} \qquad +c3 \qquad \underline{}\overline{}\underline{}$$

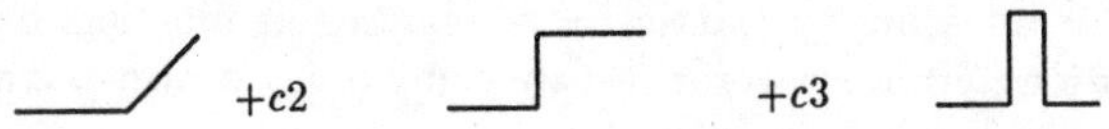

$$I(r) = c0 + c1\delta(r - r_0)^{-2} + c2\delta(r - r_0)^{-1} + c3\delta(r - r_0)$$

Figure 1: *1D-edge model, the edge is located at r_0.*

temporal derivatives will be estimated indirectly using local correspondences between two frames.

If we want to implement such operators for computing edge characteristics, we have to answer two questions:

1. What are the relationship between the characteristics of the edge curve and the intensity derivatives?

2. How to compute "good" intensity derivatives, that is, suitable to estimate edge characteristics?

2.2 Defining Edges as Level-Curve of Intensity

Authors very often implicitly consider that the intensity is locally constant along an edge and use this assumption to generalize to 1D-edge models to 2D.

Considering a general model of the intensity profile along an edge curve, let us make explicit this assumption and derive thereby some relationships about the edge characteristics.

2.2.1 1D-Edges Intensity Profiles

Edges might have different physical origins and it has been proposed, that a realistic model for an edge is a combination of a roof, a peak and a step (see Perona and Malik 1990, for a recent contribution), as shown in figure 1, ($\delta(r)$ denotes the Dirac operator, and we use the notations $\delta^{-1} = \int \delta$ and $\delta^{-2} = \int \delta^{-1}$). We will use a generalization of this model to derive our equations.

The model used in Perona and Malik 1990 is the most sophisticated model which has been seriously studied so far. In fact, in most approaches (Hildreth 1980; Deriche 1987), edge profiles are considered as simple steps ($c1 = c3 = 0$). The Canny-Deriche operator for instance has been developed assuming an edge is a simple step, although the claim has been made (Canny 1983) that it could be generalized to other profiles.

The previous analysis does not take into account the distortions caused by the imaging system to the image distribution. They correspond to different factors: motion blur, approximate focus, optical system impulse response, spatial summation by the photosensitive receptors. It is difficult to have a realistic model of these factors and it would be best to derive a theory about edge detection and analysis which *does not depend upon the intensity profile*. We will summarize all these effects in a functional $\mathcal{H}$ which is applied to the initial intensity profile I and produce a measured intensity $\mathcal{Y}$:

$$\mathcal{Y}(r) = \mathcal{H}[I(r)]$$

Moreover such a functional also represents photogrammetric effects related to the lighting conditions and the surface irradiance, and transformations related to image preprocessing (picture smoothing).

We are not going to quantitatively analyse this function, but eliminate its influence in the equations. In other words, we are going to use a **non-parametric** model of intensity profile.

2.2.2   Edge Equation and Image Intensity

Whereas 1D-edge profiles have been extensively studied, little has been done on how to generalize such profiles to 2D-edges.

A 2D-edge corresponds to a curve in the image plane, and due to the smoothing of the input data, it is realistic to assume this curve to be differentiable up to at least the second order. We thus limit our discussion to regular points of the curve, which excludes corners for instance (see Deriche and Giraudon 1990 for a justification).

In order to generalize 1D-intensity profiles to 2D-edges we are going to assume that: **In the neighbourhood of an edge, the curves of iso-intensity have the same profile as the edge curve.**

This is the basic assumption of this study[1].

From this hypothesis the following properties are easy to derive:

- The image intensity gradient is parallel to the edge normal and the latter can be computed from the intensity first order derivatives.

- The edge curvature is computable from the intensity first and second order derivatives, since the terms related to the edge profile cancel.

- It is also possible to obtain qualitative information about the kind of edge encountered: whether the edge is *locally rectilinear, locally circular* or quadratic.

- It is possible to obtain qualitative informations about the edge intensity profile. For instance, it is possible to distinguish between *step-like* and non step-like edges by comparing second and third order derivatives.

- Subpixel location of an step-edge can be computed from the intensity derivatives up to the third order. Moreover, given a location on the image, it is possible to compute the location of the nearest edges, even if the location of the edge is not known, and without any matching.

2.3   Computing Optimal Spatial Derivatives

2.3.1   Unbiased Optimal Derivative Filters

We have also to compute "good" spatial derivatives, in the vicinity of already detected edges in order to compute edge characteristics.

We consider the following two aspects for a derivative filter:

- A derivative filter is *unbiased* if it outputs only the required derivative, but not lower-order or higher-order derivatives present in the signal.

- Among these filters, a derivative filter is *optimal* if it minimizes the noise present in the signal. In our case we will minimize the output noise.

---

[1]Although the previous hypothesis is indeed not "local ", we are only going to use it only locally. In other words, we allow this level of curves to be subject to smooth deformations along a given edge.

Please note, that we are not dealing with filters which are used to detect edges, since the edges have already been detected, but rather with derivative filters used to compute edge characteristics. It is thus not relevant to consider other criteria used in optimal edge detection such as localization or false edge detection.

In fact, spatial derivatives are often computed in order to detect edges with accuracy and robustness. Performances of edge detectors are given in term of localization and signal to noise ratio (Canny 1983; Deriche 1987). Although the related operators are optimal for this task, *they might not be suitable to computing unbiased intensity derivatives on the detected edge*. Moreover it has been pointed out (Weiss 1991) that an important requirement of derivative filters, in the case where one wants to use differential equations of the intensity, is the preservation of the intensity derivatives, which is not the case with usual filters. However, this author limits its discussion to a particular set of filters, whereas we would like to determine what the general condition is for a filter to be unbiased and derive optimal unbiased filters. We have first demonstrated some properties of such filters in the continuous or discrete case and then use an equivalent formulation in the discrete case.

The resulting equations (Viéville and Faugeras 1991) have the following consequence: *the optimal derivative filter is a polynomial filter and is thus only defined on a finite window.* If not, the equations are no more defined.

The first way to compute these derivatives is to explicitly solve the obtained equations (Viéville and Faugeras 1991). There is also another way to compute these derivatives, considering the Taylor expansion of the input as a parametric model, but the two approaches are equivalent.

Considering a signal with derivatives up to a given order Q, it is thus possible to compute unbiased estimators of these derivatives with a minimum of output noise by solving a least-square problem. This result is not a surprise for someone familiar with optimization but is crucial when implementing such filters in the discrete case, as done here.

2.3.2 An Optimal Approach in the Discrete 2D-Case

Whereas most authors derive optimal continuous filters and then use a non-optimal method to obtain a discrete version of these operators, we would like to stress the fact that *the discretization of an optimal continuous filter is not necessary the optimal discrete filter.*

Moreover, this depends upon the (implicit) model used for the sampling process. For instance, in almost all implementations (Deriche 1987; Haralick 1984), the authors make the implicit assumption that the intensity measured for one pixel is related to the true intensity by a Dirac distribution, that is, corresponds to the value of the intensity at the middle point of the pixel. This is not a very realistic assumption, and in our implementation we will use another model.

The key point here is that since we have obtained a formulation of the optimal filter using any Lebesgue integration over a bounded domain, then the class of resulting filters is still valid for the discrete case.

This approach is very similar to what was proposed by Haralick (Haralick 1984), and we call these filters Haralick-like filters. In both methods the filters depend upon two integers: (1) the size of the window, (2) the order of expansion of the model. In both methods, we obtain polynomial linear filters. However it has been shown (Huertas and Medioni 1986) that Haralick filters reduce to Prewitt filters, while our filters do not correspond to already existing filters. The key point, which is – we think – the main improvement, is to consider the intensity at one pixel not as the simple value at that location, but as the integral of the intensity over the pixel surface, which is closer to reality.

Note that the sensor influence is taken into consideration at two stages of the process: the continuous smoothing of the intensity due to the optical system, for instance, has been taken into consideration when deriving a non-parametric model of the edges, while the geometry of the CCD pixels which could not be considered as a simple smoothing has been taken into account here.

Contrary to Haralick's original filters these filters are not all separable, however, this not a drawback because separable filters are only useful when the whole image is processed. In our case we only compute the derivatives in a small area along edges, and for that reason efficiency is not as much an issue[2].

2.4   Incremental Edge Analysis: Experimental Results

The algorithm   Let us now give the algorithm in detail. The original picture is smoothed using the Deriche operator, and edges are detected using the related edge detector (Deriche 1987).

The following algorithm integrates all the equations developed in this paper and has been designed to provide local information about an edge, in an image sequence:

- Smooth the picture and compute first order derivatives using the Deriche optimal operator.

- For each pixel do:

- Is the magnitude of the image gradient above a given threshold ? *If yes*:

  - Recompute first order derivatives in a finite window using unbiased derivative operators and the original picture.

  - Compute edge orientation.

  - Compute second and third order derivatives in a finite window using unbiased derivative operators and the original picture.

  - Compute edge sub-pixel location.

  - Is this location inside the present pixel or one in its neighbourhood (thus, can our local model be used) ? *If yes*:

    * Recompute derivatives at the edge location.
    * Compute edge curvature.
    * Compute edge intensity profile parameters $\mathcal{Y}2$ and $\mathcal{Y}3$, and edge geometric properties.
    * Compute edge sub-pixel location in the previous frame.
    * Is this location inside the window of the filter? *If yes*:
      · Compute the edge displacement between two frames.

---

[2]Anyway, separable filters are quicker than general filters if and only if they are used on a whole image, not a few set of points.

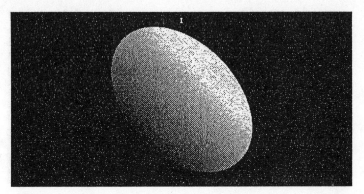

Figure 2: *A typical artificial picture used to evaluate the method.*

2.5 Experimental Results

Computation time Smoothing over space requires about 2.5 sec of cpu-time on a Sun4 workstation for each picture, and edge detection requires 2.2 sec for each picture on the same system. The rest of the computer power depends on the size of the window used, but has the same order of magnitude (from 2 to 4 sec).

2.5.1 Verification Using Noisy Artificial Pictures

In order to check the validity of on our computations, we have experimented our operators with noisy synthetic pictures, containing horizontal, vertical, or oblique edges with step and roof intensity profiles. A typical picture is shown in figure 2.

We used window sizes of 5×5 or 7×7 for the convolution kernels. In fact, we limited almost all computations to 5×5 windows in order to show the limits of the method. Obviously the higher the window size the lower the noise.

Noise has been added both to the intensity (typically 5 % of the intensity range) and to the edge location (typically 1 pixel). Noise on the intensity will be denoted "I-Noise", its unit being in percentage of the intensity range, while noise on the edge location will be denoted "P-Noise", its unit being in pixels.

Edge orientation Edge orientation is computed as the gradient orientation using Haralick-like filters. Taking a noisy circle, the orientation should vary linearly with the edge curvilinear coordinates. In fact this is not entirely the case, since the edge sampling is not homogeneous with its curvilinear coordinates, but only piecewise linear. This phenomenon is illustrated in figure 3, for a Canny-Deriche detector.

We thus analyse the orientation for part of the edge for which the sampling is linear with respect to the curvilinear coordinates and obtain the result shown in figure 4. This curve represents the edge orientation along a circular edge. Three portions of the edge for which the sampling is linear have been analysed and the theoretical values are drawn in dotted line. The theoretical orientation has been drawn as a dashed line.

In addition, we have compared the average edge orientation of a rectilinear edge, as a function of the noise. Results are given in Table 2.5.1. Error is computed as the standard deviation over a set of 20 values.

This shows that the precision of the edge orientation is about 1 degree, and is stable even if the image intensity is noisy, while it is sensitive to errors in edge localization.

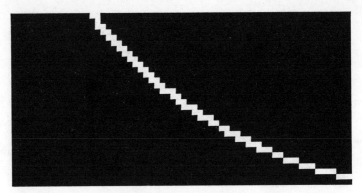

Figure 3: *A detail of the edge coordinates for a circular edge.*

However, in a realistic situation (edge localization error of 1 pixel), this computation is still robust.

Table 1: *Computation of the orientation at different levels of noise*

I-Noise	2%	5%	10%	0	0	0	10%
P-Noise	0	0	0	0.5	1	2	2
Error (in radians)	0.015	0.02	0.022	0.01	0.023	0.13	0.2

In the worst case (I-Noise=10%, P-Noise=2), we obtain the curve shown in figure 5. The true value is drawn as a dotted line. It is clear that although the estimation is noisy, the average value is not subject to a bias. Smoothing along the edge will thus increase the precision.

We have also studied the stability of this estimate when the edge is not a step-like edge but contains a roof or peak component. In this case we still obtain a robust estimate of the edge orientation with a small increase in the error ($Error \simeq 10\%$), but only if the derivatives have sufficiently large values. If not, the estimate is incoherent, and the error is huge.

We have also computed the edge orientation for edges which do not correspond to locally constant intensity and obtained high bias (more than 5 degrees).

Edge curvature   We have computed the curvature for non-rectilinear edges, either circular or elliptic. The curvature range is between 0 for a rectilinear edge and 1, since a curve with a curvature higher than 1 will be inside a pixel.

As in the case of orientation we have computed the curvature along an edge, and have compared the results with the expected values. Results are plotted in figure 6, the expected values being a dashed curve.

It is obvious that edge curvature is more sensitive to discretization errors than orientation but the result is still reliable, since these errors are not systematic but random. Again, smoothing along the edge, which has not been done here, will increase the precision of the result.

We have also computed the curvature for different circles, in the presence of noise, and evaluated the error on this estimation. Results are shown in Table 2.5.1. The results are

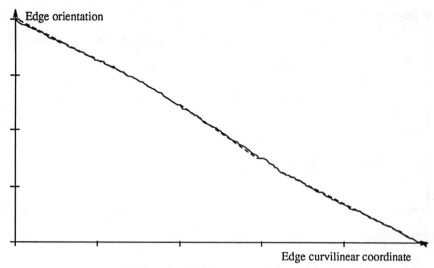

Figure 4: *Orientation along a circular edge.*

the radius of curvature, and the inverse of the curvature expressed in pixels. The circle radius was of 100 pixels.

Table 2: *Computation of the curvature at different levels of noise*

I-Noise	2%	5%	10%	0	0	0
P-Noise	0	0	0	0.5	1	2
Error (in pixels)	2.1	6.0	10.4	6.0	12.2	huge

Although the error is almost 10 %, it appears that for important edge localization errors, the edge curvature is simply not computable. This is due to the fact we use a 5×5 window, and that our model is only locally valid. In the last case, the second order derivatives are used at the border of the neighbourhood and are no longer valid.

We also have studied the stability of this estimate when the edge is not a step-like edge but contains a roof or peak component. In this case we have still obtained a robust estimate of the edge curvature with a small increase in the error ($Error \simeq 10$), but only if the derivatives have sufficiently large values. If not, the estimate is incoherent, and the error is huge.

We have finally tried to compute the edge curvature for edges which do not correspond to locally constant intensity and obtained in this case erroneous values.

Edge sub-pixel location We have also computed the edge relocalization in the direction normal to an edge. This parameter, noted d_0 is the subpixel distance from the edge to the middle of the pixel rectangule. A typical result is shown in figure 7. In order to locate the edge we have also plotted the magnitude of the gradient. The abscissa corresponds to the position of the pixel in a direction normal to the edge. The edge's exact localization is shown by an arrow. The most important values are given explicitly.

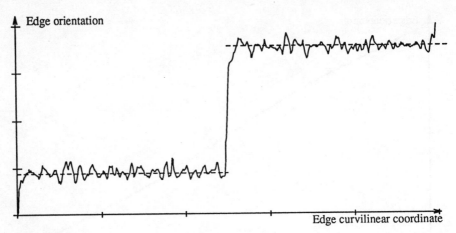

Figure 5: *Computation of the orientation along two sides of a square in the presence of noise.*

The analysis of this curve shows that we indeed obtain a good estimate of the edge sub-pixel localization in the neighbourhood of the edge, and the algorithm automatically detects if the measure is valid (please note that values at localization ±3 are very different but not taken into account by the algorithm, since they are out of window defined by the filters). Thus, this estimate is only valid in a neighbourhood of ±2 pixels, using a 5 × 5 window. Moreover, erroneous values of edge localization are observed when the gradient amplitude is small. This explains why we first perform an edge detection, and then compute the edge location in the neighbourhood of the edge only.

The computation of edge localization is a surprisingly robust primitive with respect to noise on the intensity as shown in Table 2.5.1. This is due to the robustness of the filters we used, and to the fact that we only did our computations on the edge, which corresponds to high contrast points, and is less sensitive to noise. Of course, when the edge localization itself is wrong, the parameter is also erroneous, but the error on the estimate has the same order of magnitude as the input error, which is expected. This error also obviously increases with the eccentricity of the edge location. Moreover, it is not possible to compute this parameter for such a large eccentricity.

Table 3: *Computation of the edge localization at different level of noise*

I-Noise	2%	5%	10%	0	0
P-Noise	0	0	0	0.5	1
Error (in pixels) $d_0=0$	0.08	0.11	0.07	0.6	1.2
Error $d_0=1$	0.32	0.29	0.41	0.8	none
Error $d_0=2$	0.52	0.31	0.69	1.4	none

It is not possible to apply this operator on edges with a non-step like intensity profile, since the assumption about the maximum of the gradient magnitude is no longer true.

Edge displacement   We have evaluated the performances of our operator for different horizontal displacements of a vertical edge. For a displacement of less than 2 pixels,

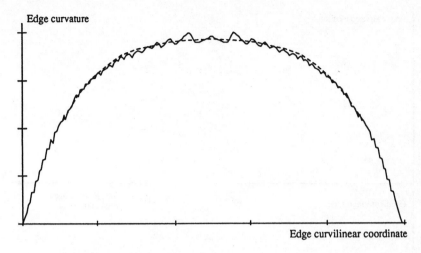

Figure 6: *Computation of the curvature along an elliptic edge.*

that is, less than half of the window size, we obtain relevant results while for higher displacement, the algorithm no longer computes the value. A nice feature is that it does not compute irrelevant values but rather rejects the measure at this point. Mean value and standard deviations are given in Table 2.5.1.

In this table we made the computation for two different sizes of window, 5 × 5 and 7 × 7. It appears, as discussed previously, that for large amplitudes, the displacement is much less biased, less than 10%, in all cases.

In fact, several other tests have been made, which are not reported here, since they yield similar results, and we obtain the following approximate experimental rule for our operator:

$$|\beta_{max}| < \frac{WindowSize}{4}$$

that is, the value of beta must not exceed a quarter of the size of the window, that is the location of the edge to be founded must be in the center part of the window, in a window with about half that size. This is an important point when designing the size of the related filters.

Table 4: *Computation of the displacement for different amplitudes*

True motion (pixels)	0.0	0.5	1.0	1.5	2.0	2.5
Estimate (pixels) (5 × 5)	-0.01±0.1	0.48±0.081	1.12±0.11	1.64±0.07	1.5±0.3	none
Estimate (pixels) (7 × 7)	0.01±0.2	0.51±0.01	1.02±0.06	1.48±0.08	2.11±0.1	2.31±0.4

We also run the test at different levels of noise. Quantitative results are given in the Table 2.5.1. Unfortunately, the algorithm rejected quite a number of measurements when there were errors in position, and this was not yet computed. However, although noise on the intensity increases the error, the value is still computable.

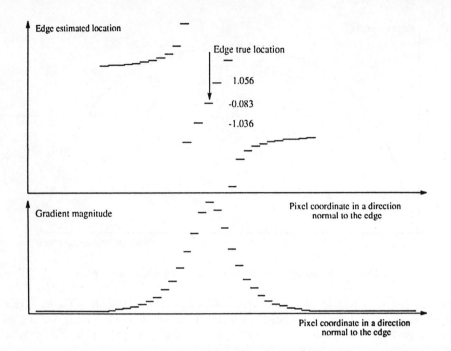

Figure 7: *An example of edge sub-pixel localization.*

We then made the computation for different edge orientation, since we might expect the method to be less reliable in this case, but we found no significant variations in the response.

The normal-motion field along two sides of this square has been also drawn in figure 8. Erroneous values have been put on the curve although *they are canceled by the algorithm.* The edge displacement is in fact underestimated, as shown in the previous tables, while distribution of erroneous values corresponds to irrelevant overestimates of the displacement. These erroneous values are related to points which are not well located, due to the edge orientation.

Table 5: *Computation of the displacement for different levels of noise*

I-Noise	2%	5%	10%	0	0	0
P-Noise	0	0	0	0.5	1	2
Error (in pixels)	0.13	0.31	0.5	0.6	none	none

2.5.2 Application to Real Scenes

Let us finally illustrate these computations with results obtained using real images. Two kinds of pictures have been used: a polyhedral object (figure 9) and a non polyhedral object (figure 10).

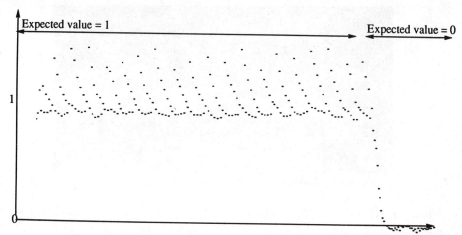

Figure 8: *Computing the normal motion field for a noisy square.*

We first have to verify if the two underlying assumptions of our study are reasonable, when considering real data. Since our discussion is based on the fact that *edges have a constant intensity profile* and that *the intensity surface can be modeled by a third order polynomial*, let us check this two points.

Is the intensity constant along an edge? This assumption is almost verified with our data. One example of the intensity variation along an edge is shown in figure 11. It appears that, with intensity levels from 0 to 255, intensity variation is of about 1%. More precisely, we have obtained a standard deviation of 0.37% for a set of 100 pixels chosen along 5 different edges.

Is the intensity well represented by a Taylor expansion? In order to verify this second assumption we have computed the quadratic mean square error between the true intensity and the intensity estimated by a third-order polynomial.

Computing this error for a set of 100 pixels over 5 different edges yields a mean square error of 1.3% of the maximum intensity. Computing the Ξ^2 square related to this sum of quadratic errors taken as a set of normalized random variables, we obtained a model rejection probability lower than $p = 0.01$. The model fits very well the true data.

Both assumptions are then relatively well verified for our data set.

Qualitative aspects of the edge characteristics We can represent the orientation around edges by an intensity value, from black for angles equal to $-\pi$ to white for angles equal to π. Values at which the gradient is too small encroached upon the white area on the left. Results are shown for a polyhedral object and a non-polyhedral object (figures 12 and 13). It is visible on these two images that the estimated orientation is qualitatively correct.

In order to observe the regularity of the edge orientation estimate we show a zoom of the obtained results in an area of the polyhedral object where the white squares lie on a dark plane (figure 14).

Similarly the absolute value of the normal distance from a pixel to the proximal edge (d_0 parameter) can be represented by an intensity value from black for $d_0 = 0$ to white for $d_0 = WindowSize$. Results are shown for a polyhedral object (figure 15).

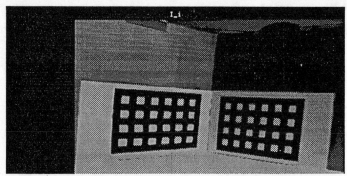

Figure 9: *A simple polyhedral object used for the experiment.*

It appears that the underlying edge local model used seems to corresponds to the edges expected in the picture, with dark values on the edges and lighter values when away from the edge, as visible on (figure 16) and (figure 17).

It is also possible to visualize the absolute value of the curvature obtained with a non polyhedral object (figure 19) or the corner of a polyhedral object (figure 18). The intensity varies from black for a null curvature to white for a curvature equal to 1 pixel. It is visible on this picture that the estimate is qualitatively correct, for instance corners (related to fingers of the operator) correspond to high curvatures (figure 18), while the curvature is almost constant along a curve (figure 19). This last picture corresponds to the part of the image where fingers are visible.

Let us now quantitatively analyse these data.

Accuracy of the parameters   In order to compare noisy synthetic pictures, we have used the rectilinear edge of a known orientation and an elliptic edge of a calculable curvature, as shown in figure 10, for three manually selected edges.

Computing the average mean square error for the orientation and the radius of curvature we have obtained a value of $0.4deg$ for the first one and of $12pixels$ for the second.

It thus appears that the parameters we compute in real images are accurate enough to be useful.

2.6   Conclusion

We have considered that the intensity profile has been smoothed by the optic and the preprocessing. We also assumed that edges correspond to level-curves of the intensity.

Using this simple but general model of the edge intensity profile, we have been able to derive a new equation for the computation of edge subpixel location. Our model also provides a theoretical justification of usual estimates of the edge orientation θ and curvature κ.

We also designed a new class of unbiased optimal filters dedicated to the computation of intensity derivatives, as required for the computation of edge characteristics. Because these filters are computed though a simple least-square minimization problem, we have been capable to implement these operators in the discrete case, taking the CCD subpixel mechanisms into account.

These filters are dedicated to the computation of edge characteristics as studied in this paper, they can be well implemented with finite windows, and correspond to unbiased

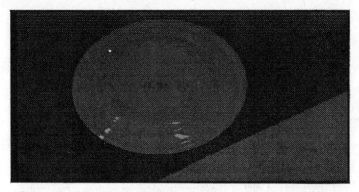

Figure 10: *A simple non-polyhedral object used for the experiment.*

derivative operators with minimum output noise. They do not corresponds to optimal filters for edge detection.

It might be surprising that we put so much energy into the calculation of the exact values of the edge parameters, while the human system can detect differences in, e.g., orientation of half a degree, which would be represented at early levels by the activity of a bank of coarse filters. One could have thought of a general front-end for the vision mechanism as developed by Koenderink and Kappers in this book, the estimation of edge characteristics being really done at "higher levels" in the system architecture. But although such an approach is satisfactory in many cases, it fails when the vision algorithms require a very high precision for the measurements. This is the case for "ill-conditioned" problems such as the well known structure from motion paradigm, or more generally, when quantitative data is to be output by the visual system.

In other words, if you consider the artificial visual system not only as a "features detector" but as a measurement device, it is crucial to obtain primitives with unbiased values, and to have an adequate model for the sensor and data flow. This is the goal of our efforts.

3 Corner and Junction Detection

Corners and vertices are strong and useful features in Computer Vision for scene analysis, stereo matching or motion analysis. This work deals with the development of a computational approach of these important features. We consider first a corner model and study analytically its behavior once it has been smoothed using the well-known Gaussian filter. This allows us to clarify completely the behavior of some well known *cornerness* measure based approaches used to detect these points of interest. Most of these classical approaches appear to detect points that do not correspond to the exact position of the corner. A new scale-space based approach that combines useful properties from the Laplacian and Beaudet's measure (Beaudet 1978) is then proposed in order to correct and detect exactly the corner position. An extension of this approach is then developed to the problem of trihedral vertex characterization and detection. In particular, it is shown that a trihedral vertex has two elliptic maxima on extremal contrast surfaces if the contrast is sufficient, and this allows us to classify trihedral vertices into 2 classes: "vertex" and "vertex as corner". The corner detection approach developed here is applied in order to

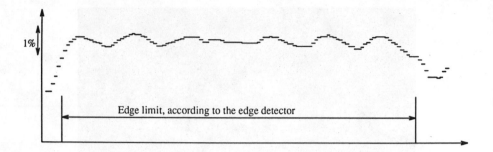

Figure 11: *A plot of the intensity, along an edge in a real scene.*

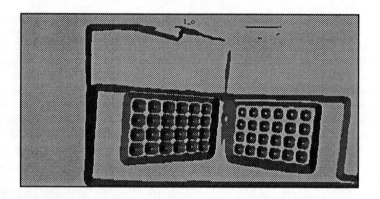

Figure 12: *Map of edge orientation for a simple polyhedral object.*

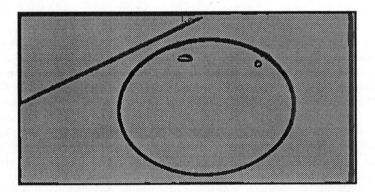

Figure 13: *Map of edge orientation for a simple non-polyhedral object.*

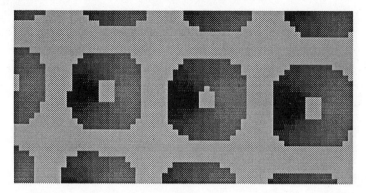

Figure 14: *Zoom on the map of edge orientation for a simple polyhedral object.*

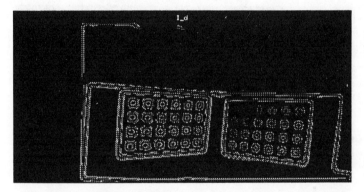

Figure 15: *Map of edge localization for a simple polyhedral object.*

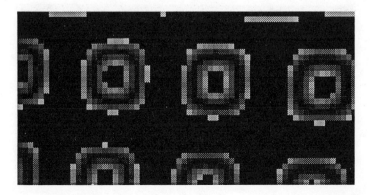

Figure 16: *Zoom on the map of edge localization for a simple polyhedral object.*

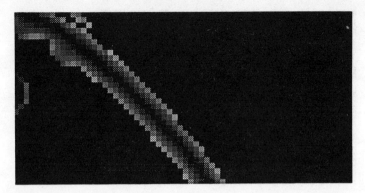

Figure 17: *Zoom on the map of edge localization for a simple non-polyhedral object.*

Figure 18: *Zoom on the map of edge curvature for a simple polyhedral object.*

Figure 19: *Zoom on the map of edge curvature for a simple non-polyhedral object.*

accurately detect trihedral vertices using an additional test to make a distinction between trihedral vertices and corners. For more details, one can refer to Deriche and Giraudon (1990) and Giraudon and Deriche (1991).

3.1 An Analytical Study for the Corner Model

We consider a corner model and study its behavior in scale space. We then show how some well known approaches used to extract corners are inadequate for an exact localization of the corner. We then propose a way to correct this drawback and get the exact position.

3.1.1 Notations and definitions

We introduce here some functions that will be widely used in the rest of the paper. Let $g(x)$ denote the zero mean Gaussian filter :

$$g(x) = \frac{1}{\sqrt{2\pi}}e^{-\frac{x^2}{2}} \qquad (1)$$

The two-dimensional Gaussian filter G can be expressed as :

$$G(x,y) = g(x)g(y) \qquad (2)$$

Following Berzins (Berzins 1984), we work in a coordinate system where the unit length is equal to the scale factor σ of the filter. In order to convert the results into a more general coordinate system (X, Y), we use the following transformation:

$$\begin{cases} x = \frac{X}{\sigma} \\ y = \frac{Y}{\sigma} \end{cases} \qquad (3)$$

Let Φ denote the error function given by :

$$\Phi(x) = \int_{t=-\infty}^{x} g(t)dt \qquad (4)$$

Let U define the unit step function

$$U(x) = \begin{cases} 1 & \text{if } x > 0 \\ 0 & \text{otherwise} \end{cases} \qquad (5)$$

The response of the 2D Gaussian filter G for a 2D input function I can be computed by evaluating the following convolution integral

$$S(x,y) = \int_{\alpha=-\infty}^{+\infty} \int_{\beta=-\infty}^{+\infty} G(\alpha,\beta)I(x-\alpha, y-\beta)d\alpha d\beta \qquad (6)$$

3.1.2 A Corner Model

An ideal corner with one edge along the x axis and an angle θ can be modeled by the following 2D step function.

$$I_\theta(x,y) = U(x)U(mx - y)U(y) \qquad (7)$$

where $m = tan(\theta)$.

If we convolve this 2D step function with the 2D Gaussian filter given by (2), we get the following filtered image $S(x, y)$:

$$S(x, y) = \Phi(x)\Phi(y) - \int_{\alpha=-\infty}^{x} g(\alpha)\Phi(y - mx + m\alpha)d\alpha \qquad (8)$$

This equation describes a Gaussian filtered corner with angle θ localized in the origin point $(0, 0)$.

Let us make the following changes of variables:

$$\begin{aligned} u &= x\sin(\theta) - y\cos(\theta) \\ v &= x\cos(\theta) + y\sin(\theta) \end{aligned} \qquad (9)$$

The components $S_x(x, y)$ and $S_y(x, y)$ of the gradient vector $\vec{\nabla} S(x, y)$ are then given to be

$$\vec{\nabla} S(x, y) = \begin{bmatrix} g(u)\Phi(v)\sin(\theta) \\ g(y)\Phi(x) - g(u)\Phi(v)\cos(theta) \end{bmatrix} \qquad (10)$$

As it can be shown, the Hessian matrix is important for the description of surfaces. The elements of such a matrix can be calculated by deriving the surface $S(x, y)$ two times in the x and y directions. Applying this to our corner model surface yields the following elements:

$$\begin{aligned} S_{xx}(x, y) &= g(u)\sin(\theta)(g(v)\cos(\theta) - u\Phi(v)\sin(\theta)) \\ S_{xy}(x, y) &= S_{yx}(x, y) = g(u)\sin(\theta)(u\Phi(v)\cos(\theta) + g(v)\sin(\theta)) \\ S_{yy}(x, y) &= -yg(y)\Phi(x) - g(u)\cos(\theta)(u\Phi(v)\cos(\theta) + g(v)\sin(\theta)) \end{aligned} \qquad (11)$$

3.1.3   About the Way to Extract Edges

It is well known that edges can be extracted from an image using the non maximum suppression scheme (Canny 1987; Deriche 1987) or the zero-crossing scheme (Bergholm 1984). In this subsection, we briefly review the main difference between both approaches in the case where corners have to be detected (Bergholm 1984, 1987).

- *Non-Maxima Suppression Method*

  Edges can be extracted as local maxima of the gradient magnitude in the gradient direction. This is equivalent to extracting points where the second directional derivative of the gradient magnitude image along the gradient direction is equal to zero. An explicit representation of the second directional derivative in the gradient direction **n** can be found to be:

  $$\frac{\partial^2 S}{\partial \mathbf{n}^2} = \frac{S_{xx}S_x^2 + 2S_xS_yS_{xy} + S_{yy}S_y^2}{(S_x^2 + S_y^2)} \qquad (12)$$

  The location of the zero-crossings of (12) corresponds to the location of the discontinuity in the 2D step filtered function (8) extracted following the non-maximum suppression scheme. Illustrating the curve that represents the set of points extracted through the use of this measure in the case of a right angle leads to an illustration of the well-known rounding effect of the edge extraction with the non-maxima suppression scheme (Bergholm 1987). In order to better appreciate the displacement of the extracted edges, let's take the particular edge point that belongs to the line bisector and look for its displacement from the its correct position (i.e. the origin point $(0,0)$). We have found the following results: for a right angle the displacement found was equal to $.7151\sigma$ while it was found to be 1.329σ and 1.893σ for angles with $\pi/4$ and $\pi/8$ respectively. The sharper the angle is, the bigger will be the displacement.

- *Zero-Crossing Method*

  Edges can also be extracted as zero-crossings in the Laplacian image given by :

  $$\nabla^2 S(x,y) = S_{xx}(x,y) + S_{yy}(x,y) \tag{13}$$

  Dealing with the corner described by (8) leads to the following expression for the Laplacian:

  $$\nabla^2 S(x,y) = -(yg(y)\Phi(x) + ug(u)\Phi(v)) \tag{14}$$

  Equation (14) can be solved numerically to find the contours where the Laplacian is equal to zero and illustrates very well the rounding effect due to the use of the Laplacian instead of the non-maxima suppression scheme. Following (Bergholm 1984), the largest deviation of the computed edge from the true edge can be found to be for $x = .839\sigma, y = -0.295\sigma$.

From these remarks, an important point to note is that the Laplacian allows one to recover exactly the position of the corner since it is equal to zero at the origin. This is not the case for the non-maximum suppression scheme where the point presenting the highest curvature is $.7151\sigma$ away from the exact position of the corner (i.e. the origin point $(0,0)$ for a right angle model. We will make use of this important point in order to correct the position of the extracted corner.)

3.1.4 Analysis of Some Classical Approaches for Corner Extraction

In this subsection, we will discuss and analyze the behavior of three classical measures that have been proposed and used in the corner detection, namely Beaudet, Nagel and Harris's approaches.

Beaudet's approach is based on the use of the determinant of the Hessian matrix associated with the intensity image and called DET. The use of this measure leads to an important local maximum in *all the directions* indicating that this approach is much more stable than the one proposed by Kitchen and Rosenfeld. Using a steepest descent based algorithm starting from the point $(x = 1, y = 1)$, we found that the local maximum in all the directions for such surface is located exactly in the bisector line at $(x = 1.17134\sigma, y = 1.17134\sigma)$ for a right angle. Therefore, detecting the corner point as the point where this local maximum occurs leads to a displacement between the true point and the one detected equal to 1.6565σ. Dealing with an angle of $\frac{\pi}{4}$ leads to displacement of the position of the local maximum to $(x = 2.58231\sigma, y = 1.06963\sigma)$ (i.e. in the bisector line of the angle). The displacement from the true corner point is in this case equal to 2.795σ. For sharper angles, the local maximum is pushed away on the bisector line. Therefore, the local extrema of measure proposed by Beaudet depend on the scale parameter σ and clearly makes the approach unstable in the scale-space. Since the displacement of the local extrema is a function of the standard deviation σ used in the filtering process, this measure does not allow us to locate the corner exactly. However, since a local maximum in all directions is much easier to extract than a maximum along a specific direction (because of the accuracy in that direction), we will show in the next subsection how we can use this measure in order to correctly extract corners.

The second approach that has been analysed is the one proposed by Dreshler and Nagel (Dreshler and Nagel 1982; Nagel 1983). It is based on the Gaussian curvature. We first detect positive as well negative extrema of the Gaussian curvature. It is worth noting that since the corner is located at a zero-crossing of one of the main curvatures, it corresponds also to a zero-crossing of the Gaussian curvature. Using a steepest descent based algorithm starting from the point $(x = 1, y = 1)$, we found that the local maximum in all the directions for the Gaussian curvature is located exactly in the bisector line at $(x = 1.216\sigma, y = 1.216\sigma)$ for a right angle. Dealing with an angle of $\frac{\pi}{4}$ leads to displacement of the position of the local maximum to $(x = 2.634\sigma, y = 1.09\sigma)$ (i.e. in the bisector line of the angle). However using the same technique we have not been able to find the local minimum in all the directions. This is because a minimum in all the directions does not exist. The positive extrema are very well detected but there are no negative extrema in all directions. This means that it will be difficult to extract this minimum. However this minimum exists in the edge direction or bisector line direction. Moreover, we have found that the zero crossings of one of the main curvature (and consequently a zero-crossing of the Gaussian curvature) does not give the correct position of the corner. Another important point to note is that the extrema locations of the Gaussian curvature as well as the zero-crossing of the Gaussian curvature on the bisector line moves in the scale space. This leads to the fact that the corners extracted following this approach are not well located. In fact, this approach suffers from the same problem than Beaudet's approach. We will show in the next subsection how we can make use of these important points in order to better locate the extracted corners.

The third approach that has been tested is the one proposed by Noble (Noble 1988; Harris 87) and known as the Plessey corner detector, and its slightly improved version as proposed by Harris. Using an angle of $\Pi/2$ and a k value of 0.04 as proposed by Harris, we have been able to find that the position of the local maximum in all the directions is located at $x_{max} = y_{max} = .7611\sigma$ and thus does not correspond to the exact position of the corner ($\sigma = 1$ was used for the Gaussian smoothing operator).

3.1.5   Combining Information to Extract Accurate Corners

Following the analysis that have been done, two important points have to be noted :

- The exact position of a corner can be detected as a stable zero-crossing in the scale-space.

- The local maxima in Beaudet and Nagel's measures moves in the scale space along the bisector line that passes through the exact position of the corner point.

We have combined these two observations in order to extract corners as follows. First a Laplacian image is calculated. Second, Beaudet's measure at two scales are calculated and an extrema detection in all directions is performed. Around each detected extrema in the image corresponding to the first scale, we look in the second image for the position of the local maxima. Once this second maxima detected, we look for the exact position of the corner as the point that belongs to the line segments joining the two positions and where a zero-crossing occurs in the Laplacian image.

With such an approach, we have combined the two important points that we have mentioned. Experimentally, this approach has been found to be reasonably stable. However some difficulties may arise. In particular for an angle less than $\pi/4$, the displacement

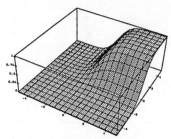

Figure 20: *A Gaussian Filtered $\pi/2$ Angle Model.*

is very important, and thus we have to look in a bigger neighborhood than the one implemented in order to have the expected results.

3.2 A General Trihedral Vertex Model and its Analytical Study

In this section, we consider a general trihedral vertex model and apply the same methodology developed for the corner case.

3.2.1 A General Trihedral Vertex Model

A general trihedral vertex as shown in Figure 21 can be modeled by the following 2D intensity function.

$$I(x,y) = \begin{cases} A & \text{if } x \leq 0 \quad \text{and } y < m'x \\ B & \text{if } x > 0 \quad \text{and } y < mx \\ C & \text{if } y \geq mx \quad \text{and } y \geq m'x \end{cases} \tag{15}$$

where $m = tan(\theta)$ and $m' = -tan(\phi)$. Note that from this model, it is easy to derive the corner model given in the previous section with A=B=0, C=1 and $\theta = 0$.

Using the step function $U(x)$ defined by (5) yields the following expression for $I(x,y)$:

$$I_{(\theta,\phi)}(x,y) = AU(-x)U(m'x - y) + BU(x)U(mx - y) + CU(y - mx)U(y - m'x) \tag{16}$$

If we convolve this 2D intensity function with the 2D Gaussian filter given by (2) we get the following filtered image $S(x,y)$:

$$S(x,y) = A + C(\int_{-\infty}^{x} g(\alpha)\Phi(y - m(x - \alpha))d\alpha + \int_{x}^{\infty} g(\alpha)\Phi(y - m'(x - \alpha))d\alpha)$$
$$-A(\int_{x}^{\infty} g(\alpha)\Phi(y - m'(x - \alpha))d\alpha) - B(\int_{-\infty}^{x} g(\alpha)\Phi(y - m(x - \alpha))d\alpha) + (B - A)\Phi(x) \tag{17}$$

In order to deal with an analytical expression for the different elements of the first fundamental form of the intensity surface described by the trihedral vertex, we have to get the components $S_x(x,y)$ and $S_y(x,y)$ of the gradient vector $\vec{\nabla}S(x,y)$.

Using the following changes of variables:

$$\begin{aligned} u &= xsin(\theta) - ycos(\theta) \\ v &= xcos(\theta) + ysin(\theta) \\ \bar{u} &= -xsin(\phi) - ycos(\phi) \\ \bar{v} &= xcos(\phi) - ysin(\phi) \end{aligned} \tag{18}$$

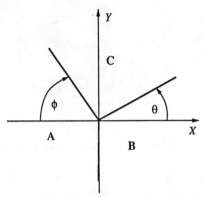

Figure 21: *A General Trihedral Vertex Model.*

The components $S_x(x,y)$ and $S_y(x,y)$ of the gradient vector $\vec{\nabla}S(x,y)$ can be calculated after some manipulations and put in the following analytical and simplified form :

$$\vec{\nabla}S(x,y) = \left[\begin{array}{c} (B-A)g(x)\Phi(-y) - (A-C)g(\bar{u})\Phi(-\bar{v})sin(\phi) + (B-C)g(u)\Phi(v)sin(\theta) \\ (C-B)g(u)\Phi(v)cos(\theta) + (C-A)g(\bar{u})\Phi(-\bar{v})cos(\phi) \end{array} \right]$$

$$(19)$$

An analytical expression for the norm of the gradient vector can then easily be calculated.

As seen in the previous section devoted to the corner modelisation, the Hessian matrix is important for the description of surfaces. The elements of such a matrix can be calculated by deriving the surface $S(x,y)$ given by our vertex model two times in the x and y direction. This yields after some manipulations the following elements:

$$S_{xx}(x,y) = (A-B)xg(x)\Phi(-y) + (C-A)g(\bar{u})\sin(\phi)(\Phi(-\bar{v})\bar{u}\sin(\phi) - g(-\bar{v})\cos(\phi))$$
$$+(B-C)g(u)\sin(\theta)(g(v)\cos(\theta) - u\Phi(v)\sin(\theta))$$
$$S_{xy}(x,y) = S_{yx}(x,y) = (A-B)g(x)g(y) + (C-A)\sin(\phi)(\bar{u}g(\bar{u})\Phi(-\bar{v})\cos(\phi)$$
$$+g(x)g(y)\sin(\phi)) + (B-C)\sin(\theta)(ug(u)\Phi(v)\cos(\theta) + g(x)g(y)\sin(\theta))$$
$$S_{yy}(x,y) = (C-B)\cos(\theta)(ug(u)\Phi(v)\cos(\theta) + g(x)g(y)\sin(\theta)) + (C-A)\cos(\phi)$$
$$(\bar{u}g(\bar{u})\Phi(-\bar{v})\cos(\phi) + g(x)g(y)\sin(\phi))$$

$$(20)$$

It is worth noting that setting $\theta = \phi$ and $C = 0$ leads to the particular results obtained for the symmetric trihedral vertex studied by Michelli *et al.* (1989) in their interesting contribution to the study of the localization in edge detection. In fact, this part of our study generalizes a part of the work addressed by Michelli (Michelli et al. 1989).

3.2.2   On the Extraction of the Edges of a Trihedral Vertex

Extracting the edges of the vertex described by (17) as zero-crossings in the Laplacian image leads to the following expression for the Laplacian:

Figure 22: *Features extracted on real image with* $\sigma = (2, 1)$.

$$
\begin{aligned}
\nabla^2 S(x,y) = & (A - B)xg(x)\Phi(-y) \\
& +(C - A)g(\bar{u})\sin(\phi)(\Phi(-\bar{v})\bar{u}\sin(\phi) - g(-\bar{v})\cos(\phi)) \\
& +(B - C)g(u)\sin(\theta)(g(v)\cos(\theta) - u\Phi(v)\sin(\theta)) \\
& +(C - B)\cos(\theta)(ug(u)\Phi(v)\cos(\theta) + g(x)g(y)\sin(\theta)) \\
& +(C - A)\cos(\phi)(\bar{u}g(\bar{u})\Phi(-\bar{v})\cos(\phi) + g(x)g(y)\sin(\phi))
\end{aligned}
\tag{21}
$$

Equation (21) can be solved numerically to find the contours where the Laplacian is equal to zero and illustrates very well the rounding dingo effect due to the use of the Laplacian instead of the non-maxima suppression scheme.

An important point to notice is that the Laplacian allows us to recover exactly the position of the vertex since it can easily be checked that $\nabla^2 S(0,0) = 0$ (i.e. the spatial position of a general vertex does not change in the scale space). This is not the case for the non-maximum suppression scheme. We will make use of this important point in order to correct the position of the extracted vertex.

3.2.3 On DET Extrema Extraction of a Trihedral Vertex

The DET equation can be written as follows :

$$
DET(x,y) = S_{xx}(x,y)S_{yy}(x,y) - S_{xy}(x,y)^2
\tag{22}
$$

Figure 23: $\pi/2$ *type angle : Corner detected with $\sigma=1$.*

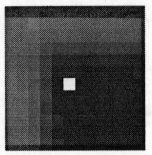

Figure 24: $\pi/2$ *type angle : Corner detected with $\sigma=2$.*

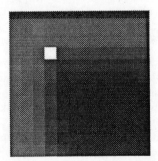

Figure 25: $\pi/2$ *type angle : Corrected position.*

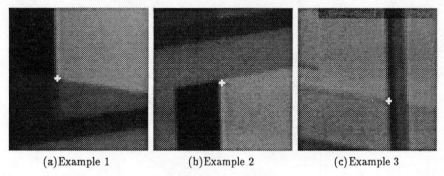

(a)Example 1   (b)Example 2   (c)Example 3

Figure 26: *Vertex localization on real image with $\sigma = (2,1)$.*

Replacing the different second derivatives that appear in the expression of the DET by their expression given in (20), we have been able to analyse the behavior of this measure. As for the corner detection we have studied the behavior of elliptic maxima in scale space. We have found that elliptic maxima localization move in scale space for the three types of contrast along a line which passes through the vertex. This means that the approach developed for the corner detection can also be applied for accurate vertex detection.

It appeared that we can class vertices into two classes function of contrast and so function of the number of elliptic maxima existing around the vertex. If two maxima exist (case 1 or 2), we have what we call "real or robust vertex" and if only one maximum exists (case 3), we have a "vertex like corner".

If a distinction must be made between a corner and a vertex, we can use a remark given above. In the case of "real or robust vertex", we have two detections of vertex location because each detection is created by a curve joined to two elliptic maxima. Thus a distinction between vertex and corner can be made easily. But for the second type of vertex ("vertex-like corner"), we have just one elliptic maximum curve, and so we will have only one detection on vertex position like a corner detection. Thus our detector cannot solve this ambiguity and another local contrast study would be needed to make a distinction.

An other possible classification is to say :

- A detected feature is named a "vertex" if its localization is created by the existence of two elliptic maxima curves

- A detected feature is a "corner" if its localization is created by the existence of only one elliptic maximum curve.

3.3   Some Experimental Results

In this section, we give some experimental results obtained by running our corner and vertex detector on real images.

To illustrate the efficiency of our detector on a real case, figure 22 shows an example of the promising result we have obtained on running the algorithm on an entire real image. In this image (indoor image of 512.512 pixels with 256 grey levels), we have corners and vertices.

Almost all the features have been well detected and corrected. Note that decreasing the threshold of maxima magnitudes would have resulted in more features points detected

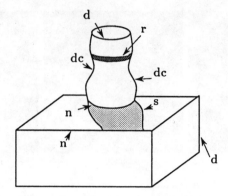

Figure 27: *The different types of edges.*

on the image. In order to illustrate some steps of the algorithm, Figures 23 to 25 show a local zoom of what happens near a corner of type $\Pi/2$. Detection of elliptic maximum and corner localization are shown using two different resolutions corresponding at $\sigma_1 = 1$ and $\sigma_2 = 2$. The corrected position derived by the algorithm seems to be quite remarkable.

In order to illustrate the detector behavior on vertices, figures 26 (a, b, c) shows a local zoom of vertix detection. As for the corner case, the corrected vertex position derived by our algorithm also seems to be quite promising. As we have indicated in the previous section, we detect as vertex, features which are created by the intersection of zero crossing and two lines (existence of two elliptic maxima). So, from a point of view of scene analysis, all vertices which are present in this image are not extracted because some of them are classified as corners.

Remark : The results obtained by running the algorithm on a triplet of trinocular stereo images are very stable and suggest that the matching process will be much more easier and more reliable using these features. Relating the width of the spiral search window to the scale space used is also a point that we are considering.

4 Shape from Occluding Edges

One of the aims of computer vision is to extract concise surface descriptions from several images of a scene. The descriptions can be used for the purpose of object recognition and for geometric reasoning (such as obstacle avoidance). Stereovision is often used for recovering the structure of the 3D world. Standard techniques can determine the depth of edges on a surface. These techniques fail with extremal boundaries as these change according to the viewpoint.

When we form an image of the environment on the retina of a camera we assume, up to a very good approximation, that it is a perspective projection of the scene and that, at every pixel, the image intensity is proportional to the scene irradiance (Horn 1986). One of the goals of computer vision is to extract from this pattern of changing intensities relevant information about the three-dimensional geometry and kinematics of the objects present. One of the key ideas to achieve this is to extract edges because they always signal important physical properties of the scene: discontinuities in the reflectance, texture, color, depth, and motion, are among the many possibilities. But even though the knowledge of the physical origin of an edge in an image is extremely important for

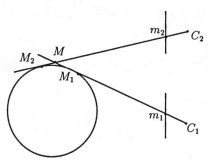

Figure 28: *Pixels m_1 and m_2 are not the images of the same physical point on the cylinder.*

further analysis, this information is usually lost, making the task of computer vision more difficult.

For example, figure 27 shows examples of several of these edges: edges labeled **r** correspond to a rapid change of the reflectance of the vase, edges labeled **s** are caused by the shadow of the vase cast on the parallelepiped that supports it, edges labeled **d** correspond to discontinuities of the distance of the object to the camera, the normal to the object being discontinuous there (peaks and troughs), edges labeled **n** signal a discontinuity of the normal to the object without a discontinuity of the distance. Finally edges labeled **dc** signal a discontinuity of the distance to the camera with no discontinuity of the normal to the object, which varies smoothly in the vicinity of the edge. The chapter by Henry and Merle discusses the topological relationships between **s** edges and **dc** edges.

Labeling correctly and robustly those edges from just one image has proved to be a formidable task. If we have several images of the same scene taken from slightly different viewpoints as in the case of a moving observer or in the case of stereo, then the task may be a little simpler.

Let us consider for example the case of stereo. Many algorithms proposed for doing stereo are edge based (Grimson 1985; Baker and Binford 1981; Ohta and Kanade 1985; Ayache and Lustman 1987,...). After matching these edges they can provide depth along the edges which can be interpolated to yield a surface representation of the scene (Grimson 1983; Faugeras *et al.* 1990). Classification of the edges can then be obtained from this surface model.

This sounds like a reasonable thing to do, but be cautious! This analysis assumes that an edge in an image which is matched to an edge in another image both correspond to the images of the **same** physical event in the 3D scene. Unfortunately, this is not true of the edges labeled **dc** in figure 27 as the reader may convince himself by looking at figure 28, which represents the stereo geometry, in a plane going through the optical centers of the two cameras. Pixels m_1 and m_2 are both on the outline of the cylinder as seen from the two cameras but because of the geometry of the object they are not the images of the same physical point: if our stereo algorithm matches them they will yield the reconstructed point M which is an error.

Such edges have received several names in the literature: some authors call them *obscuring edges* or *occluding contours*, *extremal boundaries* when referring to their images, or the *limb* or the *rim*, when referring to them in 3D. Note that the rim is not a physical marking on the object but depends both on its geometry and the viewing position.

It looks therefore as if we should pay attention to those edges and that further processing of the data is necessary in their vicinity. We may take the pessimistic view of

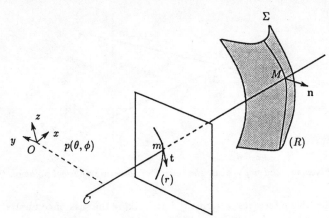

Figure 29: *A Rim (R) and its image (r).*

abandoning the hope of obtaining an accurate reconstruction of the object near those edges. As we show in this paper, this would be a big blunder since it is precisely there that we can extract the most information about the geometry of the object: unlike edges such as **d**, **r**, **s**, or **n** where only the distance to the cameras can be recovered, extremal boundaries allow us to recover the distance, the normal to the object, and the principal curvatures, or in terms of differential geometry, the differential properties of order 0, 1 and 2.

If this statement is correct (and we show that it is in the rest of the paper) obscuring edges appear more as a forgotten treasure than as a nuisance for computer vision, as a superficial analysis might conclude.

In this paper, we propose a new method for detecting extremal boundaries. We also propose an algorithm for reconstructing exactly the curves observed by each camera and computing the principal curvatures of the object surface along them.

In light of all this, our approach has several specificities which we think are interesting:

1. we identify occluding contours from triplets of images by estimating the radius of curvature of a special planar cross-section of the object surface: the radial curve (defined in section 4.2.1).

2. on the identified occluding edges, we detect the parts, if any, which are images of points on the object with zero Gaussian curvature (parabolic points).

3. we model the object as the envelope of its tangent planes and use the Gauss map (defined in section 4.1) to compute the depth, the normal and the second fundamental form of the surface along the occluding edge (except at the parabolic points where the Gauss map is degenerate).

4. we use a general camera model that uses perspective projection and is not restricted to orthographic projection.

4.1 Computing Differential Properties of the Object along the Rim

The basic idea is to consider the surface of the object as the envelope of its tangent planes.

As shown in figure 29, we consider a fixed coordinate system $(Oxyz)$; the optical center is at C. The camera looks at the rim (R) on the surface (Σ) which produces the occluding contour (r). A point m on (r) is the image of a point M on (R) at which the optical ray determined by Cm is tangent to the object surface. The tangent plane to the surface at M is defined by the optical ray and the tangent $\mathbf{t}$ to the occluding contour at m. Let $\mathbf{n}$ be the unit length normal vector to this plane, defined by its Euler angles θ and ϕ and $p(\theta, \phi)$ the distance from the origin to the tangent plane. The equation of this plane can be written as:

$$\mathbf{n}(\theta, \phi) \cdot \mathbf{X} - p(\theta, \phi) = 0 \qquad (23)$$

where $\mathbf{X}$ is the vector $(x, y, z)^T$ and $\mathbf{n} = (\cos\theta\cos\phi, \sin\theta\cos\phi, \sin\phi)^T$. With respect to image measurements,

$$\mathbf{n}(\theta, \phi) = \frac{Cm \wedge \mathbf{t}}{]]Cm \wedge \mathbf{t}]]} \text{ and } p(\theta, \phi) = \mathbf{n}(\theta, \phi) \cdot \mathbf{OC}$$

The observation of an occluding edge immediately yields the normal to the object (differential property of order 1).

In this paragraph, we state that under some hypothesis, (θ, ϕ) is a parametrization of the surface (Σ) in the neighbourhood of a point M. We consider the Gauss map:

$$\mathbf{N} : (\Sigma) \rightarrow S^2$$

where S^2 is the unit sphere of R^3. To each point M of (Σ), $\mathbf{N}$ associates the point of S^2 where the normal to (Σ) pierces the Gauss sphere.

It is known that the Gauss map is singular if and only if $\mathcal{K}_g(M)$, the Gaussian curvature of (Σ) at $M \in (\Sigma)$ is null.

A proof of this classical result can be found in Do Carmo (1976). This theorem says that at a non-parabolic point of the surface (Σ) the Gauss map is non singular.

Now consider the mapping $(\theta, \phi) \rightarrow p(\theta, \phi)$ which associates with every direction the distance from the origin to the plane tangent to the surface whose normal is in the direction (θ, ϕ).

It is possible to show that for every non-parabolic point, (θ, ϕ) is a parametrization of (Σ) and we can produce parametric equations of the surface.

This parametrization allows us to compute the two main curvatures of the surface for a point M of the rim of the surface. A remarkable point is that this computation requires only second order derivatives of the image measurements. This may come as a surprise since a superficial analysis would make us expect that the main curvatures which are second order differential properties of the surface would depend upon the third order derivatives of p.

The result is that the evaluation of the first and second fundamental quadratic forms requires an estimation of the value of θ, ϕ, $p(\theta, \phi)$, $\frac{\partial p(\theta, \phi)}{\partial \theta}$, $\frac{\partial p(\theta, \phi)}{\partial \phi}$, $\frac{\partial^2 p(\theta, \phi)}{\partial \theta^2}$, $\frac{\partial^2 p(\theta, \phi)}{\partial \phi^2}$, $\frac{\partial^2 p(\theta, \phi)}{\partial \theta \partial \phi}$. This is very interesting since these values can be estimated with sufficiently good accuracy for the points belonging to an extremal boundary, as shown later.

The problem now is that derivatives have to be estimated in the discrete case and this is known to be sensitive to noise, and especially so since we have only three images to work with.

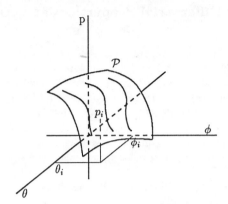

Figure 30: *The pedal surface* $(\mathcal{P})$.

In summary, there are two steps in our computation:

1. Detect the extremal boundaries.

2. Compute $p(\theta, \phi)$ and its derivatives.

Mathematically, there are no difficulties; it is in practice that they arise. Indeed we measure pieces of the surface $[\theta, \phi, p(\theta, \phi)]^T$, the pedal surface $(\mathcal{P})$, from which we have to estimate the first and second order derivatives of p which in turn yield properties of the object surface.

How is this estimation carried out? Returning to figure 29 and assuming that the curve (r) in the retinal plane has been identified as an occluding contour (the way of achieving this identification is explained later in section 4.2), if we move point m along (r), the tangent plane varies in a predictable manner and we obtain, in general, a piece of curve drawn on the pedal surface (figure 30). If we move the camera a bit and assume that we know its motion, we observe another occluding contour and generate another piece of curve on $(\mathcal{P})$.

For a given point M on the rim (R) defined by the values of θ, ϕ and $p(\theta, \phi)$, if we can obtain sufficiently many curves on the surface $(\mathcal{P})$ in the vicinity of $(\theta, \phi, p(\theta, \phi))^T$ by moving the camera with sufficient accuracy then we may hope that the first and second order derivatives of p can be accurately estimated and therefore that M can be reconstructed and the differential properties of the object surface at M computed.

In the following sections, we investigate robust ways of achieving such a goal.

4.2   Detection of Occluding Contours

We suppose that we have observed an edge from several viewpoints and we want to decide if it is an extremal boundary, and if so to compute some properties of the surface in the neighbourhood of the rim, as explained in section 4.1. We are interested in the differential properties of the surface up to the second order. Fundamental theorems of differential geometry (Do Carmo 1976) assert that these properties are sufficient to characterize the surface up to a rigid displacement.

- The zero order differential property is the simple estimation of the position of the point. It means that we have to compute the exact position of the contact point between the surface and the optical ray for each of the cameras.

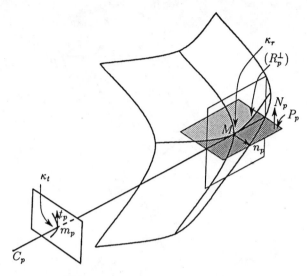

Figure 31: *The radial curve* $(R_p^\perp)$.

- The first order differential property is the estimation of the tangent plane to the surface. It is the easiest to obtain as we are observing an extremal boundary. In this case the tangent plane is the optical plane.

- The second order properties are the most difficult to obtain as they require the evaluation of second order derivatives. Such computation can be sensitive to noise.

4.2.1 Computing Differential Properties of the Object along the Rim: the Radial Curve Method

We remember that the rim (R) of a surface is a curve, and thus the image (r) of the rim must be a curve. It is always true in a generic position. So, we can assume that we have detected a curve (r_p) in each image ($p = 1, 2, 3$ in practice). For each of these curves, it is possible to compute the tangent vectors at each of their points.

Let us choose one camera p and m_p one point on the curve (r_p). We consider the curve $(R_p^\perp)$ which is the intersection of the surface of the object with the plane $P_p = (C_p, \mathbf{N_p})$, where C_p is the optical center of the camera p and $\mathbf{N_p} = (C_p m_p) \wedge \mathbf{n_p}$. $\mathbf{n_p}$ is the normal to the optical plane tangent to the surface at M_p (figure 31). This plane can be easily constructed with the tangent $\mathbf{t_p}$ to the occluding contour (r_p). The planar curve $(R_p^\perp)$ is called in the literature the radial curve and its curvature is called the **radial curvature** κ_r. The curvature of the curve (r_p) is called the apparent curvature or **transverse curvature** κ_t.

The key idea of our method is to neglect the apparent curvature and to use only the radial curvature. The objectives are to be able to decide if an edge is an extremal boundary or not, and if it is to compute the coordinates of the point M_p which belongs to the surface of the object and projects to point m_p.

We can draw another planar curve on the surface of the object which has interesting properties. Consider another camera, q, and the plane $E_{p,q} = (C_p, m_p, C_q)$ (figure 32).

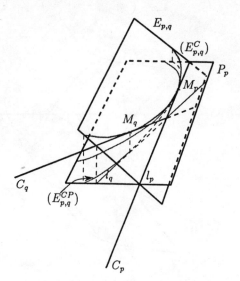

Figure 32: *The radial curve $(R_p^\perp)$ and the epipolar curve $(E_{p,q}^C)$.*

This plane is called the epipolar plane of m_p with respect to camera q[3]. Suppose that this plane intersects the image (r_q) of the rim seen from camera q at a point m_q. The intersection of the plane $E_{p,q}$ with the surface of the object is a curve $(E_{p,q}^C)$. M_q belongs to this curve. The epipolar plane $E_{p,q}$ contains the optical ray (C_p, m_p) and this implies that the point M_p belongs also to the curve $(E_{p,q}^C)$. Moreover we can prove that[4] $(R_p^\perp)$ and $(E_{p,q}^C)$ have the same tangent at the point M_p. The tangent at M_p to $(R_p^\perp)$ is the optical ray by construction. The tangent at M_p to $(E_{p,q}^C)$ is in the plane $E_{p,q}$ by construction, and in the plane tangent to the surface at M_p, since this curve is drawn on the surface. As a consequence, the tangent at M_p to $(E_{p,q}^C)$ belongs to the following two planes: the plane tangent to the surface at M_p and the plane $E_{p,q}$. The intersection of these two planes is the optical ray because the optical ray belongs to both planes. Consequently the two curves $(R_p^\perp)$ and $(E_{p,q}^C)$ have the same tangent at the point M_p. They have also the same normal curvature since they have the same tangent M_p.

Now we consider the curve $(E_{p,q}^{CP})$ obtained by projecting the curve $(E_{p,q}^C)$ orthogonally in the plane P_p. The tangent at M_p to $(E_{p,q}^{CP})$ is the optical ray. The tangent at the projection of M_q to $(E_{p,q}^{CP})$ is the projection l_q of the optical ray (C_q, m_q) on the plane P_p. If there is no apparent curvature, the two curves $(R_p^\perp)$ and $(E_{p,q}^{CP})$ will be the same. The distance between these two curves depends on the magnitude of the apparent curvature and the angle between the two planes. If the viewpoints are not too different, they will be very close. The idea of our method is to neglect the apparent curvature and to use these different projections to compute the radial curvature at M_p.

From the coordinates of the points m_p and m_q and their tangent vectors $\mathbf{t_p}$ and $\mathbf{t_q}$ and the calibration parameters, we can do the following:

- Compute the optical ray (C_p, m_p).

- Compute the radial plane $P_p = (C_p, \mathbf{N_p})$ with $\mathbf{N_p} = (C_p m_p) \wedge \mathbf{n_p}$.

---

[3]This plane is defined by the three points m_p, C_p and C_q.
[4]For simplicity, the curve $(R_p^\perp)$ has not been drawn in figure 32.

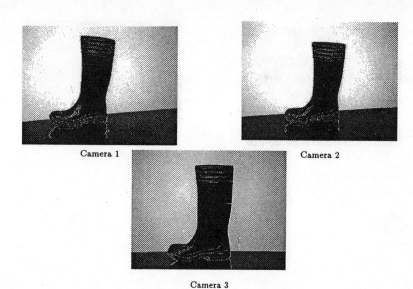

Camera 1   Camera 2

Camera 3

Figure 33: *Images of the triplet "boot"and the polygonal approximations of the edges.*

- Compute the optical ray (C_q, m_q) and its orthogonal projection l_q in the plane P_p.

We now have a set of lines l_q in the plane P_p. We have exactly one line per camera. The next question is how to find an approximation of the radial curve and especially of the point M_p. The curve $(R_p^\perp)$ is approximately tangent to all the lines l_q. As we have supposed that our viewpoints are not too different, the different contact points must be neighbours and we can use these points for estimating the osculating circle to the radial curve at M_p. This problem is that of finding the center and the radius of the cylinder and can be solved very simply. We also compute the deviation σ_r on the radius r.

At the end, we have an estimation to the osculating circle of the radial curve and an idea of the validity of this approximation through an estimation of the variance of the computed radius. We thus have a method for deciding whether or not an edge is occluding or not. This method is based on the computation of the radial curvature and its uncertainty and is very simple: if the radius and the variance are both very small, or if twice the variance is larger than the radius, we conclude that the edge is not an occluding edge, otherwise it is.

The position of the point M_p can then be easily deduced from the line l_p and the parameters of the circle.

We can then estimate the first and second derivatives of $p(\theta, \phi)$. In order to do this we consider the pedal surface $(\mathcal{P})$ (figure 30), each camera yields one curve drawn on $(\mathcal{P})$. $(\mathcal{P})$ is simple since we have assumed that locally (θ, ϕ) is an admissible parametrization of the object surface. The best way to estimate the first and second derivatives of p is to fit locally a surface on the measured data in the space (θ, ϕ, p).

4.3 Experimental Results

We have tested those ideas on synthetic and real data. We present only the real data here. We present two scenes: the first one contains a boot and the second one a bottle.

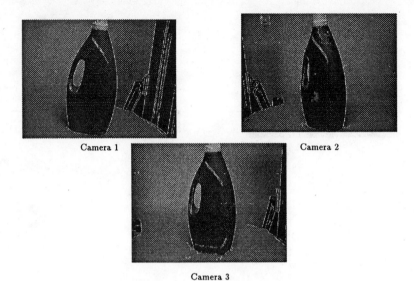

Camera 1   Camera 2

Camera 3

Figure 34: *Images of the triplet "bottle" and the polygonal approximations of the edges.*

- Figures 33 and 34 show the original images with the edges overlaid on them.

- Figures 35.a and 37.a represent the reconstructed chains. The extremal boundaries are represented with fine continuous lines. The other contours are represented with thick continuous lines.

- Figures 35.b and 37.b represent the sign of the Gaussian curvature along the extremal boundaries (thick continuous lines $\Longleftrightarrow \mathcal{K}_g > 0$, fine continuous lines $\Longleftrightarrow \mathcal{K}_g < 0$, very thick point $\Longleftrightarrow \mathcal{K}_g = 0$).

- Figures 36 and 38 represent the local surface that we have computed along the extremal boundaries. The dotted lines represent contours which are not extremal boundaries. For example, there is a marking on the bottle.

It can be seen from figures 36 and 38 that the general agreement between the results of the method and the real shape is very good.

4.4   Conclusion

We have shown that extremal boundaries can be used as a robust source of 3D information: they can be identified in images if at least three views are available and points on the rim can be accurately reconstructed. Moreover, good qualitative and quantitative estimates of the second order differential properties of the surface in the vicinity of the rim can be reliably computed.

This opens some interesting avenues for building complete models of objects shapes by active exploration as well as for the navigation of robotics systems in non-polyhedric environments.

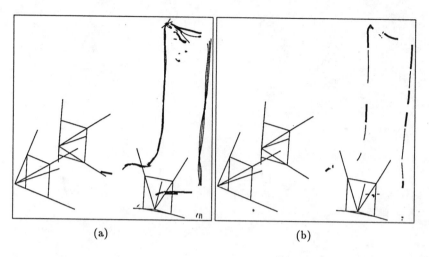

(a)   (b)

Figure 35: *A boot (a) The reconstructed contours (b) The sign of the Gaussian curvature (see text for explanation).*

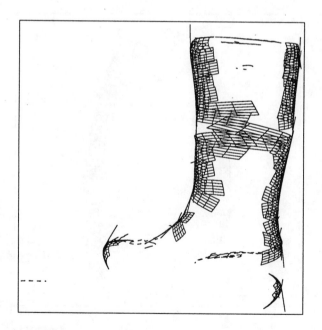

Figure 36: *The boot: the reconstructed extremal boundaries and the local surface that we have estimated from the principal curvatures and directions.*

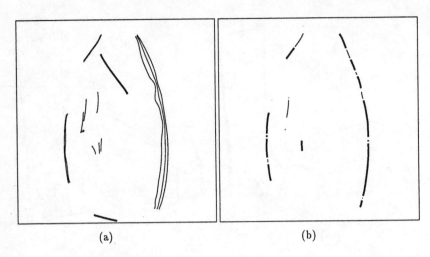

(a)   (b)

Figure 37: *A bottle (a) The reconstructed contours (b) The sign of the Gaussian curvature (see text for explanation).*

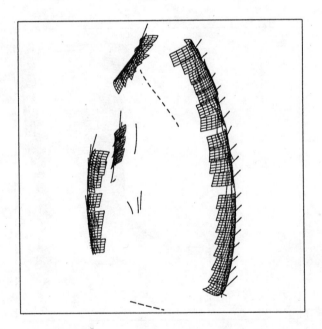

Figure 38: *The bottle: the reconstructed extremal boundaries and the local surface that we have estimated from the principal curvatures and directions.*

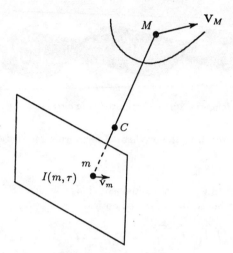

Figure 39: *Optical flow.*

5 Motion of Curves

In this part, we study the motion field generated by moving 3D curves which are observed by a camera. The emphasis will be made more on qualitative results than on equations. Interested readers can find all the proofs in (Faugeras 1990a; Faugeras 1990b, Faugeras 1991a; Faugeras and Papadopoulo 1991).

5.1 Optical Flow and Motion Field

The concept of optical flow was introduced by Gibson (1950) and is based upon the idea that there is a relationship between the temporal variations of the image intensity at one point of the retinal plane and the motion of the camera, and the motions and the shapes of objects present in the scene.

Let us consider the image intensity $I(x, y, \tau)$ at pixel m of coordinates (x, y) in the retinal plane at time τ. m is the image of a 3D point M moving in the scene with a velocity $\mathbf{V}_M$. The velocity of m is $\mathbf{v}_m = [v_x, v_y]^T$ (see figure 39).

Now let us look at the equation:

$$\nabla I \cdot \mathbf{v}_m + \frac{\partial I}{\partial \tau} = 0 \tag{24}$$

This equation has been presented by many authors as a constraint on the velocity $\mathbf{v}_m$, but it is true only under the constraint of $\dot{I} = 0$. In particular, it is discussed in the chapters by Hans-Helmut Nagel and by Verri, Straforini, and Torre.

In fact, this constraint does not hold even for non purely translating-Lambertian objects. Nevertheless, it can be used as a *definition* of optical flow:

$$\mathbf{v}_m^o = -\frac{\frac{\partial I}{\partial \tau}}{\| \nabla I \|} \frac{\nabla I}{\| \nabla I \|} \tag{25}$$

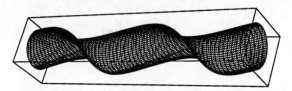

Figure 40: *The spatio-temporal surface generated by a circle rotating in front of the camera.*

So the standard approach to the problem of recovering the motion field is:

- Identify as $\mathbf{v}_m^o = \mathbf{v}_m \frac{\vec{\nabla} I}{\|\nabla I\|}$

- Invent a smooth field $\hat{\mathbf{v}}_m$ whose component along the direction of the image gradient is as close as possible to the measured optical flow $\mathbf{v}_m^o$.

$$\min_{\mathbf{v}_m} \int \int ((\mathbf{v}_m^T \frac{\nabla I}{\| \nabla I \|} - \mathbf{v}_m^o)^2 + \lambda Tr(D\mathbf{v}_m (D\mathbf{v}_m)^T))dxdy \qquad (26)$$

where $D\mathbf{v}_m$ is the derivative of the function $(x, y) \to \mathbf{v}_m(x, y)$ at the pixel m of coordinates (x, y).

A detailed analysis of the possible solutions to this minimization problem can be found in (Horn 1986). For a related approach, see (Hildreth 1984). Of course there is no guarantee that the computed field $\hat{\mathbf{v}}_m$ is close to the motion field $\mathbf{v}_m$ and in fact it is, in general, different.

Now, let us have a look at the relationship between $\mathbf{v}_m$ and $\mathbf{V}_M$ in the case of a curve.

5.2   The Motion Fields of a Curve

Suppose that we observe in a sequence of images a family (c_τ) of curves, where τ denotes time, which we assume to be the perspective projection in the retina of a 3D curve (C) that moves in space. If we consider the three-dimensional space (x, y, τ), this family of curves sweeps in that space a surface (Σ) defined as the set of points $((c_\tau), \tau)$. As an example, figure 40 shows the spatio-temporal surface generated by a circle rotating around one of its diameters in front of the camera.

At a given time instant τ, we call s the arclength of (c_τ) and S the arclength of (C). We further suppose that S is not a function of time (i.e. the motion is isometric). Now, for a point m on (c_τ), it is possible to define two different motion fields:

- The *apparent motion field* $\mathbf{v}_m^a$ (*a* for *apparent*) of $m(s, \tau)$ is the partial derivative with respect to time when s is kept constant, $\frac{\partial \mathbf{m}}{\partial \tau} = \mathbf{m}_\tau$.

- The *real motion field* $\mathbf{v}_m^r$ (*r* for *real*) is the partial derivative of $m(s, \tau)$ with respect to time when S is kept constant, or its total time derivative $\dot{\mathbf{m}}$. This field is the projection of the 3D velocity field $\mathbf{V}_M$ on the retina.

Moreover, introducing the Frenet frame $\mathbf{t}$, $\mathbf{n}$, where $\mathbf{n}$ is the unit normal vector to (c_τ) at m, we have

$$\mathbf{v}_m^a = \alpha \mathbf{t} + \beta \mathbf{n}$$
$$\mathbf{v}_m^r = w \mathbf{t} + \beta \mathbf{n}$$

where α is the tangential apparent motion field, w is the tangential real motion field and β is the normal motion field.

With these notations, under the weak assumption of *isometric* motion, we reach the following conclusions from the study of the spatio-temporal surface:

1. The normal motion field β can be recovered from the normal to the spatio-temporal surface,

2. the tangential apparent motion field can be recovered from the normal motion field,

3. the tangential real motion field *cannot* be recovered from the spatio-temporal surface.

Therefore, the full real motion field is not computable from the observation of the image of a moving curve under the isometric assumption. This can be considered as a new statement of the so-called *aperture* problem. In order to solve it, it is *necessary* to add more to the hypothesis, for example that the 3D motion is rigid. This is what will be done in the next section. Note that this is *not* what previous authors have done (Horn and Schunk 1981; Nagel 1983; Hildreth 1984; Gong 1989; Bouthemy 1989). We suspect that they actually compute the apparent motion field, which can be quite different from the real one. This of course must have some consequences on the accuracy of the 3D motion that is computed from the image flow.

5.3 Rigid Motion

Let us now assume that the curve (C) is moving rigidly. Let (Ω, V) be its kinematic screw at the optical center O of the camera. Moreover, suppose that the camera has been normalized by calibration to unit focal length.

The point is that now V_M is not just *any* 3D motion field:

$$V_M = V + \Omega \times OM \qquad (27)$$

thus by taking the total time derivative of the perspective equation (see figure 41)

$$zm = M \qquad (28)$$

we can compute w as a function of β and the kinematic screw.

The basic idea is then to combine the structural information about the geometry of (C) with purely kinematic information about the motion of its points.

In order to do this, it is necessary to introduce the accelerations $\dot{\Omega}$ and $\dot{V}$.

So, the following theorem can be stated:

Theorem 1 *At each point of (c_r) we can write two polynomial equations in the coordinates of Ω, V, $\dot{\Omega}$ and $\dot{V}$ with coefficients which are polynomials in quantities that can be measured from the spatio-temporal surface (Σ).*

The total degree is 6 for the first equation, 4 for the second. Both equations are homogeneous in $(V, \dot{V})$, of degree 4 for the first and 2 for the second. The degrees in V are 4 for the first one, 2 for the second. Both equations are linear in $\dot{V}$ and $\dot{\Omega}$. The degree in $(\Omega, \dot{\Omega})$ is 3 for the first equation, 2 for the second. The degrees in Ω are 3 for the first equation, 2 for the second.

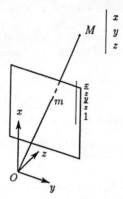

Figure 41: *The perspective projection.*

Thus, N points on (c_τ) provide $2N$ equations in the 11 unknowns[5] Ω, V, $\dot{\Omega}$, and $\dot{V}$. Therefore, we should expect to be able to find, in some cases, a finite number of solutions. Degenerate cases where this is not true do not exist can be easily found: straight lines, for example (Faugeras et al. 1989), are notorious for being degenerate from that standpoint. The crucial problem of studying the cases of degeneracy is left for further research. Note that this problem is equivalent to the following question:

For a given curve (C) and motion (V, Ω), $(\dot{V}, \dot{\Omega})$, how many algebraically independent equations can be obtained by the method just described?

As an example, in the case of a straight line the answer is 2.

Ignoring for the moment those difficulties (but not underestimating them), one major conjecture/result can be stated:

Conjecture 1 *The kinematic screw Ω, V, and its time derivative $\dot{\Omega}$, $\dot{V}$, of a rigidly moving 3D curve can, in general, be estimated from the observation of the spatio-temporal surface generated by its retinal image, by solving a system of polynomial equations. Depth can then be recovered at each point through equation up to a scale factor. The tangent to the curve can be recovered at each point.*

5.3.1   Some Simple Examples

The problem, as stated in the previous section, is difficult to solve because of the great number of unknowns. In order to validate the theory, some examples have been studied in detail. First, it can be proved that, in the case where the observed motion is in a plane parallel to the retinal plane, the apparent and real are in general identical[6].

In the following examples, we suppose that we observe a 3D conic[7]. We also consider only the one of the two polynomials with the lowest degree (because the number of solutions of a system of polynomial equations is related to the degrees of the polynomials). Two approaches are then possible:

---

[5]There are only 11 unknowns because of homogeneity of the equations in $(V, \dot{V})$. This is related to the fact that, with one camera depth can be recovered only up to a scale factor.

[6]There are some degenerate cases such as a circle whose center is on the optical axis for which it is not possible to measure the rotation in the image plane.

[7]Conics might not be a very restrictive condition since one of the two equations depends only on second derivatives on the spatio-temporal surface. Thus, locally, a conic might be a good model.

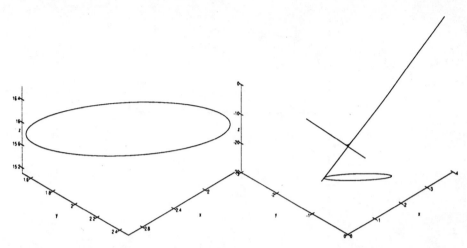

Figure 42: *The reconstructed curves corresponding to the two solutions of the motion equations: the left-hand side is the correct solution, the right-hand side shows the superposition of the correct one and the incorrect one which is a hyperbola.*

- Take a set of points along the curve, evaluate the polynom at each point and solve the system.

- Eliminate one coordinate between the polynomial and the equation of the conic in order to have all the constraints that can be expressed along the conic. For the small polynomial this gives 17 equations.

Moreover, once a solution has been found at time t then this solution can be tracked over time: this method has the advantage of avoiding the necessity of finding all the solutions of the system at each time and, given an initial set of solutions, it allows one, after a few steps, to select only the right answer.

Pure translational motion   In this case, there are only 5 unknowns, $\mathbf{V}$ and $\dot{\mathbf{V}}$[8]. We first consider the case of a circle of center $C = [\tau, 2, 2\tau^2]^T$ and radius 10. At time $\tau = 2.3$ the velocity and accelerations are $\mathbf{V} = [1, 0, 13.8]^T$ and $\dot{\mathbf{V}} = [0, 0, 6]^T$. Solving the system gives two solutions. One of them is the correct solution, the second one is incorrect and yields a reconstructed curve which is a hyperbola. Results appear in figure 42 which shows the reconstructed curves and the real one. The left-hand side figure shows a perfect superposition between the real and the reconstructed curves while the right-hand side shows the difference between the extra solution and the real one. The computed motion is in this case $\mathbf{V} = [1, 1.176, 7.845]^T$ and $\dot{\mathbf{V}} = [10.394, 7.132, 74.422]^T$.

Now add a second circle, not in the same plane as the first, and moving with the same rigid motion, and the solution to the motion then becomes unique.

It is important to note that this experiment, and the one described in the next section are only simulations, since the coefficients of the polynomial equations are not estimated from the spatio-temporal surface but computed analytically. What does it tell us then? Clearly, it does not tell us anything about the numerical stability of the problem, this

---

[8]V is normalized so that $V_x = 1$.

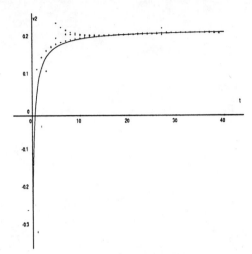

Figure 43: *The v_2 motion parameter curve over time. Here the points provided by the edge detector were directly used to compute the spatio-temporal parameters. The plain curve is the theoretical solution whereas the crosses are the computed values.*

is not the purpose of this paper, but it tells us that the number of "ghost" solutions is very small, and in the case of two circles, equals 0, even though this is the case of a very simple planar curve. Our guess is that for more general curves, the solution will always be unique.

What about numerical stability? In the following example, a synthetic sequence of ellipses was taken and the coefficients of the polynomial equations estimated from the spatio-temporal surface. In figures 43-47 we show the variation over time (the horizontal axis) of the second component of v_2 of **V**. Figures 43 and 44 show this component over time with different noise ratios.

These figures were obtained with the "with-elimination" method. Interestingly enough, as shown in figure 45, the "point" method seems more unstable.

What is happening when one tracks the first set of solution over time? Figures 46 and 47 show that after 3 or 4 steps only one solution remains.

Example of a mobile robot moving on a flat ground   The second example is the case of a mobile robot moving on a flat ground with two translational degrees of freedom (x and y) and one rotational degree of freedom (z). The first one is the correct one $V_x = 1$, $V_y = 0$, $\dot{V}_x = 0.385$, $\dot{V}_y = 0$, $\Omega_z = 1.3$, $\dot{\Omega}_z = 0.5$, the reconstructed curve is shown in the top left-hand side of figure 48, the second one is incorrect, $V_x = 1$, $V_y = -0.047$, $\dot{V}_x = -0.292$, $\dot{V}_y = -0.097$, $\Omega_z = 1.130$, $\dot{\Omega}_z = 0$, the reconstructed curve is a circle which is shown with the correct one in the top right-hand side of the same figure, and the third one is also incorrect, $V_x = 1$, $V_y = -1.400$, $\dot{V}_x = -1.242$, $\dot{V}_y = -2.281$, $\Omega_z = 0$, $\dot{\Omega}_z = 0.063$, the reconstructed curve is a hyperbola, which is shown with the correct one in the bottom part of the same figure.

In figure 49 the comparison between the apparent and real motion fields along the observed ellipse is shown. The difference between the two is clearly apparent.

Just as in the first example, a second circle, not coplanar with the first one but moving with the same motion, was added and a unique solution (the correct one) was found.

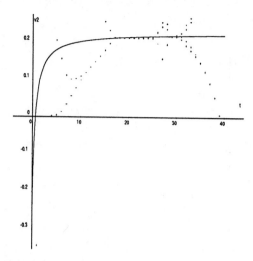

Figure 44: *The v_2 motion curve over time. Here a noise of 0.5 pixels was added to the points before computing the spatio-temporal parameters. The plain curve is the theoretical solution whereas the crosses are the computed values.*

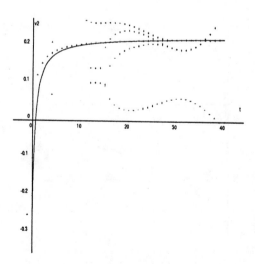

Figure 45: *The v_2 motion parameter curve over time. Here the points provided by the edge detector were directly used to compute the spatio-temporal parameters. The "point" method was used to find the motion parameters.*

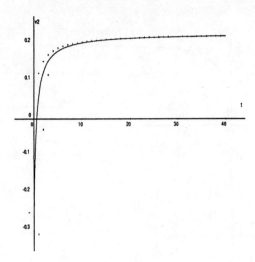

Figure 46: *Here the system was solved at time 0 and then the different solutions were tracked over time. After a few steps, all the solutions converge to one. For these results, no noise was added.*

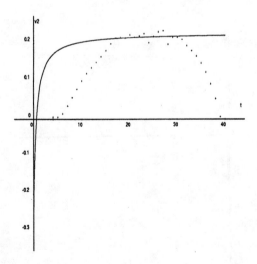

Figure 47: *Here the system was solved at time 0 and then the different solutions were tracked over time. After a few steps, all the solutions converge to one. For theses results, a noise of 0.5 pixels was added to each pixel of the initial image before computing the coefficients of the polynomials.*

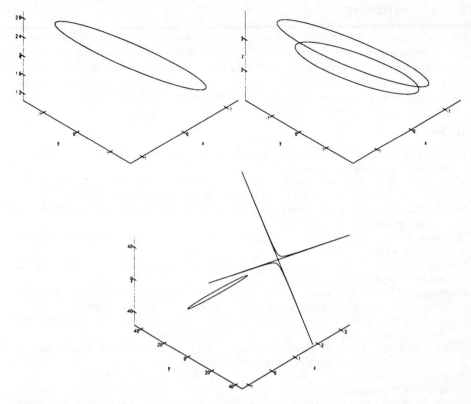

Figure 48: *The reconstructed curves corresponding to the three solutions of the motion equations: the upper left hand-side is the correct solution, the bottom and upper right-hand side show the superposition of the correct one and the incorrect ones.*

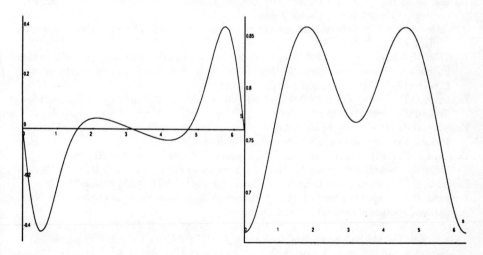

Figure 49: *A plot of the real (left) and apparent (right) motion fields along the observed ellipse.*

6 Conclusion

We have shown in this report that four major problems in machine vision, edge localization and characterization, vertexes and junctions detection, 3D shape from occluding edges, and 3D motion from moving curves could be solved using differential geometric tools. We have worked both on the theoretical side of the problems, completed computer implementations of our ideas, and tested them on synthetic and real images.

References

Ayache, N., Lustman, F. (1987) Fast and Reliable Passive Trinocular Stereovision. In: Proceedings First International Conference on Computer Vision

Baker, H., Binford, T.O. (1981) Depth from Edge and Intensity Based Stereo. In: Proceedings 7th Joint Conference on Artificial Intelligence. Vancouver, Canada, pp. 631-636

Beaudet, P.R. (1978) Rotational Invariant Operators. In: Proceedings International Conference on Pattern Recognition, pp. 579-583

Bergholm, F. (1987) Edge Focusing. IEEE Transactions on Pattern Analysis and Machine Intelligence 9(6), pp. 726-741

Berzins, V. (1984) Accuracy of Laplacian Edge Detectors. Computer Vision, Graphics and Image Processing 127, pp. 195-210

Bouthemy, P. (1989) A Maximum Likelyhood Framework for Determining Moving Edges. IEEE Transactions on Pattern Analysis and Machine Intelligence 11(5), pp. 499-511

Canny, J.F. (1983) Finding Edges and Lines in Images. Technical Report AI Memo 720, MIT Press, Cambridge MA

Canny, J.F. (1987) A Computational Approach to Edge Detection. IEEE Transactions on Pattern Analysis and Machine Intelligence 9(6), pp. 726-741

De Micheli, E., Caprile, B., Ottonello, P., Torre, V. (1989) Localisation and Noise in Edge Detection. IEEE Transactions on Pattern Analysis and Machine Intelligence 11(10), pp. 1106-1117

Deriche, R. (1987) Using Canny's Criteria to Derive an Optimal Edge Detector Recursively Implemented. The International Journal of Computer Vision 2, pp. 15-20

Deriche, R. (1990) Fast Algorithms for Low-Level Vision. IEEE Transactions on Pattern Analysis and Machine Intelligence 12

Deriche, R., Giraudon, G. (1990) Accurate Corner Detection : an Analytical Study. In: Proceedings of the 3th ICCV. Osaka, Japan

Do Carmo, M.P. (1976) Differential Geometry of Curves and Surfaces. Prentice-Hall, Englewood Cliffs, NJ

Dreschler, L., Nagel, H.-H. (1982) On the Selection of Critical Points and Local Curvature Extrema of Region Boundaries for Interframe Matching. In: Proceedings International Conference on Pattern Recognition, pp. 542-544

Faugeras, O.D. (1990a) On the Motion of 3-D Curves and its Relationship to Optical Flow. Technical Report 1183, INRIA

Faugeras, O.D. (1990b) On the Motion of 3-D Curves and its Relationship to Optical Flow. In: Proceedings First European Conference on Computer Vision.

Faugeras, O.D. (1991a) Sur le mouvement des courbes tridimensionalles et sa relation avec le flot optique. Comptes rendus de l'Academie des Sciences de Paris, t.312, Serie II, pp. 1279-1285

Faugeras, O.D. (1991b) Three-Dimensional Computer Vision. MIT Press, Boston MA

Faugeras, O.D., Papadopoulo, T. (1991) A Theory of the Motion Fields of Curves. The International Journal of Computer Vision (to appear)

Faugeras, O.D., Deriche, R., Ayache, N., Lustman, F., Giuliano, E. (1988) Depth and Motion Analysis: the Machine being developed within ESPRIT Project 940. In: Proceedings IAPR Workshop on Computer Vision (Special Hardware and Industrial Applications). Tokyo, Japan, pp. 35-44

Faugeras, O.D., Deriche, N.R., Navab, N. (1989) From Optical Flow of Lines to 3D Motion and Structure. In: Proceedings IEEE RSJ Workshop on Intelligent Robots and Systems '89. Tsukuba, Japan, pp. 646-649

Faugeras, O.D., Lebras-Mehlman, E., Boissonat, J.D. (1990) Representing Stereo Data with the Delaunay Triangulation. Artificial Intelligence Journal, 44 (1-2), pp. 41-87; also Technical Report 788, INRIA

Francois, E., Bouthemy, P. (1990) Multiframe-based Identification of Mobile Components of a Scene with a Moving Camera. Technical Report 564, IRISA, Rennes, France

Gibson, J.J. (1950) The Perception of the Visual World. Houghton Mifflin, Boston, MA

Giraudon, G., Deriche, R. (1991) On Corner and Vertex Detection. In: Proceedings International Conference on Computer Vision and Pattern Recognition. Maui, Hawaii

Gong, S. (1989) Curve Motion Constraint Equation and its Applications. In: Proceedings Workshop on Visual Motion. Irvine, CA, pp. 73-80

Grimson, W.E.L. (1983) An Implementation of a Computational Theory of Surface Interpolation. Computer Vision, Graphics, and Image Processing 22, pp. 39-69

Grimson, W.E.L. (1985) Computational Experiments with a Feature Based Stereo Algorithm. IEEE Transactions on Pattern Analysis and Machine Intelligence 7, pp. 17-34

Haralick, R.M. (1984) Digital Step Edges from Zero Crossing of Second Order Directional Derivatives. IEEE Transaction on Pattern Analysis and Machine Intelligence 6

Harris, C.G. (1987) Determination of Ego-Motion from Matched Points. In: Proceedings Alvey Vision Conference. Cambridge, U.K.

Hildreth, E. (1980) Implementation of a Theory of Edge Detection. Technical Report AI Memo 579. MIT Press, Cambridge, MA

Hildreth, E. (1984) Computation Underlying the Measurement of Visual Motion. Technical Report AI Memo 761. MIT Press, Cambridge, MA

Horn, B.K.P. (1986) Robot Vision. MIT Press, Boston, MA

Horn, B.K.P., Schunk, B.G. (1981) Determining Optical Flow. Artificial Intelligence 17

Huertas, A., Medioni, G. (1986) Detection of Intensity Changes with Subpixel Accuracy using Laplacian-Gausian Masks. IEEE Transactions on Pattern Analysis and Machine Intelligence 8, pp. 651-664

Monga, O., Rocchisani, J.M., Deriche, R. (1991) 3D Edge Detection using Recursive Filtering. CVGIP: Image Understanding 53

Nagel, H.-H. (1983) Displacement Vectors Derived from Second Order Intensity Variations in Image Sequences. Computer Vision, Graphics and Image Processing 21, pp. 85-117

Nagel, H.-H. (1988) Analyse and Interpretation von Bildfolgen. Informatik-Spektrum 8

Noble, J.A. (1988) Finding Corners. Image and Vision Computing 6, pp. 121-128

Ohta, Y., Kanade, T. (1985) Stereo by Intra- and Inter-scanline Search. IEEE Transactions on Pattern Analysis and Machine Intelligence, 7(2), pp. 139-154

Perona, P., Malik, J. (1990) Detecting and Localizing Edges Composed of Steps, Peaks and Roofs. In: Proceedings of the 3th ICCV. Osaka, Japan

Tsai, R., Huang, T.S., Zhu, W.L. (1982) Estimating Three- Dimensional Motion Parameters of a Rigid Planar Patch. II. Singular Value Decomposition. IEEE Transactions on Acoustic, Speech and Signal Processing 30, pp. 525-534

Verri, A., Poggio, T. (1986) Motion Field and Optical Flow: Differences and Qualitative Properties. Technical Report AI Memo 917. MIT Press, Cambridge, MA

Viéville, T., Faugeras, O.D. (1991) Robust and Fast Computation of High Order Spatial Derivatives in Image Sequences. In: Proceedings Insight Meeting

Weiss, I. (1991) Noise Resistant Invariants of Curves. In: Application of Invariance in Computer Vision. Dapra ESPRIT. Iceland

Geometry of Vision

*Michel Demazure, Jean-Pierre Henry, Michel Merle,
and Bernard Mourrain*

Centre de Mathématiques, Ecole Polytechnique, Palaiseau

1   General Introduction

Visual concepts are basic concepts of geometry. Among them, **Projection** is one of the most natural, one of the most intuitive, but also mathematically one of the most intricate: ambiguity is the drawback of ubiquity.

Projections are a crucial tool in several domains of mathematics : in differential and projective geometry but also in algebra (through elimination theory). The related concept of *direct image* is also fundamental (and sometimes difficult) in modern algebraic geometry.

This chapter modestly endeavours to apply these techniques to vision, using the well known fact that vision uses 2D images (*direct images*) to understand 3D surfaces. Computer vision experts are well aware that vision (by a camera) is a projection from a 3 dimension space to a 2 dimension portion of plane: the television screen.

The first part of the chapter uses differential and analytic techniques to define a system of geometric cues through projections. In fact two projections are used, because light rays from a light spot (creating shade contours) are added to visual rays (generating apparent contour) to enhance the comprehension of scenes.

The second part uses pure projective geometry to relate configurations of lines, planes and points in the *object* to configurations of points and lines in the *image*. An *algebra* of geometric objects is introduced. This has been implemented in LISP. Effective relations between geometry of image and geometry of object can be computed.

2 Shape, Shadow and Shape:

Classification of Views of Lighted Smooth Objects[1]

2.1 Introduction

We shall first motivate the use of geometric cues in intelligent vision, then explain why mathematics can be of some help to define and unify treatment of these cues. In subsection 3 of this introduction, we shall introduce a **unified mathematical treatment** of

1. a very classical cue: the apparent contour,

2. a less classical cue: the shade line separating full light from shadow,

3. a far less used cue: the cast shadow line.

They appear as a triplet of 3 curves, defined through 2 different projections, in the image. We shall then recall in section 2 what has been done in the now classical case where only the first cue is used. This includes n.⍺ny mathematical results, some ancient, that we shall only sketch and some very recent works applied to vision and inspired by the pioneering ideas of Jan Koendenrink. We shall also explain the differences from works in robotic vision which either

- do not use the special properties of the shadow lines,

- do not use the cast shadow cue,

- forget the origin of the cast shadow and do not relate it to the 2 others,

- use the grey level (or isophot) curves (usual *shape from shading*).

The next 2 sections will develop a mathematical treatment of the local situations arising from the properties of the 3 curves previously defined. We prove that, up to a change of coordinates, there is only *a finite number* of stable *"Sol y Sombra Patterns"* that can arise in generic situations:

1. a **Sol y Sombra Pattern** will be defined as a suitable *equivalence class of local singularity of the triplet of curves* .

2. a **situation** will be defined as a set-up {equation of surface, direction of light, direction of eye or camera}.

3. **generic situations** will be defined, first by impressing a *topology* onto a suitable range of applications, hence on the range of situations, then by deciding that around generic situations there must be a neighbourhood of situations where the induced Sol y Sombra Patterns must stay equivalent.

4. finally, we shall say that a Sol y Sombra Pattern is *stable towards small perturbations* (small motions of the observer or of the object or of the light), if, for nearby situations, the corresponding Sol y Sombra Patterns are in the same equivalence class.

---

[1]We are glad to thank Robert Fournier (INRIA, Sophia-Antipolis) for the use of ZICVIS, a versatile, user-friendly, software for drawing and viewing curves and lighted surfaces, a goody for geometers.

We shall also begin the classification of "Sol y Sombra Patterns" (singularity of the triplet of curves) that are not stable but can be seen when moving the "situation" (by moving the object or the observer or the light spot) along a line: these are the *singularities of codimension 1*.

We shall finally sketch how this apparatus can be used to classify, discriminate, and help in the use of the other usual cues that have already been developed in Vision theory.

2.2 A Mathematical INSIGHT in Vision?

2.2.1 Motivations

What is intelligent vision ? What do we really use to understand scenes ? What are the pertinent **features** to extract from images to detect objects or to distinguish between them ?

And second question but not less important, how do we organize these cues to classify objects ?

Why Classify?

Recognition needs a database Recognition means cognition and matching, therefore memory and a process, an *algorithm*, to match an image from the camera (the retina) with an image in the memory. Any (even the less sophisticated or the most specialized) Visual System needs some database to perform recognition. This database as well as the process of visual matching needs organisation and hierarchy, and therefore classification. Because robots as well as human beings or monkeys cannot be expected to memorize all aspects of all possible objects.

Using complexity theory to prove the need to classify Let us use a tool that J. Tsotsos emphasized (Tsotsos 1992), one of the achievements of computer science, *complexity theory*, can and must be used to discriminate between impossible and permissible procedures for vision.

A plain *complexity* consideration insures the necessity of a simplifying and classifying process:a very rough image of 512×512 pixels in black and white represents the huge amount of more than 250000 bits of information, i.e. a possibility of 2^{250000} different images.

A toy universe Let us be more clear by giving a simple example. Suppose that we restrict ourselves to an "universe" where there exists only two objects, a ball and a cube, and suppose that we get a perfect black and white camera with gives an image of 512×512 pixels, without blurring and with perfect detection of edges.

Suppose also that we always get an image of only one object at a time, and a full view of that object.

Evaluation of the total number of possible views Suppose that we want to implement a recognition algorithm based upon a database of all views of one of the object, preferably the ball. A view of the ball will be a dark disk on a white background. Therefore the number of possibilities is equal to the number of pairs (O, R) of admissible center and radius: the center has 512×512 permissible positions, the permissible radius depends on this position. By dividing the image into 8 parts and counting in each octant we arrive at a sum which looks like $\Sigma_{n=1..256}(\Sigma_{i=1...n}(i)) = \Sigma_{n=1..256}i(i+1)/2$ and is of the order of magnitude of $512^3/6$ or 20 megas.

Size of the data-base If this information is kept in the most elementary fashion (as a list of list of pixels) we would obtain as data base for this "toy universe" 20 millions of lists of 250 000 bits. Comparing it with an image (again a list of 250 000 bits) would require a ridiculous amount of time! Of course a standard image compressing algorithm, for example the popular *run length coding* (Horn 1986, **3.5**) would reduce the size of the representation of each possible image of the ball to less than $1kb$, but the size of number of entries (size of the directory) of the database would have to stay the same.

Classifying needs using neighbourhood considerations We human beings as well as artificial vision users should be able to use some sort of analogy or neighbourhood law, putting all chairs in a class and all automobiles in another one, and not mixing bananas with snakes.

How to Classify

Some usual cues Of course it seems that edges are important, and polyhedral objects provide images with lots of edges, so they should be and have been basic experimental support. More generally, for "smoothly curved surfaces", occluding contours are known to have fundamental impact on humans' perceptions (see theoretical analyses in Koenderink 1984; Koenderink and van Doorn 1984). Textures, and especially noticeable changes of textures, are also detected even in the absence of occlusions (for a discussion of the relation between these two clues, see Todd and Reichel 1990). In motion perception or in stereo vision edges or discontinuity of texture are generally used. First they are detected then matched. Optical Flow is an interesting artefact used for comprehension of motion and/or form of moving objects, especially when there are brilliant spots (due to specularity) or very dark patches on a surface. So what cues are to be used ?

Discontinuity as an origin of cues The key word appears to be discontinuity, discontinuity of texture, or discontinuity of orientation (two faces of a polyhedral object) or discontinuity of depth (an object on a background) or discontinuities in lighting.

Discontinuity out of smoothness What about smooth objects with constant texture (or slowly varying texture) ? What are the discontinuities that they generate ?

2.2.2 Mathematics to the Rescue?

Can mathematicians bring either unifying concepts and efficient tools for these two (strongly related) tasks?
We intend to show that, yes, to some extent, mathematics can help.
There are several heuristic reasons for this:

Singularities are Cornerstones in Mathematics

First of all discontinuities are no longer physical incongruities that most mathematicians try to avoid, they are mathematical objects of interest especially under the name of **singularities**. Singularity theory had always been an important part of mathematical developments and of mathematical motivations, especially in analysis, but, as more tools have been developed in recent years, especially in algebraic geometry, singularities are better understood.

Catastrophe theory penetrates all sciences   Ideas of René Thom (see Thom (1972-73, 1981) for original works, and Chaperon (1984), Bennequin (s.d.) or Gilmore (1981) for pedagogical introductions) have shown that classification of elementary singularities (Catastrophe Theory) could be used to understand many situations (see Zeeman (1976) and Hilton (1976)).

Classification

The second reason is related to the other key-word *classification*. Mathematicians are very fond of classification and they have developed numerous tools for this task.

Classification using several proximity notions   For example metrics and topology have been extensively used to give precise meaning to the usual words of "neighbourhood" or proximity and "distance". Another mathematical concept is *equivalence* which is used for classifying in *equivalence classes*, again this is a versatile concept with many avatars in several mathematical branches.   Another such concept is *isomorphism* which again is a mathematical way of saying that two (mathematical) objects are nearly the same (of the same form). To what extent they are identical depends on the "rigidity" of the isomorphism that is used. For example an analytic isomorphism is very strong and strict, a topological one is much looser, insofar as two objects can be topologically isomorphic but not analytically. We shall talk of equivalence (analytic, differentiable, topological) when 2 objects are isomorphic (analytically, differentiably, topologically).

Deformation as an intuitive and mathematical concept   Another natural idea that is related to the idea of "proximity of forms" or of " small perturbations" has been formalized by mathematicians using *deformation*. This is another way of measuring proximity of forms. Instead of looking at 2 views as 2 different points in some sort of big space of all views, and measuring their distance, we can think of them as 2 points *with a path* from one to the other. It becomes easier to measure their variation along the path, using infinitesimal calculus. If they are *neighbours* there should be a way for going smoothly (continuously or differentiably or analytically) from one to the other. Generally this procedure can be parameterized by a (small) interval of $\mathbf{R}$ for example $[0, \epsilon]$: we can find some functions depending continuously or differentiably, or in a polynomial (or even linear) way of this parameter, and such that for the value 0 the zero set is the first form, and for the value ϵ the zero set is the second one.

Specialisation and Genericity

It often happens that for t running along the path that is parameterized by the interval of $\mathbf{R}$, the objects F_t often stay isomorphic to the same object (call it F_{gen}), or even that above semi-open intervals, (which we can take to be $[0, \epsilon)$) the whole situation can be identified to a product (trivial) situation $(F_{gen} \times [0, \epsilon])$; but above some "special" points (which we can take to be 0) the *special* form F_0 is different. In that case we shall say that (for that deformation) F_ϵ is generic and specializes on F_0.

Example of specialization Consider a cylinder (with a circular base) rotating along an axis that is perpendicular to its central axis, and parallel to the plane of the camera. We can parameterize the movement or the deformation of views by the angle of the cylinder axis with the camera direction. As long as the angle is not 0 the images will look alike (with varying dimensions but same number of edges and faces): they can be explicitly described as a facet that is an elliptic disc, a surface bounded by part of the full ellipse, 2 parallel lines and a semi ellipse. But for the special value of the angle, when the cylinder is facing the camera, the image will be a circular disc. This is specialization of the generic view on the special one.

Geometry and Vision

Finally there is no need to recall that Geometry, which is evidently related to vision (through perspective, optic, cartography ...) is one of the historical sources of mathematics so we should find in it a lot of useful concepts. For example classical projective geometry (theorems of Pascal, Poncelet, Desargues, and even Thalès) has some vivid applications in robot vision especially in stereo matching (see the article of Bernard Mourrain). In the present chapter we shall focus on the use of some differential and mainly algebraic geometry.

Let us now proceed from these heuristic reasons to applications of math to vision.

2.2.3   Towards a Mathematical Formulation

We shall now proceed to define the cues that we intend to use in the rest of this paper as mathematical objects. We shall begin with a first very classical cue in vision studies.

The Apparent or Occluding Contour

We have seen that discontinuities were good cues. We can be a little more precise and use very simple mathematical terms. What is a view ? A view is a 2D representation of a 3D object. This can be seen as a projection either from a near point (conical projection) or an infinite point. The center of projection is the eye or the camera.

The apparent contour and the edges are lines of points where this projection is singular: either the tangent plane is seen as a line by the projection point or there is none (or two or even three secant tangent planes, that is too many). This is a singularity for the projection: at all those points there is no "isomorphism" between the tangent plane and the image plane (the television screen for example) even at a very low level, because they do not have the same dimension.

On a smooth object, which from a mathematical viewpoint can be seen as a smooth surface, other discontinuities can be seen: lines separating lighted parts from shaded ones. These can be seen again as lines of points which are singular for another projection, projection from the sun or from a light spot (at least when there is one point-like light source). The shade line is made of points where the sun rays are tangent to the surface, i.e. included in the tangent plane.

Now we have several lines on smooth surfaces, and we can now proceed to classify them and relate their "forms" to the forms of the projected surfaces.

Our aim is to introduce a description and a coarse classification of all stable Sol y Sombra Patterns of objects which are lighted by a one point source at infinite distance.

In fact by Sol y Sombra Pattern we will mean the apparent contour **and** the shadow edges of a smooth surface.

The Shade and Cast Shadow Curves

We look at a smooth object, i.e. a smooth generic surface defined by polynomial equations in 3-space, in Monge form we can define the surface M as the image of a 2-plane by

$$(x, y) \longrightarrow (x, y, f(x, y))$$

and take a first orthogonal projection λ of this surface onto a plane ; the critical locus for this projection is a curve on the surface, the "shade curve" S, which in most situations separates shaded parts from lighted parts on the surface (critical points for λ are points where light rays are tangent to the surface). Now if we take the set of all points where this light ray intersects again for the first time the smooth surface M we obtain a second curve, which we will call the "cast shadow curve", (or sometimes for short, the "shadow curve") O, which, together with the "shade curve", bounds the shaded part on the surface M.

The Sol y Sombra Pattern

Take another projection π of the surface orthogonally onto a plane, which will be the camera image, and look at the critical locus of this projection C and at the projections of this curve Γ, the *outline* (apparent contour of the surface which is the set of critical values of π since the surface is smooth), and of the *shade and shadow curves*.

By this process we get three plane curves:

1. the image Σ of the shade curve, which we will call the shade contour,

2. the image Ω of the shadow curve, which we will call the shadow contour,

3. the apparent contour Γ of the surface M (intersection of visual rays tangent to M with the image plane).

This apparent contour (which is sometimes called outline or profile of the surface) is also the image by π of what is called by computer-vision authors the "rim" or the contour generator.

The singularities of apparent contours of smooth surfaces have been studied by several authors (Whitney (1955), Mather (1973a and 1976), Gaffney and Ruas (Gaffney 1983), Bruce and Giblin (1985), Arnold and his school) and classified (for a survey see Bennequin (s.d.))

In this paper we go two steps forward, by adding the shade and cast shadow curves to the apparent contour and we call this artefact the Sol y Sombra Pattern.

Classifying The Singularities of the Contour, Shade and Shadow Curves

We intend to classify the types of their common local singularities, i.e. what we have called "Sol y Sombra Patterns" in the introduction. This study needs three steps:

1. Relate the singularity of this triple of curves to a convenient map, and construct a space of maps that model the "Sol y Sombra Patterns".

2. Define one (or several) equivalence relations within this space in a compatible way with analytic equivalence ("isomorphism") of local singularities.

3. Compute and describe the local singularities that remain isomorphic by small changes of the map.

We shall determine these types and show that their number is finite.

2.3   Mathematical Recalls and Notations

Our task is a little more complex than, but related to, the usual task of classifying the singularities of the sole apparent contour curve. We recall, in this section, the usual treatment of this problem, and its solution: there are only a finite number of stable (resp. codimension 1) singularity types for the apparent contour of generic surfaces.

Don't forget that in our application to vision, a View is a *projection* of a surface to a plane (a portion of a plane: the TV set on which we monitor the view of the camera). We will study local features precisely and we will say that two views are equivalent (see 2.3.1) if and only if there exists (local) diffeomorphisms of the surface and its image which identify the two projections. In particular the two apparent contours will be (locally) diffeomorphic as embedded curves in the plane.

We are interested in finding cases when the projection is stable to small perturbations (the view looks alike, when moving the camera or the object a little). We postulate, after Thom (Catastrophe theory, Thom 1972-73) that any feature that is not stable either is not seen by the camera, or is not detected because of noise and blurring. This postulation should be discussed in relation to scaling problems.

When one wants to study smooth objects (surfaces), it is possible to *classify* all projections of generic surfaces, or to classify generic projections of all surfaces (notice that the two cases overlap for generic projections of generic surfaces (Whitney 1955)).

In this paper we shall restrict ourselves to the first case. Some differences may appear in the classification according to the choice of the type of equivalence and topology on the set of maps to classify.

In the following subsections, we shall give very few historical milestones, and then we shall recall the mathematical tools to give precise meanings to the previous sentences.

2.3.1   Some Historical Ramblings

Previous and related work in **transversality, stability and preparation theorems,** can be coarsely classified in three groups.

Fundamental Theorems

The seminal task was Hassler Whitney's (Whitney 1955). The main mathematical work of defining concepts, relating geometric intuitions to precise statements, finding the good set-up for separating necessary hypothesis from uncertain grounds, and finally proving very hard theorems, is mainly due to René Thom and John Mather (Mather 1969, 1970, 1973a and 1976), with collateral work of J.-C. Tougeron *et al.*

Specialised Results

The papers of Arnold and his school (Arnold 1972, 1984; Platonova 1981, 1984; see also Bennequin (s.d.) for a survey) or C.T.C. Wall, Bruce *et al.* (Wall 1977; Bruce and

Giblin 1985; Gaffney 1983), *on this subject* is rather on the R&D side, using Mather's and Thom's theorems in specified contexts, adding lemmas and computations in specialised set-ups.

Applied Results

There is now a third firesquad using again Thom's and Mather's results and generalisations of the specialised lemmas of the second group, and aiming toward situations approximating more and more the real world (Rieger 1987 and 1987b; Tari 1990 and 1992; Bruce and Giblin). The present work can be included in that group.

Equivalence Classes in C^∞ Mappings

We want now to give a meaning to sentences of the type: "function f and function g define *similar forms*."

In other words we intend to define several equivalence relations in the space of infinitely derivable functions from one space to another.

A function g that can be obtained as composition of f with diffeomorphisms at the source space and at the target space is said to be diffeomorphically equivalent to f. As required, if two maps f and g are diffeomorphically equivalent, they have diffeomorphic critical locuses and diffeomorphic discriminant locuses[2]. In other words they have diffeomorphic apparent contours. If instead of diffeomorphisms at the source and at the target space, we take homeomorphisms, we shall only speak of Topological equivalence. If two maps f and g are topologically equivalent, their discriminant locuses are homeomorphic as *embedded* spaces in the target space (this consequence is not so easy to see as for the diffeomorphic equivalence).

Definition 2.3.2 *Stability*
We shall say that f is differentiably stable if the equivalence class of f for diffeomorphic equivalence contains all sufficiently near functions of f.

Infinitesimal Stability and Stability

We sketch a result, essentially due to Mather (see Chenciner 1972-1973 for a survey, and Mather 1969, 1973a and 1973b) that makes effective computations possible.

Instead of looking at the space of diffeomorphisms g in a neighbourhood of f Mather considers the tangent vector space. He compares it to the sum of 2 vector spaces: those corresponding to composition of f with diffeomorphisms on the source space (resp. the target space). The property that these 2 spaces are equal is also called "f is infinitesimally stable," and a theorem of Mather can be stated as "infinitesimally stable $\implies$ stable." More generally the codimension of the later space in the former space (when it is finite but not 0) characterizes the "defect" of stability.

Stable Contours

As a special case of the preceding result of John Mather we get that if we define

$$\mathcal{K} \; := \; \mathbf{R}\{y, f\} + \mathbf{R}\{x, y\}\partial f(x, y)/\partial x + \mathbf{R}\{y, f\}\partial f(x, y)/\partial y$$

---

[2]the discriminant locus is the locus of critical values

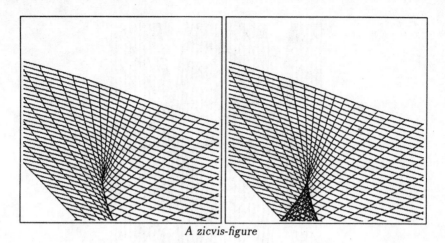

A zicvis-figure

Figure 1: *"The Whitney cusp, without/with hidden parts."* On the left image, the hidden parts do not appear, the apparent contour (a semi-cusp) has a terminal point, on the right image a full cusp is seen.

the equality

$$\mathcal{K} \;=\; \mathbf{R}\{x,y\}$$

implies that the analytic type of the apparent contour is infinitesimally stable, and by Mather's result is stable: any small deformation will give an equivalent germ with a locally diffeomorphic apparent contour.

Stable singularities of the apparent contour are unavoidable (inescapable), i.e. they are seen from every viewpoint (perhaps after a small perturbation). They are known since works by H. Whitney (see Whitney 1955) to be

the **etale** map

$$(x,y) \longrightarrow (x,y,z=x)$$

the **fold** map

$$(x,y) \longrightarrow (x,y,z=x^2)$$

the **cusp** map

$$(x,y) \longrightarrow (x,y,z=xy-x^3)$$

Codimension 1

If we allow the observer or the camera (or the object) to move along a curve, we will get more singularity types, all those with orbit of codimension 1 in their deformation space. These are obtained if $\mathcal{K}$ is of codimension 1 in $\mathbf{R}\{x,y\}$.

We get 3 other situations:

the **lips**

$$z = x^3 + xy^2$$

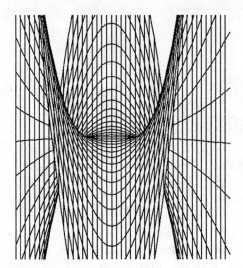

Figure 2: *"Beak to beak." Two semi-cusps (terminal point singularities) facing each other. The "beak to beak" singularity is the "collision" of the 2 cusps. It is unstable, but very near it, presence of 2 semi cusps allows its detection.*

the **beaks** (see figure 2.3.1)

$$z = x^3 - xy^2$$

and the **camel**

$$z = x^4 + xy \quad .$$

If the camera can be moved in all directions, i.e. in a projective space of dimension 2, it is necessary to look at singularities with orbits of codimension 2 in their deformation space. If the camera can be moved from infinity within a short distance of the object then we get one parameter more, namely the distance to the object and classification has to be carried on to codimension 3 (Platonova, Sherback (Platonova 1984)).

2.4   Classifying Views of Lighted Smooth Surfaces

2.4.1   The Mathematical Set-Up

Notations and Hypothesis

As in the classical case we can assume that our object is a smooth surface M locally embedded in $\mathbf{R}^3$ and that, in a neighbourhood of the origin, it is the image of $\mathbf{R}^2$ by the morphism:

$$\begin{array}{ccc} \mathbf{R}^2 & \stackrel{f}{\longrightarrow} & \mathbf{R}^3 \\ (x, y) & \longrightarrow & (x, y, z = f(x, y)) \end{array}$$

We assume that the camera is at the infinity in the direction of the x-axis (projection on the (y,z)-plane) and that the sun direction is the y-axis, (with the sun on the $y > 0$ side), the only restriction we have thus made is that the camera is not in line with the sun.

In our coarse manner of taking images, what we call the image of the lighted surface is the union of three plane curves:

1. the image Σ of the shade curve, which we call the shade contour; this is the image of points on M where the sun ray is tangent to M

2. the image Ω of the shadow curve, which we call the shadow contour; it is the image of the points on M where a tangential sun ray cuts the surface M out of the shade curve

3. the apparent contour Γ of the surface M; it is the image of points on M where the visual ray is tangent to M.

We want to classify the local singularity types of the *union* of 3 analytic curves:

1. The curve of critical values of the x-axis projection π on M i.e. the critical value curve of the morphism $\bar{f}$,

$$
\begin{array}{ccc}
\mathbf{R}^2 & \xrightarrow{\bar{f}} & \mathbf{R}^2 \\
(x,y) & \longrightarrow & (y, f(x,y))
\end{array}
$$

will be called $\Gamma(\bar{f})$ or Γ if no confusion is possible.

2. The shade contour Σ, which is the image by π of the visible part of the shade curve S. The shade curve S has been defined as the critical value curve S of the projection λ of y-axis.

   The equations in $\mathbf{R}^3$ of S will be:

$$
\begin{array}{rcl}
z & = & f(x,y) \\
0 & = & \partial f(x,y)/\partial y
\end{array}
$$

3. The shadow contour Ω, image by π of the shadow curve O.

   The equations in $\mathbf{R}^3$ of $O \cup S$ will be given by a projection of an analytic set of $\mathbf{R}^4$:

$$
\left\{ (x,y,y',z) \; ; \;
\begin{array}{rcl}
z & = & f(x,y') \\
z & = & f(x,y) \\
0 & = & \partial f(x,y')/\partial y
\end{array}
\right\}
$$

   We can also define O as the inverse image of the locus of critical values of the projection λ of kernel the y-axis.

Classification Principles

We would like to use as equivalence relation, analytic equivalence, i.e. existence of an embedded analytic isomorphism between the union of 3 curves, sending apparent contour on apparent contour, shade curve on shade curve and shadow curve on shadow curve. Of course, analytic (or even topological) equivalence of the union of the three contours would be a better criterion, but harder to deal with. Moreover, stability with respect to analytic equivalence is a sufficient condition for topological stability .

$$
\begin{array}{ccc}
\mathbf{R}^2 & \xrightarrow{\sim} & \mathbf{R}^2 \\
\uparrow & & \uparrow \\
\Gamma_1 \cup \Sigma_1 \cup \Omega_1 & \xrightarrow{\sim} & \Gamma_2 \cup \Sigma_2 \cup \Omega_2
\end{array}
$$

The Right-Left equivalence on the morphism $\bar{f}$ defined by

$$\bar{f} \simeq L \circ \bar{f} \circ R$$

where R and L are analytic isomorphisms of $\mathbf{R}^2$, will of course induce an embedded isomorphism of the critical curve of $\bar{f}$ with the critical curve of $L \circ \bar{f} \circ R$.

As $\Gamma(\bar{f})$ can also be defined as the locus of critical values of the morphism $\bar{f}$, we get a sufficient condition for the equivalence of $\Gamma(\bar{f})$ with $\Gamma(L \circ \bar{f} \circ R)$, that is, of the two apparent contours. What can we say about the other two curves ?

A Criterion for Invariance of Shade and Shadow

We need extra conditions to insure that the shade and shadow curves are invariant by a combined change of coordinates on the source space and on the target space. A first necessary condition is that they are such by an *infinitesimal* change of coordinates.

First of all, we remark that the ideal S (the "**shadow ideal**") defining the union of the shade and shadow curves is given by:

$$\begin{aligned} S &\subset \mathbf{R}\{x, y, y', z\} \\ S &= (z - f(x, y'),\ z - f(x, y),\ \partial f(x, y')/\partial y\,)\mathbf{R}\{x, y, y', z\} \end{aligned}$$

Remark If the sun is in the $y > 0$ direction, then the shadow curve is better defined by adding the inequality $y' \geq y$ which says that the point (x, y, z) is in the shadow of the point (x, y', z). In the following, we will study both parts of the shadow curve, as we can switch from one to the other by reversing the propagation of the light.

We want to compare, for a change of coordinates on the source and on the target, the transform of the *shadow ideal* with the shadow ideal of the transform.

To simplify notations we will write P' instead of $P(x, y')$ for an element of the algebra $\mathbf{R}\{x, y\}$ and write ϕ' instead of $\phi(y, f(x, y'))$, for an element of the subalgebra $\mathbf{R}\{y, f\} \subset \mathbf{R}\{x, y\}$.

For ε small, that is computing $(mod\ \varepsilon^2)$, we have to compare the following ideals in $\mathbf{R}\{x, y, y', z\}$:

$$\mathcal{I}_1 = \left\{ \begin{aligned} &\frac{\partial f}{\partial y}(x + \varepsilon P', y' + \varepsilon \phi') \\ &\frac{1}{y' - y}(f(x + \varepsilon P', y' + \varepsilon \phi') - f(x + \varepsilon P, y + \varepsilon \phi)) \end{aligned} \right.$$

and the shadow ideal of the embedding associated to $g(x, y)$, where

$$g(x, y) = f((x + \varepsilon P(x, y), y + \varepsilon \phi(y, f)) + \varepsilon \psi(y, f)$$

which is

$$\mathcal{I}_2 = \left\{ \begin{aligned} &\frac{\partial g}{\partial y}(x, y') \\ &\Delta_y g \end{aligned} \right.$$

where Δ_y is the partial difference operator on y:

$$\Delta_y(g(x, y)) := \frac{g(x, y) - g(x, y')}{y - y'} = \frac{g - g'}{y - y'}.$$

$$\mathcal{I}_2 = \begin{cases} \partial f/\partial x(x + \varepsilon P', y' + \varepsilon\phi')\varepsilon\partial P/\partial y + \varepsilon\partial\psi/\partial y(y', f') + \varepsilon\partial\psi/\partial z\partial f/\partial y \\ \qquad\qquad + (\partial f/\partial y(x + \varepsilon P', y' + \varepsilon\phi'))(1 - \varepsilon\partial\phi/\partial y) \\ \Delta_y(f(x + \varepsilon P, y + \varepsilon\phi) + \varepsilon\psi(y, f)) \end{cases}$$

Comparing these two ideals give two conditions insuring invariance of shade and shadow.

$$\frac{\partial f}{\partial x}\frac{\partial P}{\partial y} + \frac{\partial \psi}{\partial y} \ \in \ \frac{\partial f}{\partial y}\mathbf{R}\{x, y\}$$

$$\Delta_y\psi \ \in \ (\frac{\partial f}{\partial y}(x, y'), \Delta_y f)\,\mathbf{R}\{x, y, y'\}$$

Let's observe that, using the Taylor formula, the second condition implies

$$\frac{\partial \psi}{\partial y} \in \frac{\partial f}{\partial y}\mathbf{R}\{x, y\}.$$

Hence we get a new formulation of the invariance of the shadow ideal by an infinitesimal change of coordinates. Gluing together with 2.4.1 we get the following

Proposition 2.4.2 *An infinitesimally small deformation of the function f given by g = f + εh induces an infinitesimally small trivial deformation of the shade, shadow and apparent contour curves if and only if there exists functions $P \in \mathbf{R}\{x, y\}$, $\phi \in \mathbf{R}\{y, f\}$ and $\psi \in \mathbf{R}\{y, f\}$ such that*

$$(o) \qquad h \ = \ P\frac{\partial f}{\partial x} + \phi\frac{\partial f}{\partial y} + \psi$$

$$(i) \qquad \frac{\partial f}{\partial x}\frac{\partial P}{\partial y} \ \in \ \frac{\partial f}{\partial y}\mathbf{R}\{x, y\}$$

$$(ii) \qquad \Delta_y\psi \ \in \ (\frac{\partial f}{\partial y}(x, y'), \Delta_y f)\,\mathbf{R}\{x, y, y'\}$$

Using this result we want to give (at least sufficient) conditions for a function f to give a stable image, that is, conditions which insure that *every* infinitesimal perturbation of f induces an infinitesimally small trivial deformation of the shade, shadow and apparent contour curves

Lemma 2.4.3 *The two conditions (i) and (ii) are naturally satisfied if*

1. *$P(x, y)$ depends of y only through powers of $f(x, y)$, that is belongs to the subalgebra $\mathbf{R}\{x, f(x, y)\}$ of $\mathbf{R}\{x, y\}$*

2. *$\psi \in \mathbf{R}\{f(x, y)\}$.*

Proof: If

$$P(x, y) = \mathcal{P}(x, f(x, y))$$

we get

$$\frac{\partial P}{\partial y} = \frac{\partial \mathcal{P}}{\partial z}\frac{\partial f}{\partial y}$$

and if ψ only depends of $z = f(x, y)$:

$$\frac{\partial \psi}{\partial y} = \frac{\partial \psi}{\partial z}\frac{\partial f}{\partial y}$$

condition (i) is easily verified, so is (ii) because it is satisfied by all powers of f. ∎

For the study of stable shade, shadows and contours we will need the following consequence:

Proposition 2.4.4 *If the module*

$$\mathcal{K}' = \mathbf{R}\{x,f\}\frac{\partial f}{\partial x} + \mathbf{R}\{y,f\}\frac{\partial f}{\partial y} + \mathbf{R}\{f\}$$

is equal to the whole ring $\mathbf{R}\{x,y\}$ *all infinitesimal perturbations of the mapping* $(x,y) \longrightarrow (y,f(x,y)$ *will not change the shade, shadow and apparent contours. Hence the singularity associated to* f *is infinitesimally stable.*

2.4.5 Stable Patterns

From now on, we will talk of **Sol y Sombra Patterns** as a short-term for the analytic embedded type of the union of the three curves, apparent contour, shade and shadow.

We will define **stable Sol y Sombra Patterns** to be Sol y Sombra Patterns such that there exists an equivalent Sol y Sombra Pattern lying in any small deformation of it.

Computation Method

A necessary condition for a surface M to provide a stable Sol y Sombra Pattern, is that its associated apparent contour must be stable. So we have first to consider

Stable Contours

As a special case of the preceding remark we get that if we define

the **etale** map

$$(x,y) \longrightarrow (x,y)$$

the **fold** map

$$(x,y) \longrightarrow (x,y^2)$$

the **cusp** map

$$(x,y) \longrightarrow (x,xy-y^3)$$

We compute for each type the shade and shadow. If it is stable we have a stable Sol y Sombra Pattern, if it is not, we look at small perturbations of this surface until we have enumerated all stable Sol y Sombra Patterns.

We will first look for stable Sol y Sombra Patterns with contour isomorphic to the contour of the first type (the "etale"), i.e. empty (not going through the point).

Far from the Rim

Let us first consider the embedding of M, given by

$$(x, y, z = x)$$

in that case 0 is not on the apparent contour of M for the projection π which is generic, but this surface is tangent to the sun rays, which is a very instable situation (gliding rays). We look for perturbations of the morphism $(x, y) \longrightarrow (y, x)$ which are stable Sol y Sombra Patterns. We get those perturbations by adding to x terms of increasing degree.

We obtain three different stable "Sol y Sombra Patterns"

"Sunny side" (or full shade)   If the initial form of f is linear and not divisible by x, the pattern is stable and is equivalent to the mapping

$$(x, y, z = x + y)$$

the point is not on the rim,
the point is in full light (or in full shadow).

"Shade line"   If the initial form of $f - x$ is a quadratic form which doesn't vanish along $x = 0$, then the pattern is stable and equivalent to the mapping

$$(x, y, z = x + y^2)$$

the point is not on the rim,
the point is on the shade line, which is a fold for the shading, the shadow line is not going through this point.

Shade crease or "shaded box pleat"   If the initial form of $f - x$ is a quadratic form vanishing on $x = 0$, we first notice that the mapping $(x, y) \longrightarrow (y, x + xy)$ is not a stable one (2.4.6). Hence assume that the initial form of $f - x - xy$ is a cubic form not divisible by x. The module

$$\mathcal{K}' = \mathbf{R}\{x, f\}\frac{\partial f}{\partial x} + \mathbf{R}\{y, f\}\frac{\partial f}{\partial y} + \mathbf{R}\{f\}$$

is equal to the whole ring $\mathbf{R}\{x, y\}$ (just notice first that $\partial f/\partial y \mathbf{R}\{y, f\}$ is the ideal $\partial f/\partial y \mathbf{R}\{x, y\}$ and choose a basis of the quotient $\mathbf{R}\{x, y\}/\partial f/\partial y$ generated by powers of y)

Then by 2.4.4 we get a stable pattern, equivalent to the mapping

$$(x, y) \longrightarrow (y, x + xy + y^3).$$

The rim

$$\frac{\partial f}{\partial x} = 1 + y = 0$$

does not contain the point, the shade is given by

$$S = \left\{ (x, y, z) \; ; \quad \begin{array}{rl} z & = \; x + xy + y^3 \\ \frac{\partial f}{\partial y} & = \; x + 3y^2 = 0 \end{array} \right\}$$

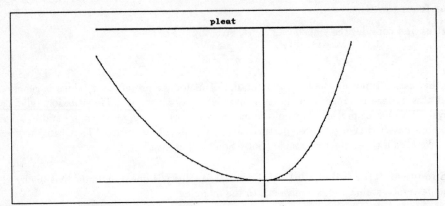

Figure 3: *A graph of the pleat in* $\mathbf{R}^2$*. The cast shadow is on the left, the shade is the right semi parabola. The sun rays come from the right. The z-axis is vertical, the y-axis is horizontal. Note the difference in curvatures of the 2 branches. The graph was created using the AXIOM computer algebra system*

which gives
$$\Sigma = \left\{(y,z) \; ; \; z = -3y^2 - 2y^3\right\}$$
the shadow curve Ω is the projection of the part of
$$\left\{(x,y,y',z) \; ; \; \begin{aligned} z &= x + xy + y^3 \\ z &= x + xy' + y'^3 \\ \frac{\partial f}{\partial y}(x,y') &= x + 3y'^2 = 0 \end{aligned} \right\}$$
not contained in S, on the (y,z) plan.
$$\Omega = \left\{(y,z) \; ; \; z = -\frac{3}{4}y^2 + \frac{1}{4}y^3\right\}$$

We obtain two half parabolas with the same tangent but different curvatures, (there is no symmetry with respect to the normal line).

Remark   If the initial form of $f - x - xy$ is a cubic form divisible by x, or if it is in the ideal $\mathbf{m}^4$, then f is not a stable pattern.

On the Apparent Contour

The simplest equation for the fold of the apparent contour is:
$$(x,y,z = x^2)$$
But this surface is not stable for shade.

To have a fold on the apparent contour, the derivative $\partial f/\partial x$ must vanish at the origin.

The apparent contour fold   If the initial form of f is linear and not divisible by x (in order to have $(\partial f/\partial x)(0,0) = 0$, the pattern is stable and is equivalent to the mapping
$$(x,y) \longrightarrow (y, x^2 + y)$$
gives a stable Sol y Sombra Pattern (the shade curve is empty).

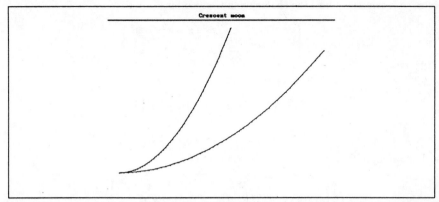

Figure 4: *Crescent moon, projection to* $\mathbf{R}^2$. *The shade curve is the upper semi-parabola, the apparent contour is the lower semi parabola, with smaller curvature. The z-axis is vertical, the y-axis is horizontal. The lighted part lies between the 2 curves. The sun rays come from the right. Graph created by AXIOM.*

"The crescent moon" If the initial form of f is a quadratic form not divisible by x, then we get an unstable pattern (cf 2.4.6). If not, we get a stable one, equivalent to the mapping

$$(x, y) \longrightarrow (y, x^2 \pm xy)$$

The apparent contour is a smooth line, with a shadow curve which is half a parabola, with the same tangent line, and different curvature, beginning at the summit.

The cusp fold of the apparent contour Another stable local Sol y Sombra Pattern is of course a cusp fold of the apparent contour which is far from the shade edges, with a local equation which is given by:

$$f(x, y) = y + xy + x^3$$

2.4.6 The Codimension 1 Singularities

Far From the Rim, Shade-Shadow Singularities

To get an unstable pattern in the "far from the rim" case (cf 2.4.5) (that is, 0 not on the apparent contour), we must take an f such that

1. $f(x, y) - x$ is a cubic form

2. $f(x, y) - x - xy$ is a cubic form divisible by x.

Shade folds confluence or the "Diabolo singularity" If $f(x, y) - x$ is a cubic form non divisible by x, we get a codimension 1 pattern, equivalent to

$$(x, y, z = x + x^2 y - y^3)$$

the rim

$$\frac{\partial f}{\partial x} = 1 + 2xy = 0$$

does not contain the point, the shading rim is given by

$$S = \left\{ (x,y,z) \; ; \; \begin{array}{rcl} z & = & x + x^2 y - y^3 \\ \dfrac{\partial f}{\partial y} & = & x^2 - 3y^2 = 0 \end{array} \right\}$$

which gives two branches for the shade contour

$$\Sigma_1 = \left\{ (y,z) \; ; \; z = +\sqrt{3}\,y + 2y^3 \right\}$$

$$\Sigma_2 = \left\{ (y,z) \; ; \; z = -\sqrt{3}\,y - 2y^3 \right\}$$

the shadow curve Ω is the projection of the part of

$$\left\{ (x,y,y',z) \; ; \; \begin{array}{rcl} z & = & x + x^2 y - y^3 \\ \dfrac{\partial f}{\partial y}(x,y') & = & x^2 - 3y'^2 = 0 \end{array} \right\}$$

not contained in S, on the (y,z) plan.

$$\Omega_1 = \left\{ (y,z) \; ; \; z = -\frac{\sqrt{3}}{2}y - \frac{1}{4}y^3 \right\}$$

$$\Omega_2 = \left\{ (y,z) \; ; \; z = +\frac{\sqrt{3}}{2}y - \frac{1}{4}y^3 \right\}$$

In the equivalence module y is missing and therefore this situation can be deformed by a one parameter family

$$(x,y,z = x - u^2 y + x^2 y - y^3)$$

on a scene with 2 "shade crease" points $(x = u, z = 0)$ and $(x = -u, z = 0)$ of type

$$(X,y,z = X + Xy + y^3)$$

The Grazed Camel   If $f(x,y) - x - xy$ is a cubic form divisible by x, the function f can be written down

$$x + xy + xm^2 + m^4$$

A typical form is the "shade" analog of the *camel* singularity of the apparent contour:

$$(x,y,z = x + xy + y^4)$$

The shading rim is given by

$$S = \left\{ (x,y,z) \; ; \; \begin{array}{rcl} x & = & -4y^3 \\ z & = & -4y^3 - 3y^4 \end{array} \right\}$$

On the Rim, Rim-Shade Singularities

The half-moon singularity In codimension 1 by deformation of the "fold of the apparent contour" case given by

$$(x, y, z = x^2)$$

(in that case 0 is on the apparent contour of M) we can get an unstable (codimension 1) situation:

$$(x, y, z = x^2 - y^2)$$

the rim

$$\frac{\partial f}{\partial x} = 2x = 0$$

contains the point 0, and is the smooth curve $z = -y^2$. The shade curve is given by:

$$S = \left\{ (x, y, z) \; ; \; \begin{array}{rcl} z & = & x^2 \\ \dfrac{\partial f}{\partial y} & = & -2y = 0. \end{array} \right\}$$

this is the z axis, another smooth curve orthogonal to the apparent contour.

Computation of the tangent space of the deformation shows that xy is the only missing term. The module

$$\mathcal{K}' = \mathbf{R}\{x, f\}\frac{\partial f}{\partial x} + \mathbf{R}\{y, f\}\frac{\partial f}{\partial y} + \mathbf{R}\{f\}$$

is equal to the sum of the two algebras $\mathbf{R}\{x, y^2\}$ and $\mathbf{R}\{x^2, y\}$.

2.5   Application to Vision

We can now look forward using our new tools for intelligent vision.

We propose to add some preprocessing or eventually post processing in a number of situations where they could help, which we shall now attempt to enumerate.

2.5.1   Inadequate and Adequate Set-Ups

Binary or Ternary Images?

Let us begin by trivial considerations. Can we use our Sol y Sombra Pattern in the world of what are called *binary images* ? By binary images we mean (Rosenfeld 1982, Horn 1986, chapter 3 and 4) black-and-white (two-valued) images. This means that the only known information about the objects is their *silhouette* against the background. This is very poor information but recognition process can be efficient in specialized environment, especially in the case of robots manipulating a finite number of objects, one at a time ; and of course the economic implications are tremendous. The usual artefacts or features extracted are area, position and orientation, through definitions of center-of-mass, pseudo-axis (axis of inertia), and mostly by adequate (discrete) integration methods (summations).

What Singularities?

In that frame it is clear that there are no shade or shadow lines and that there are no local singularities of the contour (no terminal point). However, even with a single smooth object, there can be normal crossings (semi-local singularity of the apparent contour).

Pluri-thresholding   As binary images are usually obtained from ordinary images by thresholding the multiple brightness levels, one could think of combining all the usual features of binary images with the features obtained from a *ternary thresholding*. Instead of two-valued images, it is possible to obtain three-valued images, black-grey-white images with black background, one-intensity- grey shadows and white full-light patches. In this set-up our classification of Sol y Sombra Pattern would become pertinent, either as a discriminating refinement in recognition, after processing the binary image, or as an added cue in orientation and position interpretation.

Texture and Sol y Sombra Patterns?

Another situation where Sol y Sombra Patterns seem to be of no use, is in the opposite direction, when the visual structure is too rich, for example if there is a very regular pattern such as the grid used to represent surfaces in computer programs. In fact even in that case we can exhibit examples where the texture suggests a plane or quasi-plane gentle surface but where the sole Sol y Sombra Pattern detects a much more complex form.

Of course it is possible by using special textures to make the detection of singularities rather difficult: see a discussion in (Todd and Reichel 1990) where T-patterns are projected as a texture on a surface, embedding Terminal points singularities (cusps in real life) in a sea of T-patterns.

2.5.2   Use of Sol y Sombra Patterns in Adjunction to Classical Methods

Shape from Grey Levels and Sol y Sombra Patterns

The most natural motivation for determinating the Sol y Sombra Patterns is to use them in relation with usual procedures of *Shape from Shading*. The apparent contour is generally used as lines of determinacy of the normals, which gives limit conditions for integration of vector fields on the plane image of the surface. The curves of the contour are not only solutions of ordinary differential equations related to the computation of the *image irradiance equation* (Horn 1986 ch. 11 and Horn 1990), i.e. base characteristics, but even more they are *characteristic strips*, which means that the surface orientation is known along this curve.

The shade curves provide information of the same type, since we know that along points of the shade curve the sun rays are tangent to the surface, which means that the normal to the surface is orthogonal to the sun direction. The tangent to the curve gives other information on the normal to the surface at this point. Therefore they can also be used as an initial curve, with a major advantage at occluding contours, because the slope of the surface is not infinite at these points. But the cast shadow curve does not provide the same sort of information, and discriminating them is of the upmost necessity. This question introduces our last consideration, labelling.

Labelling Sol y Sombra Patterns

We shall not develop this point. It will be the subject of a following paper. Let us only say we shall need classification of singularities of the triplet of curves, not only local singularities, as in this article, but also semi-local ones: they are due to projections of 2 or 3 different points of the surface or of different surfaces on the *same point* of the image.

After this again mathematical task, it is possible to discriminate between permissible patterns and impossible ones. This allows us to label the three curves in a very similar way to that initiated by Clowes and Huffman for edges of polyhedral objects, (world of children's blocks), extended by Waltz (Waltz 1975) to polyhedral objects with shadows, and by Malik (Malik 1987) to opaque objects with piecewise smooth surfaces.

2.6 Conclusion

We would like to point out that we only introduced a new feature, and that we hope that implementation in computer vision[3] and experimentation in psycho-physiology will show its relevance in at least one of the 2 complementary domains of vision. If that is the case, additional works in geometry will certainly be necessary but they should be exciting, both from a mathematical and a vision view point !

3 Projective Configurations of Points and Vision

3.1 Introduction

Projective geometry and the study of perspectives appeared with the works of painters who tried to describe the real world. Then, mathematicians formalized these ideas and extended them to any dimension. Generalization of this geometry and works in invariant theory were the continuation, but also the apogee, of studies in this branch of mathematics which slowly fell into oblivion. Now, artificial vision is trying to "reconstruct" the real world from images. The aim of this article is to show how projective geometry is more than ever a subject of the day. It gives a good mathematical frame-work with which to manipulate geometric objects and to describe mechanisms of vision, but is also convenient for analysis of (automatically) the structure of images.

Consider the following example: we are looking at 4 points A, B, C, D on a square (figure 11). The images a, b, c, d are defined by the projection from the plane of the square to the plane of the image from a point O which is the center of the camera (or the eye).

In the "real plane" of A, B, C, D, the lines $(A, B), (C, D)$ have no common point. However in the image, common points X, Y exist. This new line (X, Y) which does not appear in the "real plane" is called the line at infinity. All relations such as collinearity, parallelism of lines, middle points correspond in the image plane to "projective" relations with respect to this line at infinity: parallelism becomes concurrency with the line at infinity, a middle point becomes a conjugate point with respect to the intersection point with the line at infinity ... Euclidean relations such as orthogonality can also be described in this language with respect to two special points I, J on (X, Y) which are associated with the scalar product in the real plane. For instance, "(A, D) and (B, C) are orthogonal" means that the intersection points of (a, d) and (b, c) with the line at infinity are conjugated with respect to I, J (cross ratio equal to -1). With this language, we can naturally give the relations between the real plane and the image plane.

Moreover, this language describes what is intrinsic in the image of a configuration: going from one view to another corresponds to a projective transformation of the plane of

[3]We thank Joel Marchand and Eric Boix for the constant help in debugging problems of interaction among graphical software, X utilities and laser printers. Without them, most figures would have been hand drawn or would have stayed Postscript files on a disk.

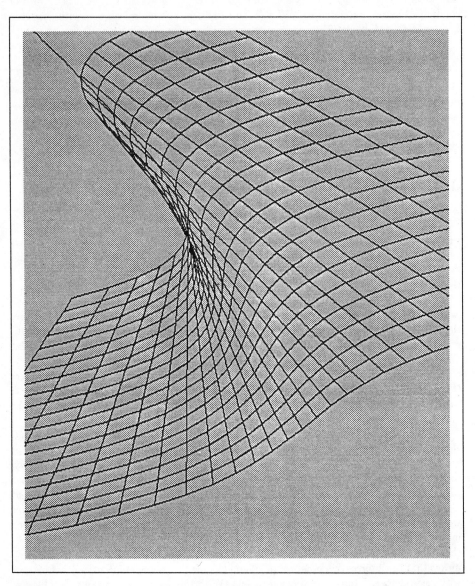

Figure 5: *The stable cusp singularity of the apparent contour. This is a view of the surface called the "Whitney Cusp" featuring the curve of the apparent contour (a curve with terminal point) for an x-projection, seen from another projection view point. This figure can also be interpreted as featuring the shade curve (but not the cast shadow curve) for a sun direction.*

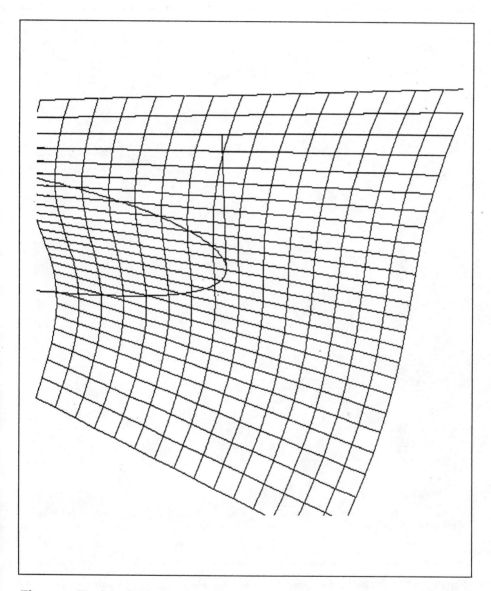

Figure 6: *The Sol y Sombra Pattern of the stable "shaded box pleat". This is again a view of the surface called the "Whitney Cusp" featuring the curve of the shade contour (a semi parabola with terminal point) and the cast shadow curve (a semi parabola with smaller curvature, with same terminal point). The y axis (sun direction) is here vertical, and the patch between the two semi parabolas (the inside) is in the shade: it is a "shade box".*

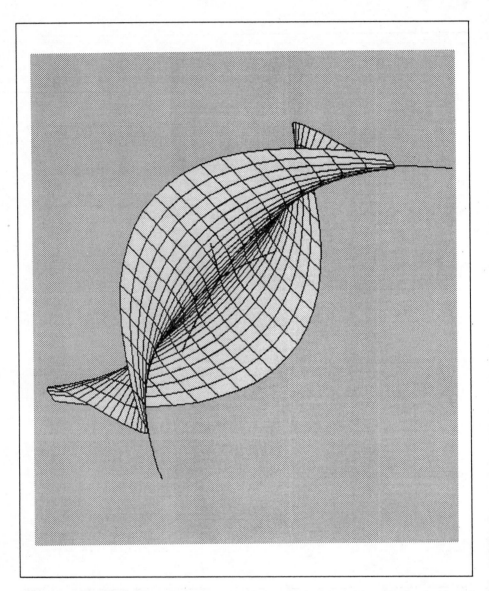

Figure 7: *The "diabolo", codimension 1 singularity, the junction of 2 stable singularities. A half y axis has been drawn, pointing towards the sun, which is above the camera, approximately in the same vertical plane (slightly on the left). This axis also marks the special point O. There are 4 curve branches departing from this point: two shade curves meet in O, they have been prolongated outside the surface grid. Contrarily, the 2 cast shadow branches have been shortened, to make distinction easier for the reader. The two Whitney cusps are clearly visible, compare with figure.*

Figure 8: *The "diabolo", a shaded view, the junction of two shaded box pleats. Although this singularity is unstable, it is nevertheless made noticeable by the presence, in the vicinity, of 2 stable singularities. Blurring and scale effect "stabilize" and "reveal" a mathematically unstable singularity: there is very little chance of the spotlight being exactly in a position to give rise to this confluence, but poor definition results in a sort of "blowing down" of a patch into a few pixels spot understood as a point. This singularity seems evident to human eyes, the question of its detection by artificial vision means remains a task to experiment on.*

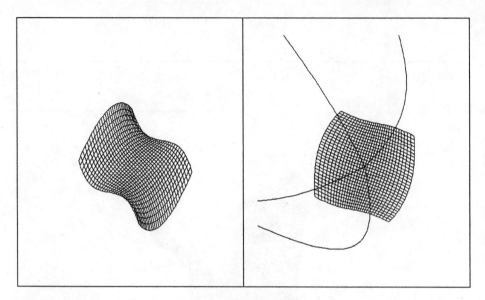

Figure 9: *The double box-pleat, the gentle views, comparison of cues. Two views of the same surface, from nearby view points, one with only grid texture, the right one with Sol y Sombra Pattern. On the right picture the two shade branches are on the right, the two cast shadow branches on the left. The sun comes from the right.*

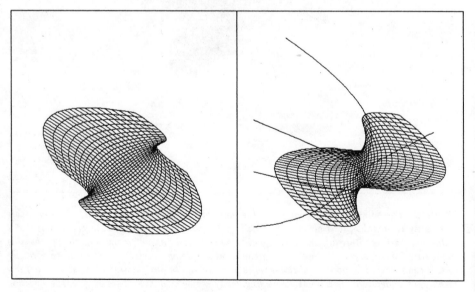

Figure 10: *The double box pleat, view without and with Sol y Sombra Pattern. Figure on the left and on the right are nearly symmetric except for the Sol y Sombra Pattern present on the right picture. The y- axis has also been drawn pointing towards the sun. The cast shadow curves are the leftmost branches. The shade contour curve has 2 branches.*

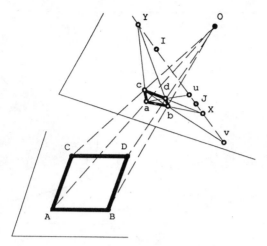

Figure 11.

the image and common informations from all these images are called projective invariants. So, we want to manipulate intrinsic functions, which have a meaning in all views and to find a way to compute in this algebra of invariants.

What is true for a projective plane can be generalized to a space of dimension three (the line at infinity becomes the plane at infinity, the points I, J become a conic in the plane at infinity), to dimension four, ...

So a good context for mathematical descriptions of vision mechanisms is a (projective) space of dimension 3, projected onto a (projective) plane from a point, considering only projective invariants or directly dealing with geometric objects such as points lines, conics, ...We assume that simple relations (coming from a first analysis) such as alignments, collinearity, concurrency ... of points, lines are known. But to better understand what is seen, new properties often have to be deduced from the first projective relations. One may want to consider on possible configurations, verify that they fit with the rest of the geometry ...

Here is an artificial example where simple properties deduced from the first analysis help us to analyze the view. Suppose we know that the points $a_1, b_1, \ldots, f_1$ (resp. $a_2, b_2, \ldots, f_2$) (resp. a_1, b_1, a_2, b_2) (resp. c_1, c_2, d_1, d_2) (resp. e_1, e_2, f_1, f_2) are on a same plane (figure 12). Then, we wonder if these views are the possible projections of a 3D-object or simply of a 2D-object. We have to check whether the dotted lines are concurrent or not (in the first figure, this is the case but not in the second). This task can be achieved by a systematic method and the purpose of this article is to show how to generalize it. We propose to consider the following situation: given configurations of points, described by projective relations, show how we can automatically check properties satisfied by these configurations. The approach we adopt is to work as much as possible with intrinsic quantities. By this, we mean that we do not take any basis, but compute directly with the geometric objects involved in the figure and with invariant coefficients.

3.2   Geometric Tools

We suppose that the configurations we are looking at are described by hypotheses of collinearity, conjugation, concurrency of points, lines, planes ...

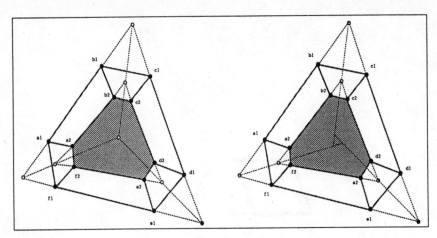

Figure 12.

The first objective is to be able to manipulate these geometric objects without taking a basis or coordinates in this basis, but work directly with geometric objects in an intrinsic way.

Let us call $\mathbf{E}$ a vector space of dimension $n + 1$ (for instance $E = \mathbf{k}^{n+1}$ with $\mathbf{k} = \mathbf{Q}$ or $\mathbf{R}$ or $\mathbf{C}$) and we note $e_i = (0, \ldots, 0, 1, 0, \ldots, 0)$ the ith vector of the canonical basis of this vector space.

Points are elements of the projective space $P(\mathbf{E})$ associated with E. An affine point will be represented by a list of coordinates of the form $(x_1, \ldots, x_n, 1)$ and a point at infinity by $(x_1, \ldots, x_n, 0)$.

To manipulate linear subspaces of $\mathbf{E}$, we introduce the following tool: let $\wedge \mathbf{E}$ be the exterior algebra of $\mathbf{E}$ over k. An element of this k-vector space is a linear combination of the basis vectors $(e_{j_1} \wedge \cdots \wedge e_{j_l})_{1 \leq j_1 < \cdots < j_l \leq n+1, 1 \leq l \leq n+1}$ where $\wedge$ denotes the exterior product.

So, for all vectors $u_1 = (u_{1,1}, \ldots, u_{n+1,1}), \ldots, u_m = (u_{1,m}, \ldots, u_{n+1,m})$ of $\mathbf{E}$ the coordinates of $u_1 \wedge \cdots \wedge u_m$ in the basis $(e_{i_1} \wedge \cdots \wedge e_{i_m})_{1 \leq i_1 < \cdots < i_m \leq n+1}$ of $\wedge^m E$ are the determinants $|u_{i_k,j}|_{1 \leq i,k \leq m}$. Writing "$u_1 \wedge \cdots \wedge u_m = 0$" is equivalent to saying that $\{u_1, \ldots, u_m\}$ are linearly dependent.

Every vector v of the linear space $\overrightarrow{L}$ generated by the independent vectors $v_1, \ldots, v_d$ satisfies

$$v \wedge v_1 \wedge \cdots \wedge v_d = 0.$$

To any linear space $\overrightarrow{L} = \langle v_1, \ldots, v_d \rangle$, we associate the element $L = v_1 \wedge \cdots \wedge v_d$ of $\wedge \mathbf{E}$. Conversely, to an element L of the form $L = v_1 \wedge \cdots \wedge v_d$, we associate the linear space $\overrightarrow{L} = \langle v_1, \ldots, v_d \rangle$.

Definition 3.1 *Let $\mathcal{G}$ be the set of elements of $\wedge E - \{0\}$ of the form $v_1 \wedge \cdots \wedge v_d$ with $v_i \in \mathbf{E}$. To each element $L = v_1 \wedge \cdots \wedge v_d$ of $\mathcal{G}$, is associated the linear subspace $\overrightarrow{L}$ generated by $v_1, \ldots, v_d$.*

Remark that this definition is not ambiguous (it does not depend on the choice of the vectors $v_1, \ldots, v_d$), for $\overrightarrow{L}$ is just the set of vectors $v \in E$ such that $v \wedge L = 0$.

Beware that an element of $\wedge E$ is not necessarily an element of $\mathcal{G}$, as we can see with

$$L' = e_1 \wedge e_2 + e_3 \wedge e_4$$

where $e_1, \ldots, e_4$ are independent vectors of a space of dimension 4. This element cannot be of the form $u \wedge v$ with $u, v \in E$. If this were the case, we would have $L' \wedge L' = 0$ but here $L' \wedge L' = 2e_1 \wedge e_2 \wedge e_3 \wedge e_4$ In fact, the subset $\mathcal{G}$ of $\wedge E$ is in the algebraic variety of $P(\wedge E)$, called *the grassmannian* whose equations are the well-known "Plücker Relations". The previous element $e_1 \wedge e_2 + e_3 \wedge e_4$ of $\wedge E$ does not satisfy these equations.

The elements of $\mathcal{G}$ are in correspondence with the linear subspaces of E. We could say that they are the elements of $\wedge E$ which have a geometric interpretation. We are going to manipulate them and use operators which preserve this form.

Intersections and Sums

To handle configurations of linear spaces, we need to construct intersections and sums.

We consider linear combinations of elements of $\wedge E$ with coefficients in C. This rough description corresponds to a precise mathematical construction that we note $\wedge E \otimes C$ or $\wedge_C E$.

The elements of C will be denoted

$$\sum_I \lambda_I (v_1^1, \ldots, v_{k_1}^1) \cdots (v_1^l, \ldots, v_{k_l}^l)$$

where $v_i^j \in E$. The element $(v_1, \ldots, v_{k_1})$ is here the determinant

$$]v_1, \ldots, v_l, t_{l+1}, \ldots, t_n]$$

where $(t_1, \ldots, t_n)$ is any basis of E. These elements are invariants by a change of coordinates and so are intrinsic tools.

An element of the space $\wedge_C E$ will be a linear combination of elements of the form:

$$u_1 \wedge \cdots \wedge u_l.(v_1^1, \ldots, v_{k_1}^1) \cdots (v_1^l, \ldots, v_{k_l}^l)$$

where $u_i, v_i^j \in E$.

Consider two lines $X_1 \wedge X_2$ and $X_3 \wedge X_4 \in \wedge^2 E$ which are coplanar (i.e. $X_1 \wedge X_2 \wedge X_3 \wedge X_4 = 0$). Then, the intersection exists and can be written as:

$$X_1 \wedge X_2 \cap X_3 \wedge X_4 = X_1 (X_2, X_3, X_4) - X_2 (X_1, X_3, X_4).$$

which is an element of $\wedge_C E$. This is a linear combination of vectors X_1, X_2 of $\wedge E$ with coefficients $(X_2, X_3, X_4), (X_1, X_3, X_4)$ in C. If we want to compute the coordinates of this point, we take any basis $(t_1, \ldots, t_n)$ of E and up to a scalar, these are the coordinates of:

$$X_1]X_2, X_3, X_4, t_4, \ldots, t_n] - X_2]X_1, X_3, X_4, t_4, \ldots, t_n]$$

This formalism allows us to work "as if" we were taking an arbitrary basis so that we may ignore it.

We introduce now the following operation: for all elements $L_1 = x_1 \wedge \cdots \wedge x_p$, $L_2 = y_1 \wedge \cdots \wedge y_q \in \mathcal{G}$, we define a new operator

$$D(L_1, L_2) = D^0(L_1, L_2) + D^1(L_1, L_2) + \cdots + D^p(L_1, L_2)$$

which is an element of $\wedge_C E$, $D^i(L_1, L_2)$ being the component of $D(L_1, L_2)$ which is in $\wedge_C^{p-i} E$. Let us give a glimpse of this operation on the following example: take two planes $X_1 \wedge X_2 \wedge X_3$ and $X_4 \wedge X_5 \wedge X_6$ (elements of $\wedge^3 E$) of a space E of dimension $n > 3$. Then,

$$
\begin{aligned}
D(X_1 & \wedge X_2 \wedge X_3, X_4 \wedge X_5 \wedge X_6) \\
= ~ & X_1 \wedge X_2 \wedge X_3 (X_4, X_5, X_6) \\
+ ~ & X_1 \wedge X_2 (X_3, X_4, X_5, X_6) - X_1 \wedge X_3 (X_2, X_4, X_5, X_6) + X_2 \wedge X_3 (X_1, X_4, X_5, X_6) \\
+ ~ & X_3 (X_1, X_2, X_4, X_5, X_6) - X_2 (X_1, X_3, X_4, X_5, X_6) + X_1 (X_2, X_3, X_4, X_5, X_6) \\
+ ~ & 1 (X_1, X_2, X_3, X_4, X_5, X_6)
\end{aligned}
$$

We define another application D_* just by exchanging the terms between parentheses with the others, in the previous development. The definition of D is extended by C-linearity to $\wedge_C \mathbf{E}$ so that it could apply to any elements of $\wedge_C E$.

Proposition 3.2 – *Let L_1, L_2 be elements of $\mathcal{G}$. We note $\overrightarrow{L_1}, \overrightarrow{L_2}$ the corresponding linear subspace. Let $L_1 \triangle L_2$ be the last non-null term of the development:*

$$
D(L_1, L_2) = D_0(L_1, L_2) + \cdots + D_l(L_1, L_2) + \cdots
$$

Then $L_1 \triangle L_2 = I.(S)$ with $\overrightarrow{I} = \overrightarrow{L_1} \cap \overrightarrow{L_2}$ and $\overrightarrow{S} = \overrightarrow{L_1} + \overrightarrow{L_2}$. In the same way, the last non-null term (noted $L_1 \triangledown L_2$) of the development of $D_*(L_1, L_2)$ is $S.(I)$.

(for the proof, see Mourrain 1991). These operations define elements in $\wedge_C \mathbf{E}$ which correspond to simple geometric operations on the two space $\overrightarrow{L_1}, \overrightarrow{L_2}$. To construct the element associated to the sum of $\langle v_1 \rangle$ and $\langle v_2 \rangle$, we just take $v_1 \wedge v_2$. Here is a generalisation of this product, taking into account the possible intersection of the two spaces. Given two spaces of dimension d_1, d_2, the dimension of the sum can vary from d_1 to $d_1 + d_2$ and this explains the possible choice of the terms D^i.

On the other hand, intersection can also be viewed by duality as an operation of this type: the intersection is the set of vectors which are in the orthogonal of the sum of the orthogonal of the two vector spaces $(L_1 \cap L_2)^\perp = L_1^\perp + L_2^\perp$. This duality appears naturally when we exchange the two factors between D and D_*. If the last non-null term is D^l then the dimension of $\overrightarrow{I}$ is $\dim(L_1) - l$ and the dimension of $\overrightarrow{S}$ is $\dim(L_2) + l$. It corresponds to the degree of the two terms in $\wedge \mathbf{E} \otimes \wedge \mathbf{E}$. We have the relation $\dim(L_1) + \dim(L_2) = \dim(I) + \dim(S)$.

Let's come back to the previous example: if the last non-null term is

$$
X_1 \wedge X_2 \wedge X_3 (X_4, X_5, X_6),
$$

this means that the two planes are equal and the intersection is $X_1 \wedge X_2 \wedge X_3$. If the last non-null term is the second one, then the intersection is a line given in our formalism by :

$$
X_1 \wedge X_2 (X_3, X_4, X_5, X_6) - X_1 \wedge X_3 (X_2, X_4, X_5, X_6) + X_2 \wedge X_3 (X_1, X_4, X_5, X_6)
$$

and so on. In a space of dimension 4 (or in a projective space of dimension 3), the last two terms vanish but for greater dimension of E, this is not necessarily the case.

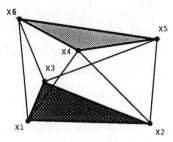

Figure 13.

An Example

We now consider in detail the following case and compare our approach with the approach of analytical geometry. Let $X_1, X_2, X_3, X_4, X_5, X_6$ be six arbitrary points of a space of dimension 3. Suppose we want to know the equation which says that the four planes

$$(X_1, X_2, X_4), (X_4, X_5, X_6), (X_2, X_3, X_5), (X_3, X_1, X_6)$$

have a common point (figure 13). This corresponds to a concrete problem of instability for a robot which consists of two platforms X_1, X_2, X_3 and X_4, X_5, X_6, linked by arms $(X_1, X_4)(X_4, X_2)\ldots(X_6, X_1)$ (see Mourrain 1991 for more details). We compute the first intersection:

$$
\begin{aligned}
&D(X_1 \wedge X_2 \wedge X_4, X_4 \wedge X_5 \wedge X_6) \\
&= X_1 \wedge X_2 \wedge X_4 (X_4, X_5, X_6) \\
&- X_1 \wedge X_4 (X_2, X_4, X_5, X_6) + X_2 \wedge X_4 (X_1, X_4, X_5, X_6) \\
&+ X_4 (X_1, X_2, X_4, X_5, X_6)
\end{aligned}
$$

As we are in a space of dimension 3 and because the points are arbitrary, we find that

$$
\begin{aligned}
&X_1 \wedge X_2 \wedge X_4 \,\triangle\, X_4 \wedge X_5 \wedge X_6 \\
(0.0.0^*.1) \quad &= -X_1 \wedge X_4 (X_2, X_4, X_5, X_6) + X_2 \wedge X_4 (X_1, X_4, X_5, X_6) \\
&X_1 \wedge X_2 \wedge X_4 \,\triangledown\, X_1 \wedge X_2 \wedge X_3 \\
&= -X_2 \wedge X_4 \wedge X_5 \wedge X_6 (X_1, X_4) + X_1 \wedge X_4 \wedge X_5 \wedge X_6 (X_2, X_4)
\end{aligned}
$$

In the same way, we found that the intersection of the three planes

$$(X_1, X_2, X_4), (X_4, X_5, X_6), (X_2, X_3, X_5)$$

is

$$
\begin{aligned}
&(X_1 \wedge X_2 \wedge X_4 \,\triangle\, X_4 \wedge X_5 \wedge X_6) \,\triangle\, X_2 \wedge X_3 \wedge X_5 \\
&= -X_1 (X_2, X_4, X_5, X_6)(X_4, X_2, X_3, X_5) + X_4 (X_2, X_4, X_5, X_6)(X_1, X_2, X_3, X_5) \\
&+ X_2 (X_1, X_4, X_5, X_6)(X_4, X_2, X_3, X_5)
\end{aligned}
$$

So, the equation which says that the 4 planes

$$(X_1, X_2, X_4), (X_4, X_5, X_6), (X_2, X_3, X_5), (X_3, X_1, X_6)$$

are concurrent is:

$$(X_1, X_3, X_4, X_6)(X_2, X_4, X_5, X_6)(X_1, X_2, X_3, X_5)$$
$$- (X_1, X_2, X_3, X_6)(X_1, X_4, X_5, X_6)(X_4, X_2, X_3, X_5)$$

The translation of this equation in analytical geometry is

$$-x_{4,1}^2 x_{6,2}\, x_{5,3}\, x_{1,2}\, x_{6,3}\, x_{5,1}\, x_{3,2}\, x_{2,3} - x_{4,1}^2 x_{6,2}\, x_{5,3}\, x_{1,2}\, x_{6,3}\, x_{5,1}\, x_{2,2}\, x_{3,3}$$
$$+x_{4,1}\, x_{3,2}^3 x_{5,3}^2 x_{6,1}\, x_{1,2}\, x_{1,1}\, x_{4,3} + x_{2,1}\, x_{5,2}\, x_{6,3}\, x_{6,1}\, x_{3,2}\, x_{4,3}^2 x_{3,1}\, x_{1,2}$$
$$-x_{4,1}^2 x_{6,2}\, x_{5,3}\, x_{3,2}\, x_{6,3}\, x_{5,1}\, x_{2,2}\, x_{3,3} - x_{4,1}\, x_{6,2}\, x_{5,3}\, x_{3,1}\, x_{4,2}\, x_{6,3}\, x_{5,1}\, x_{3,2}\, x_{2,3}$$
$$+ \cdots 10938 \text{ other terms} \cdots = 0$$

Iterations of Intersection Operator

We remark that if two elements H_1, H_2 are of the form $L_1.c_1$ (resp. $L_2.c_2$) with $L_1, L_2 \in \mathcal{G}$ and $c_1, c_2 \in C$, then $H_1 \triangle H_2 = I.(S)c_1c_2$ is also of the form $I.c$ with $I \in \mathcal{G}$ and $c = (S)c_1c_2 \in C$. we sum up the situation using the following definition:

Definition 3.3 *Let $\mathcal{G}_C$ be the set of elements of $\wedge_C E - \{0\}$ of the form $g.c$ with $g \in \mathcal{G}$ and $c \in C$. and $\overrightarrow{g.c}$, the linear subspace $\overrightarrow{g}$ associated to g.*

(The linear space $\overrightarrow{H}$ associated to an element $H \in \mathcal{G}_C$ does not depend on the representation of H, for it is the set of vectors v such that $v \wedge H = 0$). This, again, will be the set of elements of the algebra $\wedge_C E$ to which we will naturally associate a geometric object (a linear subspace of E).

The two operations $\triangle$ and $\triangledown$ are defined on $\wedge_C E$ but we remark that **if the two elements H_1, H_2 are in $\mathcal{G}_C$ then their images are also in $\mathcal{G}_C$**. In other words, these operations transform two elements of $\wedge_C E$ which have a geometric interpretation in another one which also corresponds to a geometric object. In the following, we will apply this operations on elements which are not necessarily written as products of vectors (see for instance (3.2.0*.1)) but which, however, correspond to linear subspace.

These operations are generalizations of the Cayley algebra operators "meet" and "join" (see Barnabei *et al.* 1985 or Doubilet *et al.* 1974), so that we are able to compute the two operations whatever the dimension of spaces.

The fact that the two application transform two elements of $\mathcal{G}_C$ in an element of $\mathcal{G}_C$ is fundamental to iterate the operation. This allows us to simplify a system of geometric conditions of the form

$$\begin{cases} v \wedge L_1 = 0 \\ \vdots \\ v \wedge L_s = 0 \end{cases}$$

where $L_i \in \mathcal{G}$, in the equivalent system

$$v \wedge I = 0$$

with $I_2 = L_1 \triangle L_2$, $I_3 = I_2 \triangle L_3$, $\ldots I = I_s = I_{s-1} \triangle L_s \in \mathcal{G}_C$. This remark will be the key of the algorithm which will formally check properties of a configuration of points.

We can go a little further and also simplify an equation of the form $v \wedge L = 0$ where $L = v_1 \wedge \cdots \wedge v_d\, c$ is an element of $\mathcal{G}_C$ of degree d and $v_i \in E$. This equation means that v is in the linear space $\overrightarrow{L}$ associated to L. As $v \wedge L = 0$, we could use, what we call the *Cramer Rule*

$$v \rightarrow v_1.(v, v_2, \ldots, v_d)\, c - v_2.(v, v_1, v_3, \ldots, v_d)\, c + \cdots \pm v_d.(v, v_1, \ldots, v_{d-1})\, c$$

to transform up to a scalar a vector v in a combination of vector v_i. This operation can also be described with the function D so that we can compute it for any representation of L.

Translation of Geometric Properties

With these tools, we can now easily translate geometric relations, as follows:

- $v \in \langle v_1, v_2, \ldots, v_d \rangle$ means that $v \wedge v_1 \wedge \cdots \wedge v_d = 0$ or

$$v \rightarrow v_1.(v, v_2, \ldots, v_d) - v_2.(v, v_1, v_3, \ldots, v_d) + \cdots \pm v_d.(v, v_1, \ldots, v_{d-1})$$

- The points (a, b) (c, d) on a same line are conjugate if $a \wedge c.(b, d) + b \wedge c.(a, d) = 0$ in $\wedge_C \mathbf{E}$ or

$$(a, c)(b, d) + (b, c)(a, d) = 0$$

in C.

- m is the middle of the points a, b if $m \wedge a \wedge b = 0$ and

$$(m, a)(b, \%) + (m, b)(a, \%) = 0.$$

Here the new letter $\%$ plays the part of the hyperplane at infinity and the previous equation is saying that (a, b) are conjugate with respect to m and the intersection of (a, b) with this hyperplane at infinity.

- In the plane, the lines (a, b) (c, d) (e, f) are concurrent if

$$(a, b, c)(d, e, f) - (a, b, d)(c, e, f) = 0.$$

- The lines (a, b) (c, d) are parallel if they are concurrent with the line at infinity :

$$(a, b, c)(d, \%) - (a, b, d)(c, \%) = 0.$$

3.3   The Algebraic Point of View

We want to study configurations of m points defined by projective relations such as collinearity, conjugation ... and check automatically properties of these configurations. We are going to describe now what we mean by checking a property of a configuration.

We adopt an algebraic point of view and consider m generic vectors $X_j = (x_{1,j}, \ldots, x_{n,j})$ where $1 \leq j \leq m$ and $x_{i,j}$ are variables. These vectors are associated to the points of the involved configuration. We note $X = \{x_{i,j}\}_{1 \leq i \leq n, 1 \leq j \leq m}$ and $R = k[X] = k[x_{i,j}]$.

Suppose that the configuration of points is described by a system of equations

$$\begin{cases} h_1(x_{i,j}) = 0 \\ \vdots \\ h_l(x_{i,j}) = 0 \end{cases}$$

We call $\mathcal{I} = \mathcal{I}(h_1, \ldots, h_l)$ the ideal of R generated by $h_1, \ldots, h_l$. This is the set of all elements of the form $\sum_i p_i h_i$ with $p_i \in R$ and is the algebraic set that we are really able to manipulate. We aim at the set of solutions of the equations $(h_i = 0)_i$ (or equivalently

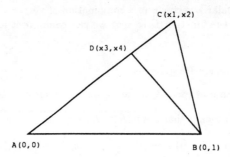

Figure 14.

of all the elements of I). This set, called V, is an algebraic variety of E^m and is the set of points $(a_1, \ldots, a_m)$ which satisfy the hypotheses.

Using classical results (see Atiyah-Mac Donald 1969) on algebraic varieties, we know that V can be decomposed into irreducible components (of a variety which is not the reunion of other varieties) as:

$$V = V_1 \cup \ldots \cup V_q.$$

Now, we want to go from this geometric description to the underlying algebra. We could associate to the variety V the set of polynomials which vanish on this set. This is an ideal which is called the radical of $\mathcal{I}$ and is noted $\sqrt{\mathcal{I}}$. We easily see that if $f^n \in \mathcal{I}$ for some integer n then the polynomial f vanish on V and is in $\sqrt{\mathcal{I}}$. This is exactly the definition of $\sqrt{\mathcal{I}}$.

Suppose, we want to check a property of the points, corresponding to a polynomial g in R. This property g is true if and only if

- under the condition $h_1 = 0, \ldots, h_l = 0$ and if **some conditions of non-degeneration** $s \neq 0$ (There are always such conditions) are satisfied, we have $g = 0$.

which is equivalent to the fact that

- $s.g \in \sqrt{\mathcal{I}}$ but $s \notin \sqrt{\mathcal{I}}$.

or

- g vanish on a component of V.

The crucial point here is to separate the "good components" from the other. We do not want to check a property "$s = 0$" under the condition $g \neq 0$ if $s = 0$ is a condition of degeneration.

To see how careful we must be, let consider the following example (figure 14): we take three points A, B, C of the plan and D a point of (A, C) such that (D, B) and (A, C) are orthogonal. Choose as coordinates $A(0,0)$, $B(0,1)$, $C(x_1, x_2)$, $D(x_3, x_4)$.

The hypotheses are translated by the following equations

$$\begin{cases} x_4 x_1 - x_2 x_3 = h_1 = 0, \\ x_4 x_2 + x_1 x_3 - x_2 = h_2 = 0 \end{cases}$$

Suppose we want to check, with algebraic tools, the simple following conclusion $AB = BC$, which is translated by the equation $c = x_1^2 + (x_2 - 1)^2 - 1$. As we have

$$(2x_3^2 + 2x_4^2 - 2x_4).c = (2h_1 - 2x_1 + 4x_3)h_1 + (2h_2 + 2x_2 - 4x_4)h_2$$

the conclusion $AB = BC$ is true under the condition $(2x_3^2 + 2x_4^2 - 2x_4) \neq 0$ (i.e. $AD^2 + DB^2 \neq AB^2$ and in this case $A = B$).

A technique used by Wu W.T. and C. C. Chou (see Wu Wen Tsün 1984), consists in working with the localization of $\sqrt{\mathcal{I}}$ by a multiplicative set $\mathcal{M}$, for instance the non-null polynomial in the variables we take as parameters: the localization by $\mathcal{M}$ is an algebraic construction which allows us to divide by elements of $\mathcal{M}$ in the ring of polynomials. The geometric interpretation of this operation consists to restrain the variety V to an open subset defined by the points where the elements of $\mathcal{M}$ do not vanish. By this method the geometric structure completely disappears.

We show that in the case of projective configurations of points, we can describe explicitly "geometric conditions" of non-degeneration and obtain only one "good" component.

We introduce n other generic vectors $T_k = (t_{1,k}, \ldots, t_{n,k})$ where the $t_{k,l}$ are new variables. These vectors will play the part of a generic basis of E. We note $T = \{t_{k,l}\}_{1 \leq k,l \leq n}$ and $S = \mathbf{k}[X, T] = R[T]$.

We extend the vector space E by taking $F = E \otimes S$ and $\wedge_S F = \wedge E \otimes S$. For all vectors $f_1, \ldots, f_l \in F$, define

$$(f_1, \ldots, f_l) =]f_1, \ldots, f_l, T_{l+1}, \ldots, T_n]$$

We are going to look at points $a_1, \ldots, a_m$ that cancel elements of S (resp. $\wedge_S F$). This means that these elements of S are null when we substitute the variables $x_{i,j}$ by the corresponding coordinates of the vectors a_j. So by definition, *the ideal associated with a set of elements of S (resp. $\wedge_S F$) is the ideal in R generated by the coefficients (polynomial in $x_{i,j}$) of these elements seen as polynomials (resp. vectors of polynomials) in the variables $T = \{t_{k,l}\}$.*

In the statement of a geometric theorem, the hypothesis is often in a natural triangular form: a new point is constructed with respect to the previous ones, so that the hypotheses which describe the point X_i only depend on the points $X_1, \ldots, X_{i-1}$. Then, X_i is chosen as general as allowed: X_i is "generic" above the configuration of the points $X_1, \ldots, X_{i-1}$.

3.4   Algorithm and Examples

Here is the algorithm, which follows the ideas of the previous sections and detailed on the configuration of a "projective cube" (figure 15).

Inputs:
1) First the list of points written in the order they are constructed.

`(a b c d e f g h i j k)`

2) Then the hypotheses corresponding to position of each point:

`(e d b a) colinear (i c f) colinear`
`(f b c a) colinear (i a b) colinear`
`(g d a c) colinear (j c g) colinear`

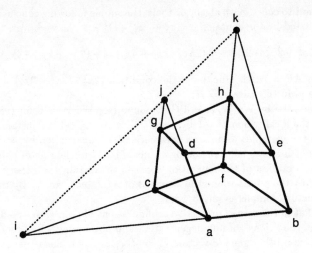

Figure 15.

```
(h f c g) colinear (j a d) colinear
(h f e b) colinear (k f h) colinear
(h g e d) colinear (k b e) colinear
```

These hypotheses correspond to equations on elements of $\wedge E$:

$$e \wedge d \wedge b \wedge d = 0, \ldots, k \wedge b \wedge e = 0.$$

3) The conclusion we want to check:

(i j k) aligned

Outputs:
This algorithm will construct a system of rules such that any element of S which vanishes with the generic component is reduced to zero with this system of rules. Conversely, all polynomial which is reduced to zero vanishes with the generic component. It works in the following way:

Construction:
We suppose here that the vector space E in which we are working is of dimension 4 (projective space of dimension 3). Consider the first point e which really depends on the previous point. Here, there is only one equation at this point so that we construct the Cramer rule associated with this equation

```
e -> + d.(e b a) - b.(e d a) + a.(e d b)
```

We do the same thing for f and g:

```
f -> - c.(f b a) + b.(f c a) - a.(f c b)
g -> - d.(g c a) + c.(g d a) - a.(g d c)
```

Now consider the equations on the point h:

$$\begin{cases} h \wedge f \wedge e \wedge b = 0 \\ h \wedge f \wedge c \wedge g = 0 \\ h \wedge g \wedge e \wedge d = 0 \end{cases}$$

We compute the intersection of $f \wedge e \wedge b$ and $f \wedge c \wedge g$:

$$D(f \wedge e \wedge b, f \wedge c \wedge g)$$
$$= f \wedge e \wedge b(f, c, g)$$
$$+ f \wedge e(b, f, c, g) - f \wedge b(b, f, c, g)$$

The last term does not vanish when we apply the rules (in this order)

```
g -> - d.(g c a) + c.(g d a) - a.(g d c)
f -> - c.(f b a) + b.(f c a) - a.(f c b)
e -> + d.(e b a) - b.(e d a) + a.(e d b)
```

so that we take

$$L = f \wedge e \wedge b \triangle f \wedge c \wedge g = f \wedge e(b, f, c, g) - f \wedge b(e, f, c, g)$$

and the two first equations are equivalent to

$$h \wedge L = 0$$

Now we compute $D(L, g \wedge d \wedge e)$

$$D(L, g \wedge d \wedge e)$$
$$= f \wedge e(b, f, c, g)(g, d, e) - f \wedge b(e, f, c, g)(g, d, e)$$
$$- e(b, f, c, g)(f, g, d, e) - f(e, f, c, g)(b, g, d, e) + b(e, f, c, g)(f, g, d, e)$$

Here also, the last term does not vanish when we apply the previous rules and we take

$$L \triangle g \wedge d \wedge e = -e(b, f, c, g)(f, g, d, e) - f(e, f, c, g)(b, g, d, e) + b(e, f, c, g)(f, g, d, e)$$

The corresponding Cramer rule is

```
h -> - e.(b,f,c, g)(f,g,d,e) - f.(e,f,c,g)(b,g,d,e)
+ b.(e,f,c,g)(f,g,d,e)
```

We go on with the point i:

$$D(f \wedge c, b \wedge a) = f \wedge c(b, a)$$
$$+ f(c, b, a) - c(f, b, a)$$
$$+ 1.(f, c, b, a)$$

The last term of this expression vanishes when we reduce it by the rules already constructed but not the second one, so that we have a new rule:

```
i -> + f.(c,b,a) - c.(f,b,a)
```

and in the same way, we obtain the rules on j, k:

```
j -> + g.(d c a) + c.(g d a)
k -> - e.(h f b) + b.(h f e)
```

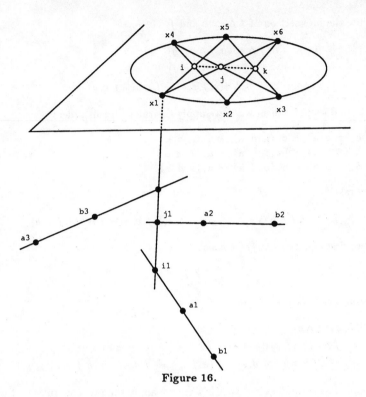

Figure 16.

Reduction:
Now, we can reduce the conclusion by this rules starting from the last constructed and developing the terms between parentheses by multilinearity:

```
(i j k)
-> - (i j e)(h f b) + (i j b)(h f e)
-> - (i g e)(h f b)(d c a) + (i g b)(h f e)(d c a)
+ (i e c)(h f b)(g d a) + (i c b)(h f e)(g d a)
-> ...
-> 0
```

which means that the conclusion is true with the "generic" component. This method, described with an example can be generalized to any problem where the hypotheses are of required form (elements of $\mathcal{G}_C$) in a space E of any dimension.

We now give outputs of a program, written in LE LISP, which computes the reduction rules and reduces the conclusion.

Pascal Theorem for a Conic (figure16)

Let **x1, x2, x3, x4, x5, x6** be six point on a conic. Then i = (x1 x5) ∩ (x2 x4), j = (x1 x6) ∩ (x3 x4), k = (x2 x6) ∩ (x3 x5) are on a same line.

Variables :
?(a1 b1 a2 b2 a3 b3 i1 j1 i2 j2 i3 j3 i4 j4 i5 j5 i6 j6 x1 x2 x3 x4 x5 x6 i j k)

```
Hypotheses :
h1 : (a3 b2 a2 b1 a1) colinear
h2 : (b3 b2 a2 b1 a1) colinear
h3 : (i1 b1 a1) aligned
h4 : (j1 i1 b3 a3) coplanars
h5 : (j1 b2 a2) aligned
 ...
h18 : (i6 b1 a1) aligned
h19 : (j6 i6 b3 a3) coplanars
h20 : (j6 b2 a2) aligned
 ...
h28 : (x6 x3 x2 x1) coplanars
h29 : (x6 j6 i6) aligned
h30 : (i x4 x2) aligned   h31 : (i x5 x1) aligned
h32 : (j x4 x3) aligned   h33 : (j x6 x1) aligned
h34 : (k x5 x3) aligned   h35 : (k x6 x2) aligned

Conclusion :
c1 : (i j k) aligned

Derivation :
r1 := k --> x5.(x6 x3 x2) - x3.(x6 x5 x2)
r2 := j --> - x4.(x6 x3 x1) + x3.(x6 x4 x1)
 ...
r22 := b3 --> b2.(b3 a2 b1 a1) - a2.(b3 b2 b1 a1) + b1.(b3 b2 a2 a1) - a1.(b3 b2
 a2 b1)
r23 := a3 --> b2.(a3 a2 b1 a1) - a2.(a3 b2 b1 a1) + b1.(a3 b2 a2 a1) - a1.(a3 b2
 a2 b1)

Result: t1 := ()
```

Here the six lines (i_l, j_l) touch the lines (a_1, b_1), (a_2, b_2), (a_3, b_3) and are on a hyperboloid. The intersection with the plane (x_1, x_2, x_3) is a conic and the cross intersections i, j, k of $x_1, x_2, x_3, x_4, x_5, x_6$ are on the same line.

3.5 Conclusion

Projective geometry is a natural language for vision. It provides a good framework to describe the intrinsic mechanisms of projection from a space of dimension 3 to the plane of the image. In the case of configurations of points defined by relations of alignment, coplanarity, . . . , conjugation, we have shown how to treat directly the geometric objects, computing with invariant coefficients.

The intrinsic techniques have to be enlarged to test more complicated geometric relationships where for instance, quadratic forms appears: tools such as factorization, resultants, etc. will be needed.

We prefer to adopt an intrinsic point of view and not to follow analytical methods. Here, computations are carried out in an algebra of invariants and often lead to smaller polynomials. The advantage of this method is that geometric properties can be translated and manipulated more easily and that the intermediate results during a computation still have a geometric meaning.

So, considering Vision as "concrete Projective Geometry", we can bet that it will raise more and more problems on points, lines, conics, etc., using results of one the oldest branch of Mathematics.

References

Abraham, R., Robbin, J. (1967) Transversal Mappings and Flows. Benjamin, New-York

Arnold, R. (1972) Lectures on Bifurcations in Versal Families. Russian Math. Surveys 27

Arnold, V. (1974) Critical Points of Smooth Functions. Proc. Intern. Congr. Math. Vancouver 1974, pp. 19-39

Arnold, V.I. (1976) Wave front evolution and equivariant Morse lemma. Comm. Pure Appl. Math. 29, 557-582

Arnold, V.I. (1983) Singularities of systems of rays. Russian Math. Surveys 38

Arnold, V. (1984) Catastrophe Theory. Springer Verlag

Arnold, V., Varchenko, A., Goussein-Zade, S. (s.d.) Singularités des applications différentiables. Traduit du russe, Editions Mir

Atiyah-Mac Donald, (1969) Introduction to Commutative Algebra. Addison-Wesley

Barnabei, M., Brini, A., Rota, G.C. (1985) On the Exterior Calculus of Invariant Theory. J. of Alg. 96, pp. 120-160

Bennequin, D. (s.d.) Caustique mystique d'après Arnold *et al.*. Séminaire Bourbaki 84-85, n° 634

Boardman, J.M. (1967) Singularities of differentiable mappings. Publ. I.H.E.S. 33, 21-57

Bruce, J.W., Giblin, P.J. (1984) Curves and Singularities. Cambridge University Press

Bruce, J.W., Giblin, P.J. (1985) Outlines and their duals. Proc. London Math. Soc. 50, 552-570

Bruce, J.W., Giblin, P.J. (1990) Projections of Surfaces with boundary. Proc. London Math. Soc. (3) 60, 392-416

Chenciner, A. (1972-1973) Travaux de Thom et Mather sur la stabilité topologique. Séminaire Bourbaki, 25 ème année, (1972-1973), exposé 424

Demazure, M. (1975) Classification des germes à point critique isolé et à nombre de modules 0 ou 1 (d'après V.I. Arnold). Séminaire Bourbaki 1973-1974, exposé 443, Springer Lecture Notes in Math. 431, 124-142

Demazure, M. (1987) Géométrie, Catastrophes et bifurcations. Ecole Polytechnique, Cours, Édition 1987, published under the title:Catastrophes et bifurcations. Éditions Ellipses, Paris

Doubilet, P., C. Rota, G.C., Stein, J. (1974) Foundations of Combinatorics IX : Combinatorial methods in invariant Theory. Studies in Appl. Math. 53, p 185-216

Dufour, J.P. (1983) Famille de courbes planes différentiables. Topology, vol. 22/4, 449-474, Printed in Great Britain, Pergamon Press Ltd.

Dufour, J.P. (1989) Modules pour les familles de courbes planes. Annales de l'Institut Fourier 39/1, 225-238

Gaffney, T. (1983) The structure of $T\mathcal{A}(f)$, classification and application to differential geometry. Orlik P. (ed.) Singularities Proc. Symp. Pure Math. 40 (part 1), 409-428.

Golubitsky, M., Guillemin, V. (1980) Stable Mappings and Their Singularities, Graduate Texts in Mathematics, vol. 14, Springer-Verlag, 1974, second corrected printing 1980

Gorjunov, (1990) in "Singularitiy theory and its applications" Advances in Soviet Math. Vol 1 (Arnold ed.) AMS

Henry, J.P., Merle, M. (1990) Fronces et doubles plis. In: M. Merle, Thèse d'État, Paris, Février 1990

Henry, J.P., Teissier, B. (1979) Suffisance des familles de jets et Équisingularité. In J.P. Henry, Thèse d'État. Paris, Septembre 1979, et Séminaire F. Norguet (1974-75), Springer Lecture Notes in Math. 482, (1976), 351-357

Hilton, P.J. (ed.) (1976) Structural Stability, the Theory of Catastrophes, and Applications in the Sciences. Proceedings of the conference held at Battelle Seattle Researh Center 1975, Springer Lecture Notes in Math. 525 (1976)

Horn, B.K.P. (1975) Obtaining shape from shading information. In P.H. Winston (ed.) The Psychology of Computer Vision. McGrawHill: New-York, 115-155

Horn, B.K.P. (1986) Robot Vision. MIT Press, MA ., and McGrawHill: New-York

Horn, B.K.P. (1990) Height and Gradient from Shading. International Journal of Computer Vision, 5:1, 37-75, Kluwer Academic Publishers

Iooss, G., Joseph, D. (1980) Elementary Stability and Bifurcation Theory, Undergraduate Texts in Mathematics. Springer-Verlag

Koenderink, J.J. (1984) What does the occluding contour tell us about solid shape. Perception, 13, 321-330

Koenderink, J.J., van Doorn, A.J. (1976) The singularities of the visual mapping. Bio. Cybernet. 24, Springer-Verlag, 51-59

Koenderink, J.J., van Doorn, A.J. (1984) The shape of smooth objects and the way contours end. Perception 11, 129-137

Malik, J. (1987) Interpreting Line Drawings of Curved Objects. International Journal of Computer Vision 1, 73-103, Kluwer Academic Publishers, Boston

Mather, J. (1969) Stability of C^∞ mappings III. Publ. I.H.E.S. 35, 127-156

Mather, J.N. (1970) Stability of C^∞ mappings IV. Publ. I.H.E.S. 37, 223-248

Mather, J.N. (1973a) Generic projections. Annals of Math., 98 no. 2, September 1973, 226-245

Mather, J. (1973b) Stratifications and Mappings. Dynamical Systems, Academic Press, Inc., New-York and London

Mourrain, B. (1991) Approche effective de la théorie des invariants des groupes classiques. Ph.D. (Ecole Polytechnique), September (1991)

Platonova, O.A. (1981) Singularities of the mutual disposition of a surface and a line. Russian Math. Surveys 36:1, 248-249

Platonova, O.A. (1984) Singularities of projections of smooth surfaces. Russian Math. Surveys 39:1, 177-178

Rieger, J.H. (1987a) Families of maps from the plane to the plane. J. London Math. Soc. 2, 36, 351-369

Rieger, J.H. (1987b) On the classification of views of piecewise smooth objects. Image and vision computing, vol. 5, no. 2, 91-97

Rosenfeld, A., Kak, A.C. (1982) Digital Picture Processing. Vols. 1 & 2, second edition, Academic Press, New-York

Shang Shing Chou (1988) Automatic prover in Geometry. Reidel

Tari, F. (1990) Some applications of Singularity Theory to the Geometry of Curves and Surfaces. Ph. D. Thesis, Liverpool

Tari, F. (1992) Projections of piecewise smooth surfaces. to appear in J. London Math. Soc.

Thom, R. (1972-73) Stabilité structurelle et morphogénèse: essai d'une théorie générale des modèles, Benjamin, Reading (Mass), 1972 or Addison-Wesley, Reading, Mass. 1973

Thom, R. (1981) Modèles mathématiques de la morphogénèse, Ch. Bourgois, Paris

Tougeron, J.C. (1967-68) Stabilité des applications différentiables. d'après Mather, Séminaire Bourbaki, 1967-1968, exposé 336, Addison Wesley/Benjamin

Todd, J.T., Reichel, F.D. (1990) Visual Perception of smoothly curved surfaces from double-projected contour patterns. Journal of experimental Psychology: Human Perception and Performance., Vol. 16, No 3, 665-674

Tsotsos, J. (1992) Locating and Localising Stimuli in the Visual Field. Invited Lecture, Final Meeting of the INSIGHT project, EEC Basic Research Action 3001, February 1992, Leuven

Tueno, J.P. (1989) Thèse de 3 ème Cycle. Université de Montpellier, November 1989

Waltz, D. (1975) Understanding line drawings of scenes with shadows. In P.H. Winston (ed.) The Psychology of Computer Vision. McGraw Hill: New-York, 19-91

Whitney, H. (1955) On singularities of mappings of Euclidean spaces. I. Mappings of the plane into the plane. Ann. of Math. 62, 374-410

Wu Wen Tsün (1984) Some Recent Advances in Mechanical Theorem Proving: After 25 years. American Mathematical Society, Contemporary Mathematics 29, p 235-242

A Method of Obtaining the Relative Positions of 4 Points from 3 Perspective Projections

H. Christopher Longuet-Higgins

Laboratory of Experimental Psychology, University of Sussex,
and Department of Engineering Science, University of Oxford

Abstract

According to Ullman's Structure-from-Motion Theorem (Ullman 1979), three orthogonal projections of four points in a non-planar configuration uniquely determine their structure, and the relative orientations of the three projections, up to a reflection in the image plane. This paper describes a direct method, based on quaternion algebra, for computing the alternative structures, and shows that it can be simply generalized from orthogonal to para-perspective projections, in which the viewing distances are large but finite. This leads on to an iterative method of computing structure from motion in the fully perspective case – a method that makes it possible, in favourable circumstances, to resolve the Necker ambiguity and arrive at an accurate structure after a small number of computational cycles.

1 The Orthogonal Case

We adopt one of the points P_0 as origin and denote the 3D coordinates of the others by $(X_n, Y_n, Z_n), (X'_n, Y'_n, Z'_n)$ and (X''_n, Y''_n, Z''_n) in the three projection frames, (X_n, Y_n), (X'_n, Y'_n) and (X''_n, Y''_n) being the (relative) image coordinates and Z_n, Z'_n and Z''_n the (relative) depth coordinates. Then there will exist unique rotation matrices $U = [u_{ij}]$ and $V = [v_{ij}]$ such that

$$X'_n = u_{11}X_n + u_{12}Y_n + u_{13}Z_n, \tag{1}$$

$$Y'_n = u_{21}X_n + u_{22}Y_n + u_{23}Z_n, \tag{2}$$

$$Z'_n = u_{31}X_n + u_{32}Y_n + u_{33}Z_n \tag{3}$$

and

$$X''_n = v_{11}X_n + v_{12}Y_n + v_{13}Z_n, \tag{4}$$

$$Y''_n = v_{21}X_n + v_{22}Y_n + v_{23}Z_n, \tag{5}$$

$$Z_n'' = v_{31}X_n + v_{32}Y_n + v_{33}Z_n. \tag{6}$$

The problem is to compute U and V, and the depth coordinates, from the three sets of image coordinates. As each of these rotations involves 3 unknown parameters, and there are 9 depth coordinates, we have 18 equation for only 15 unknowns, and may expect to encounter 3 consistency conditions, useful for checking purposes.

Elimination of Z_n between (1) and (2) gives (see Appendix A)

$$X_n u_{32} - Y_n u_{31} + X_n' u_{23} - Y_n' u_{13} = 0, (n = 1, 2, 3) \tag{7}$$

from which one can obtain the ratios of the "border elements" $u_{32} : u_{31} : u_{23} : u_{13}$.

The computation fails if either (i) the four points are coplanar, in which case equations (7) are no longer independent, or (ii) the Z and Z' axes coincide, in which case $u_{33} = \pm 1$ and all four border elements vanish. Otherwise, because U is a rigid rotation, its border elements must satisfy

$$u_{32}^2 + u_{31}^2 (= 1 - u_{33}^2) = u_{23}^2 + u_{13}^2. \tag{8}$$

This is one of the three consistency conditions mentioned above, and can be checked as soon as the ratios $u_{32} : u_{31} : u_{23} : u_{13}$ have been obtained from (7).

Introducing the normalized quaternion

$$\mathbf{Q} = \mathbf{i}p + \mathbf{j}q + \mathbf{k}r + s, \quad p^2 + q^2 + r^2 + s^2 = 1 \tag{9}$$

related to U by the equations

$$U(\mathbf{Q}) = \begin{bmatrix} u_{11} & u_{12} & u_{13} \\ u_{21} & u_{22} & u_{23} \\ u_{31} & u_{32} & u_{33} \end{bmatrix} = \begin{bmatrix} p^2 - q^2 - r^2 + s^2, & 2(pq - rs), & 2(pr + qs) \\ 2(pq + rs), & -p^2 + q^2 - r^2 + s^2, & 2(qr - ps) \\ 2(pr - qs), & 2(qr + ps), & -p^2 - q^2 + r^2 + s^2 \end{bmatrix} \tag{10}$$

we see (Appendix B) that the ratios $u_{32} : u_{31} : u_{23} : u_{13}$ determine the ratio of p to q and the ratio of r to s, but not that of p to r. It follows that if $\mathbf{Q}$ is written in the form

$$\mathbf{Q} = (\mathbf{i}sinA + \mathbf{j}cosA)sinC + (\mathbf{k}sinB + cosB)cosC \tag{11}$$

then the two images (X_n, Y_n) and (X_n', Y_n') yield the values of the two parameters A and B, but not the "vergence" parameter C (equal to half of the angle between the Z and Z' axes – see Appendix C). To compute C we need all three images, and the A and B parameters of the rotations connecting them, which we now denote (see Figure 1) by $U_1(= U), U_2(= V^{-1})$ and $U_3(= VU^{-1})$, satisfying

$$U_1 U_2 = U_3^{-1}. \tag{12}$$

Armed with these parameters we substitute them in the parallel equation

$$\mathbf{Q}_1 \mathbf{Q}_2 = \mathbf{Q}_3^{-1} \tag{13}$$

where

$$\mathbf{Q}_1 = (\mathbf{i}sinA_1 + \mathbf{j}cosA_1)sinC_1 + (\mathbf{k}sinB_1 + cosB_1)cosC_1, \tag{14}$$
$$\mathbf{Q}_2 = (\mathbf{i}sinA_2 + \mathbf{j}cosA_2)sinC_2 + (\mathbf{k}sinB_2 + cosB_2)cosC_2, \tag{15}$$
$$\mathbf{Q}_3^{-1} = (-\mathbf{i}sinA_3 - \mathbf{j}cosA_3)sinC_3 + (-\mathbf{k}sinB_3 + cosB_3)cosC_3. \tag{16}$$

Using the rules of quaternion multiplication ($\mathbf{i}^2 = \mathbf{j}^2 = \mathbf{k}^2 = -1, \mathbf{ij} = \mathbf{k} = -\mathbf{ji}$, etc.) we equate coefficients of $\mathbf{i}, \mathbf{j}, \mathbf{k}$ and unity on the two sides of (13), to obtain four equations

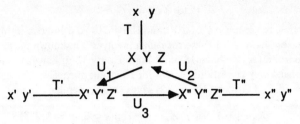

Figure 1: *Key to Algebraic Notation Adopted in the Text.*

which we may call I, J, K and L (see Appendix D). Elimination of $sinC_3$ from I and J and of $cosC_3$ from K and L gives, after a certain amount of algebra,

$$tanC_1 = \pm\sqrt{(S_0 S_1 / S_2 S_3)}, \tag{17}$$

$$tanC_2 = (S_2 / S_1) tanC_1, \tag{18}$$

$$tanC_3 = (S_3 / S_1) tanC_1 \tag{19}$$

where

$$S_0 = sin(B_1 + B_2 + B_3), \tag{20}$$

$$S_1 = sin(B_1 - A_2 + A_3), \tag{21}$$

$$S_2 = sin(B_2 - A_3 + A_1), \tag{22}$$

$$S_3 = sin(B_3 - A_1 + A_2). \tag{23}$$

These expressions for C_1, C_2 and C_3 in terms of the A's and B's enable us to determine $\mathbf{Q}_1$, $\mathbf{Q}_2$ and $\mathbf{Q}_3$, and from them the relative orientations U_1, U_2 and U_3. The relative depths Z_n then follow from (1) or (2), though their absolute signs are subject to an overall "Necker" ambiguity associated with the undetermined sign of $tanC_1$ in (17).

2   The Para-Perspective Case

Whereas in orthogonal projection the image coordinates are obtained by projecting 3D coordinates directly on to the (X, Y) plane, in perspective and para-perspective projection the equations for the image coordinates involve the distances T, T' and T'' of the three viewpoints from the reference point P_0. Without serious loss of generality we now assume that in each image P_0 lies on the optic axis of the camera (if it does not, the image coordinates must be transformed appropriately); the plane projective coordinates of the other 3 points are then

$$x_n = X_n / (T + Z_n), \qquad y_n = Y_n / (T + Z_n), \tag{24}$$

$$x'_n = X'_n / (T' + Z'_n), \qquad y'_n = Y'_n / (T' + Z'_n), \tag{25}$$

$$x''_n = X''_n / (T'' + Z''_n), \qquad y''_n = Y''_n / (T'' + Z''_n), \tag{26}$$

where, as before, $X_n, \ldots, Z''_n$ are the 3D coordinates of P_n (relative to P_0) in the three frames. In the para-perspective or "small object" approximation one neglects the relative depths Z_n in comparison with the viewing distances T; this amounts to identifying the image coordinates x_n and y_n with the "reduced widths" ξ_n and η_n defined by equations (27) etc., and neglecting the "reduced depths" ζ_n altogether.

$$\xi_n = X_n/T, \quad \eta_n = Y_n/T, \quad \zeta_n = Z_n/T, \tag{27}$$

$$\xi_n' = X_n'/T', \quad \eta_n' = Y_n'/T', \quad \zeta_n' = Z_n'/T', \tag{28}$$

$$\xi_n'' = X_n''/T'', \quad \eta_n'' = Y_n''/T'', \quad \zeta_n'' = Z_n''/T''. \tag{29}$$

Substituting from (27) and (28) into (7) we obtain, for the frames connected by U_1,

$$\xi_n T u_{32} - \eta_n T u_{31} + \xi_n' T' u_{23} - \eta_n' T' u_{13} = 0, (n = 1, 2, 3) \tag{30}$$

and these three equations yield the ratios $T u_{32} : T u_{31} : T' u_{23} : T' u_{13}$. *Assuming* the rigidity condition

$$u_{32}^2 + u_{31}^2 = u_{23}^2 + u_{13}^2 \tag{31}$$

makes it possible to calculate *both* $T : T'$ *and* $u_{32} : u_{31} : u_{23} : u_{13}$, from which the values of A and B follow. The gain in generality over the orthogonal case has been bought at the expense of a consistency check at this stage; but when a similar computation has been carried out for the other two rotations, U_2 and U_3, the ratios $T : T', T' : T''$ and $T'' : T$ may be compared, to see whether their product is unity, as consistency demands. Thereafter, from the A and B parameters of U_1, U_2 and U_3 we can compute their C parameters, and hence the matrices themselves, as explained in the previous section.

The final step is to calculate the reduced depths. The method most in keeping with the para-perspective approximation is to replace the 3D widths X and Y in equations (1) and (2) by appropriately scaled image coordinates Tx and Ty, obtaining equations such as (32) and (33)

$$T'x_n' = T(u_{11}x_n + u_{12}y_n) + u_{13}Z_n, \tag{32}$$

$$T'y_n' = T(u_{21}x_n + u_{22}y_n) + u_{23}Z_n. \tag{33}$$

Multiplying (32) by u_{13}, (33) by u_{23} and adding the results we obtain eventually

$$\zeta_n = Z_n/T = (1 - u_{33}^2)^{-1}(u_{33}(u_{31}x_n + u_{32}y_n) + (T'/T)(u_{13}x_n' + u_{23}y_n')). \tag{34}$$

In the next section, however, we shall give reasons for preferring an alternative course.

3   The Perspective Case

The fact that (34) supplies values for the very quantities that are initially neglected in the para-perspective approximation raises the hope that one might be able, in favourable circumstances, to bootstrap one's way from the para-perspective to the fully perspective case. Having obtained provisional values of the elements u_{ij}, why persist with an approximation that is known to be inexact? Would it not be better to take full account of perspective effects in obtaining the reduced coordinates from the image coordinates? All we need do is to replace (32) and (33) by (35) and (36), obtained by substituting from (24) and (25) into (1) and (2):

$$(T' + Z_n')x_n' = (T + Z_n)(u_{11}x_n + u_{12}y_n) + u_{13}Z_n, \tag{35}$$

$$(T' + Z_n')y_n' = (T + Z_n)(u_{21}x_n + u_{22}y_n) + u_{23}Z_n. \tag{36}$$

Equations (35) and (36) immediately give the ratios of $(T' + Z_n')$, $(T + Z_n)$ and Z_n, and the reduced widths and depths may then be computed from the relations

$$\xi_n = x_n(T + Z_n)/T, \; \eta_n = y_n(T + Z_n)/T \text{ and } \zeta_n = Z_n/T. \tag{37}$$

With these new reduced coordinates, one set for each Necker alternative, one can return to equation (30) and recompute, first the A and B parameters and then the C parameters of the three rotation matrices. This time, however, the earlier choice of Necker alternative will affect, not only the absolute signs of the depth coordinates but their relative magnitudes as well, and also the magnitudes of the ratios $T : T'$ etc. The product of these three ratios serves as a measure of the consistency of the chosen alternative with the three sets of image coordinates, and one will naturally prefer that alternative for which the product is closer to unity. Thereafter one may iterate cyclically through the various steps described until the process either converges or evidently fails to do so – because the viewpoints are too close to the object, the images too noisy or the views too similar.

4   Results

As yet the only results available are some that have been obtained by computer simulation. Since the application of the method demands that the 4 points are non-coplanar, P_0 is placed at one vertex of a unit cube, with P_1, P_2 and P_3 at the three neighbouring vertices, so that the triple product of the vectors $P_0 P_n$ (a convenient measure of non-planarity) is unity. The 3D coordinates are then evaluated in three randomly oriented frames centred on P_0, and perspective image coordinates are evaluated for three viewpoints located at specified distances T, T' and T'' along the Z, Z' and Z'' axes respectively. These image coordinates, corrupted with gaussian noise of specified standard deviation, comprise the sole information available to the structural computation. It is, of course, necessary to maintain a constant value of the Necker parameter – the sign of $tanC_1$ – throughout each computation, and the alternative choices lead to distinct structures, of differing consistency.

Tables 1 and 2 show, for two different sets of 3D coordinates $X_1, \ldots, Z_3$, (i) the chosen values of T, T' and T'', (ii) the choice of sign for $tanC_1$, (iii) the standard deviation of the gaussian noise added to the image coordinates, (iv) the final value of the consistency measure (when it ceased to change in the fourth decimal place), (v) the number of iterations needed to achieve this value, and the final values of (vi) the non-planarity measure and (vii) the cosines of the "vergence" angles between the three optic axes. In Table 2 the word "fail" in column (iv) indicates that the theoretical expression for tan^2C_1 acquired a negative value in the course of the iteration process.

The conditions for the application of the method – that the tetrahedron be sufficiently non-planar and the viewing directions sufficiently distinct – make its convergence less than certain in any given case, especially when the image coordinates are subject to substantial measurement errors. In Table 2, for example, where one of the vergence angles has a cosine dangerously close to unity, the method assigned a negative value to tan^2C_1 on the very first iteration at even quite moderate viewing distances. More typical is the case documented in Table 1, where the vergence cosines are substantially less than unity, and both Necker structures emerge from the computation, one with a distinctly better consistency measure than the other. But, as the Table shows, at fairly large viewing distances there is little to choose between the Necker alternatives on the basis of the consistency measure, and even quite a small amount of noise in the data leads to a rapid decline in the accuracy of the final solution.

Table 1: *The performance of the method on images derived from a typical set of 3D coordinates.*

(i)			(ii)	(iii)	(iv)	(v)	(vi)	(vii)		
Correct	values		-1		1.000		1.000	0.117	-0.835	-0.461
3	4	5	1	0	1.074	13	-0.764	-0.447	-0.454	-0.253
3	4	5	-1	0	1.000	10	0.999	0.117	-0.835	-0.461
3	4	5	1	0.001	1.069	13	-0.756	-0.449	-0.451	-0.252
3	4	5	-1	0.001	0.989	10	0.966	0.095	-0.831	-0.445
3	4	5	1	0.003	1.092	13	-0.763	-0.450	-0.435	-0.275
3	4	5	-1	0.003	0.981	12	0.982	0.107	-0.845	-0.456
7	8	9	1	0	1.006	6	-0.821	-0.256	-0.651	-0.284
7	8	9	-1	0	1.000	6	1.000	0.117	-0.835	-0.461
7	8	9	1	0.001	0.995	6	-0.799	-0.232	-0.657	-0.303
7	8	9	-1	0.001	1.023	6	1.019	0.108	-0.843	-0.441
7	8	9	1	0.003	1.088	5	-0.831	-0.327	-0.656	-0.173
7	8	9	-1	0.003	0.979	6	1.001	0.049	-0.793	-0.450
30	40	50	1	0	0.990	4	-0.930	0.012	-0.798	-0.402
30	40	50	-1	0	1.000	4	1.000	0.117	-0.835	-0.461
30	40	50	1	0.001	0.966	4	-0.978	-0.010	-0.828	-0.307
30	40	50	-1	0.001	1.099	4	0.959	-0.143	-0.760	-0.283
30	40	50	1	0.003	1.121	4	-1.374	0.097	-0.778	-0.252
30	40	50	-1	0.003	1.422	4	0.976	-0.353	-0.433	-0.529

5 Discussion

These results, such as they are, show that it is possible to derive accurate structural parameters from sets of images without relying on the delicate "perspective effects" exploited in, for example, (Longuet-Higgins 1981). To achieve this it is necessary, as Ullman (Ullman 1979) and others (Huang and Lee 1989) have realized, to compare the images obtained from at least three sufficiently distinct viewpoints, although, as Koenderink and van Doorn (Koenderink and van Doorn 1992) point out, two orthogonal projections suffice to determine those features of an object's structure that are invariant under an affine transformation. Within the orthogonal and paraperspective approximations the present work may be regarded as complementary to that of Tomasi and Kanade (Tomasi and Kanade 1991), who show how to decompose a matrix of essentially orthogonal images into one matrix specifying the camera orientations and another encapsulating the structure of the object. In particular, the present procedure for determining U and V from three orthogonal projections supplies a simple and straightforward way of computing their 3×3 matrix **A** from three representative members of the sequence.

The present work breaks new ground, however, in showing that one consistency constraint survives the generalization from 3 orthogonal to 3 perspective projections, and that this makes it possible in favourable cases to transcend Ullman's theorem and select the correct Necker alternative in a principled fashion; having done so one may proceed to compute the structure with an accuracy limited only by that of the image coordinates. It is, however, necessary to emphasize that this method of obtaining the structure from

Table 2: *The performance of the method on images derived from a different set of 3D coordinates.*

(i)			(ii)	(iii)	(iv)	(v)	(vi)		(vii)	
Correct values:			-1		1.000		1.000	-0.550	0.676	-0.944
30	40	50	1	0	0.988	4	-1.238	0.717	0.781	-0.974
30	40	50	-1	0	1.000	5	1.000	-0.550	0.676	-0.944
7	8	9	1	0	fail					
7	8	9	-1	0	fail					

three perspective images is still unsupported by any formally precise conditions for its convergence, so that it is impossible to be quite sure that it will not go off the rails in any novel situation.

Acknowledgments

Warmest thanks to Mike Brady for his invaluable encouragement and support and to Andrew Blake, Bernard Buxton, Chris Harris, Steve Maybank, Larry Shapiro and Andrew Zisserman for valuable comments and discussions.

The work was funded by the European Community, through the Esprit "Insight" project.

References

Ullman, S. (1979) The interpretation of visual motion. MIT Press, Cambridge MA, USA

Tomasi, C., Kanade, T. (1991) Shape and motion from image streams: a factorization method. 2. Point features in 3D motion. School of Computer Science, Carnegie-Mellon University, Pittsburgh

Longuet-Higgins, H.C. (1981) The reconstruction of a scene from two projections. Nature Vol. 293, pp. 133-135

Koenderink, J.J., van Doorn, A.J. (1992) Affine Structure from Motion. J. Opt. Soc. Amer. (to appear)

Huang, T.S., Lee, C.H. (May 1989) Motion/structure from orthographic projections. IEEE Trans. on Pattern Analysis and Machine Intelligence

Kanatani, K. (1988) Group theoretical methods in image understanding. p. 273, Proposition 6.13

Appendices

Appendix A.
To obtain (7) from (1) and (2) it is necessary to use the relations

$$u_{21}u_{13} - u_{11}u_{23} = u_{32}, \qquad u_{22}u_{13} - u_{12}u_{23} = u_{31} \tag{38}$$

which follow from the fact that in U every element u_{ij} equals its cofactor U_{ij}.

Appendix B.
From equation (10) we deduce that

$$p/q = (u_{13} + u_{31})/(u_{23} + u_{32}), r/s = (u_{23} + u_{32})/(u_{13}u_{31}) \tag{39}$$

Appendix C.
As shown by Kanatani (Kanatani 1988) the angles A, B and C in the representation (11) are closely related to the Euler angles of the rotation $U(\mathbf{Q})$. In particular, the angle C is one-half of the second Euler angle, which is none other than the angle between the old and the new Z axes.

Appendix D.
[In this section the indices 1, 2 and 3 are printed on the line, for clarity, rather than as subscripts.]

The equations I, J, K and L are as follows:

$$I: \quad sinA1sinC1cosB2cosC2 + sinA2sinC2cosB1cosC1 +$$
$$cosA1sinC1sinB2cosC2 - cosA2sinC2sinB1cosC1 = -sinA3sinC3$$

or

$$(sinC1cosC2)sin(A1 + B2) + (sinC2cosC1)sin(A2 - B1) + sinA3sinC3 = 0. \tag{40}$$

$$J: \quad cosA1sinC1cosB2cosC2 + cosA2sinC2cosB1cosC1 +$$
$$sinB1cosC1sinA2sinC2 - sinA1sinC1sinB2cosC2 = cosA3sinC3$$

or

$$(sinC1cosC2)cos(A1 + B2) + (cosC1sinC2)cos(A2 - B1) + cosA3sinC3 = 0 \tag{41}$$

From (40) and (41) it follows that

$$-(sinC1cosC2/sinC2cosC1) = \quad (sin(A2 - B1)cosA3 - cos(A2 - B1)sinA3)/$$
$$(sin(A1 + B2)cosA3 - cos(A1 + B2)sinA3)$$

or

$$tanC1/tanC2 = sin(A2 - B1 - A3)/sin(A1 + B2 - A3). \tag{42}$$

$$K: \quad sinB1cosC1cosB2cosC2 + sinB2cosC2cosB1cosC1 +$$
$$sinA1sinC1cosA2sinC2 - sinA2sinC2cosA1sinC1 = -sinB3cosC3$$

or

$$(cosC1cosC2)sin(B1 + B2) + (sinC1sinC2)sin(A1 - A2) + sinB3cosC3 = 0. \quad (43)$$

$$L: \quad cosB1cosC1cosB2cosC2 \; - \; sinA1sinC1sinA2sinC2 \; -$$
$$cosA1sinC1cosA2sinC2 \; - \; sinB1cosC1sinB2cosC2 = cosB3cosC3$$

or

$$-(cosC1cosC2)cos(B1 + B2) + (sinC1sinC2)cos(A1 - A2) + cosB3cosC3 = 0. \quad (44)$$

From (43) and (44) it follows that

$$-(sinC1sinC2/cosC1cosC2) \; = \; (sin(B1 + B2)cosB3 + cos(B1 + B2)sinB3)/$$
$$(sin(A1 - A2)cosB3 - cos(A1 - A2)sinB3)$$

or

$$-tanC1tanC2 = sin(B1 + B2 + B3)/sin(A1 - A2 - B3). \quad (45)$$

From (42) and (45), equations (17) and (18) follow immediately; (19) then follows by symmetry.

Direct Estimation of Optical Flow and of Its Derivatives

Hans-Hellmut Nagel

Institut für Algorithmen und Cognitive Systeme, Fakultät für Informatik,
Universität Karlsruhe (TH), und
Fraunhofer-Institut für Informations- und Datenverarbeitung IITB

Abstract

The notion of *optical flow* has become somewhat controversial during the past years, due to the fact that it does not coincide with the precisely defined motion or displacement rate field – a geometrical and kinematic concept – and due to the fact that different approaches for the estimation of optical flow frequently imply different definitions of it. A particular approach based on spatiotemporal derivatives of the gray value pattern is discussed here which not only estimates optical flow itself, but in addition its spatiotemporal derivatives. Local smoothness of optical flow is introduced implicitly by restricting the functional form for its spatiotemporal variation to a first order Taylor expansion. Given sufficient spatiotemporal variation of the local gray value pattern in the image plane, optical flow and its derivatives can be estimated locally. Explicit expressions for these estimates are derived and explored to show that neglecting the spatiotemporal derivatives in the estimation process may introduce a bias. The relations between this and alternative approaches reported in the literature are discussed.

A literature survey reveals that in most cases estimates for optical flow have been assessed only qualitatively. Recent publications describing quantitative error estimates are discussed. An experimental environment is described which has been realized and is expected to facilitate a quantitative assessment of optical flow estimates. Such comparisons contribute to an improved understanding of the implications of different assumptions and thus should consolidate the concept of optical flow as a background for related neurophysiological investigations.

1 Introduction

Whoever has driven on a free highway – and therefore was not forced to consciously scan the entire environment for imminent collision threats – at a speed well beyond 50 km/h will remember the impression that objects close to the roadside seemed to rush towards the boundary of the visual field almost like dragging whiskers along them. Analogous experiences may be associated with observations of the environment from a moving train or plane.

The notion of 'optical flow' thus appears intuitively plausible. Initially, attempts to analyze this phenomenon related it merely to geometric and kinematic considerations. If, for example, a camera translates into a depicted scene, images of selected scene points move in the image plane along trajectories which are defined by the distance between scene points and the camera as well as by the relative motion. The tangent field to such trajectories for a specific instant exhibits a structure analogous to that perceived while observing the environment from the driver's seat in a moving car. Geometry and kinematics thus provide a precise explanation for the structure of a field which is at first glance equated to the optical flow field perceived intuitively.

This state of affairs caused expected questions as well as some unexpected ones. Obvious questions are the ones about how the perception of an optical flow field can be explained in neurophysiological terms – see, e. g., Orban (1992), Verri *et al.* (1992). Less expected was the difficulty to define the notion of optical flow in a manner which makes it amenable to algorithmic analysis of recorded signals, in particular of digitized image sequences. By using moving dot displays, one can finesse for a while the question how a primate isolates and tracks the image of a mathematical point moving in space, based on the energy flux impinging on its retina. The lack of a satisfactory procedure to perform the analogous miracle for digitized image sequences – moreover, at a much lower spatial resolution – resulted in two developments.

First, a distinction was made between the geometric notion of a tangent to the image plane trajectory of a moving (mathematical) point in space on the one hand and an optical flow vector to be extracted from a continuous spatiotemporal intensity distribution on the other hand. The former has been denoted as a vector from a motion field (Horn 1986, Chapter 12.1) or from a displacement rate field (Nagel 1989). In contradistinction to these purely geometric concepts, optical flow is understood to describe the apparent motion of gray value – or intensity or brightness – patterns in the image plane. The concept of optical flow, therefore, comprises radiometric aspects in addition to the geometric aspects covered by the motion or displacement rate field concept. Once it has been realized that these two concepts describe two different fields, one should not expect that estimation results for an optical flow field can be substituted for a motion or displacement rate field without the danger of making – sometimes gross – approximation errors (Verri and Poggio 1987).

The second consequence consisted in that a multitude of prescriptions for the extraction of optical flow from digitized image sequences have been proposed. It is not easy, in general, to find out whether two such prescriptions relate to the same entity or boil down to different definitions of optical flow. It appears useful, therefore, to concentrate on those approaches which can be related to each other in a manner amenable to detailed analysis. This excludes most of the so-called token-based approaches if tokens are defined and extracted from digitized images by heuristics. Moreover, it has already been shown some time ago (Nagel 1984) that the most elementary tokens – gray value extrema, corners, and line elements – can be understood as special cases of a continuous approach towards the extraction of optical flow.

Following some notational conventions, a set of alternative approaches to the definition of optical flow will be narrowed down in section 2 to a subset which will subsequently be analyzed in more detail (section 3). Section 4, a review of recent approaches towards the quantitative estimation of optical flow from real world image sequences, sets the stage for the discussion, in section 5, of an experimental setup which should allow to explore in a quantitative manner some of the alternatives mentioned earlier.

A vector in the 2D image plane is given by a lower case bold character, for example the image position $x = (x, y)^T$. The motion or displacement rate vector will be denoted by $v = (v_1, v_2)^T$ and the optical flow by $u = (u_1, u_2)^T$. The subscript indices $_1$, $_2$ of the components of a vector refer to the x and y component, respectively. Scalar entities such as the gray value function $g(x)$ at image plane position x are indicated by regular lower case characters. Matrices are denoted by regular upper case characters. The subscripts $_x$, $_y$, $_t$ specify the partial derivative with respect to the image plane coordinates x and y or with respect to time t, for example $g_x = \partial_g/\partial x$.

2 Non-Token-Based Approaches towards the Definition of Optical Flow

A thorough investigation of an approach based on interframe matching of edge elements has recently been published by Murray and Buxton (1990) who discuss earlier literature. One possibility to avoid the extraction of a token for interframe matching consists in tesselating the image frame recorded at frame time t and to select the entire gray value distribution in each 'tile' or 'window' as a reference template. A detailed analysis concerning the algorithmic selection of the appropriate window size can be found in Okutomi and Kanade (1990). Given a similarity or scoring function, one has to search in the image recorded at frame time $t + \Delta t$ for the matching position where the similarity between this reference template and the recorded gray value pattern is optimized. Cross correlation or the sum of squared differences (see, e.g., Anandan 1989) are frequently employed scoring functions. The search yields a (finite) shift estimate in the image plane which is taken to be an approximation to optical flow, provided the time difference Δt between frames is small enough. Without specific assumptions about how the gray values within a reference window depend on the pixel position, only coarse statements are possible about the number of candidate positions and the precision of optic flow estimates. General statements can be based, e.g., on the shape of the autocorrelation function of the recorded gray value distribution within the reference window.

The aforementioned approaches are based on the assumption that the tile is merely translated from frame time t to frame time $t + \Delta t$ without changing its shape, and that the gray value distribution is convected together with the tile without structural change of the gray values ('convected invariance of intensity' – see, e. g., Wu and Wohn 1991), although small changes due to spatially and temporally uncorrelated noise are usually admitted.

The subsequent discussion emphasizes continuous approaches towards the definition of optical flow in order to facilitate the exploration of relations between different such approaches. Moreover, it concentrates on the spatiotemporal domain and will not be extended to approaches based on a spatiotemporal frequency analysis. Recent references to relations between local spatiotemporal and spatiotemporal frequency approaches can be found, e.g., in Bigün *et al.* (1991).

An infinitesimal form of the 'convected invariance of intensity' assumption, namely the local stationarity of the intensity $g(x, t)$ as a function of time, is represented by the well-known 'Optical Flow Constraint Equation (OFCE)' (Horn and Schunck 1981):

$$\frac{dg}{dt} = \nabla g^T u + g_t = 0 \tag{1}$$

Since this equation does not allow to estimate both components u_1 and u_2 of the optical flow $\boldsymbol{u}$, it has to be supplemented by various assumptions. The simplest assumption is, of course, that the optical flow is locally constant within some small subsection of the image plane. It will be seen that this assumption turns up later on in unsuspected disguises. Alternatively, one postulates smoothness for the spatial variation of the optical flow field, for example by minimizing the squared gradient of both components over an image plane area (Horn and Schunck 1981) :

$$\iint \left\{ \left(\nabla g^{\mathsf{T}} \boldsymbol{u} \, + \, g_t \right)^2 \, + \, \lambda^2 \left(\parallel \nabla u_1 \parallel^2 \, + \, \parallel \nabla u_2 \parallel^2 \right) \right\} \mathrm{d}x \mathrm{d}y \tag{2}$$

Variations of this approach have been proposed in different forms. One can extend the integral only along a curve, which implies that at least one component of the optical flow can be estimated along some curve (Hildreth 1984; see also Gong 1989, Gong and Brady 1990 who use a different integrand, but compare their results to the approach of Hildreth). Another curve-based approach proceeds on the hypothesis that a curve extracted from the image at frametime t is shifted and possibly deformed into another curve at frametime $t + \Delta t$. A mapping between two such curves is taken to establish a one-dimensional sampling of the optical flow field. This is then extended into the image plane environment of the curve obtained at frametime t by assuming some specific form for the spatial variation of the optical flow, for example that it can be described by a Taylor expansion up to second order (see, e.g., Waxman and Wohn 1988). If the curves are chosen to be the zero-crossing curves of the image convolved by the Laplacian of a Gaussian, it is implied that the deformation of the zero-crossing curve as a function of time is entirely due to the motion of light reflecting surfaces in 3D space. As has been shown recently (Wu and Wohn 1991), this hypothesis is in general incompatible with the hypothesis of 'convected invariance of intensity'. Although the differences between curves estimated on the basis of these two different hypotheses will be small along stretches with strong intensity gradients, the implications for optical flow estimates have not yet been evaluated quantitatively.

A substantially different curve-based approach has been proposed and investigated by Faugeras (1990). The perspective image of an inelastic curve which moves in 3D scene space can be conceived to trace a so-called 'spatiotemporal surface' in a three-dimensional space spanned by the image coordinates $\boldsymbol{x}$ and time t. Faugeras studies the differential properties of this spatiotemporal surface and shows that the full optical flow of the image curve can not be recovered, although the stronger assumption of a rigid curve motion in scene space allows him to recover structure and motion of the scene curve without explicitly computing the tangential component of optical flow. Details can be found elsewhere in this volume (Deriche *et al.* 1992). One has to realize that what is called optical flow in Faugeras (1990) is a purely geometric and kinematic concept so far. It thus is equivalent to the motion or displacement rate vector field. Its relationship to optical flow can only be evaluated after it has been specified by which procedure the image of a real curve associated with some surface moving in the scene will be extracted from the spatiotemporal gray value variation recorded by an image sequence.

Although curve-based approaches towards the estimation of optical flow appear attractive in view of many prominent curve segments in images, an area-based approach should be explored, too. A supplementary smoothness requirement can be introduced not only explicitly as in equ. (2), but implicitly through some specific form for admissible spatiotemporal variations of optical flow. Such approaches exploit significant spatiotemporal gray value variations not only at image locations declared to be part of a curve segment.

This is expected to be advantageous in image areas where curve segments are usually less well defined, for example near junctions. Part of the price to be paid for this alternative is the dependence on a particular functional form relating spatiotemporal variations of the gray value pattern to spatiotemporal variations of optical flow. Whether this price is worthwhile paying must be decided on the basis of quantitative experimental results.

3 Area-Based Direct Estimation Approaches for Optical Flow

Let us consider the spatiotemporal gray value pattern $g(\boldsymbol{x}, t)$ as some entity filling the three-dimensional space spanned by the two components x and y of the image plane coordinate vector $\boldsymbol{x}$ and by the time t. It appears plausible that in general the gray value of a pixel representing the image of a scene surface element will not change abruptly with time if the camera moves infinitesimally relative to this scene surface element. Moreover, the same assumption appears plausible – apart from locations on occluding contours – if we consider the spatial neighborhood of the pixel in question. In a first approximation, therefore, the trajectory of the image of a scene surface element should be confined to the tangent plane of the intensity pattern $g(\boldsymbol{x}, t)$ – as expressed by the OFCE of equ. (1). The area-based approach towards the estimation of optical flow, therefore, starts from the OFCE similarly to many curve-based approaches. It differs from other approaches by the supplementary assumption introduced.

3.1 Derivation of the System of Equations

In distinction to approaches which require local constancy or minimal spatial derivatives of optical flow, it is now postulated that the OFCE represents an acceptable approximation not only at the point $\boldsymbol{x}$, but also in some environment $(\boldsymbol{x} + \delta\boldsymbol{x}, t + \delta t)$ around it, provided first order Taylor expansions for $\nabla g(\boldsymbol{x}, t)$ as well as for the components of $\boldsymbol{u}$ are used to express their spatiotemporal variation in this environment. In order to obtain a compact notation, four-dimensional vectors denoted by doubly underlined bold characters will be introduced: $\underline{\underline{\boldsymbol{x}}} = (1, \delta x, \delta y, \delta t)^{\mathrm{T}}$. Analogously, vectors denoting the extension of the gradient components $\underline{\underline{\boldsymbol{g}_x}} = (g_x, g_{xx}, g_{xy}, g_{xt})^{\mathrm{T}}$, $\underline{\underline{\boldsymbol{g}_y}} = (g_y, g_{yx}, g_{yy}, g_{yt})^{\mathrm{T}}$, $\underline{\underline{\boldsymbol{g}_t}} = (g_t, g_{tx}, g_{ty}, g_{tt})^{\mathrm{T}}$, and of the optical flow components $\underline{\underline{\boldsymbol{u}_1}} = (u_1, u_{1_x}, u_{1_y}, u_{1_t})^{\mathrm{T}}$ and $\underline{\underline{\boldsymbol{u}_2}} = (u_2, u_{2_x}, u_{2_y}, u_{2_t})^{\mathrm{T}}$ are introduced. Using these conventions, the extended OFCE can be written in the form

$$\left(\underline{\underline{\boldsymbol{g}_x}}^{\mathrm{T}} \underline{\underline{\boldsymbol{x}}}\right)\left(\underline{\underline{\boldsymbol{u}_1}}^{\mathrm{T}} \underline{\underline{\boldsymbol{x}}}\right) + \left(\underline{\underline{\boldsymbol{g}_y}}^{\mathrm{T}} \underline{\underline{\boldsymbol{x}}}\right)\left(\underline{\underline{\boldsymbol{u}_2}}^{\mathrm{T}} \underline{\underline{\boldsymbol{x}}}\right) + \left(\underline{\underline{\boldsymbol{g}_t}}^{\mathrm{T}} \underline{\underline{\boldsymbol{x}}}\right) = 0 \qquad (3)$$

The left hand side represents a polynomial of up to second order in the components of $\underline{\underline{\boldsymbol{x}}}$. Since this should vanish identically, all coefficients of this polynomial must be zero. One obtains the following 10 linear equations in the 8 unknown components of $\underline{\underline{\boldsymbol{u}_1}}$ and $\underline{\underline{\boldsymbol{u}_2}}$ where for simplicity it has been assumed that the coordinate system in the image plane has been rotated such that the mixed partial derivative g_{xy} vanishes (alignment with the projection of the two principal component axes corresponding to x and y, i.e. $g_{xy} = 0$) :

$$\begin{pmatrix}
g_x & 0 & 0 & 0 & g_y & 0 & 0 & 0 \\
g_{xx} & g_x & 0 & 0 & 0 & g_y & 0 & 0 \\
0 & 0 & g_x & 0 & g_{yy} & 0 & g_y & 0 \\
g_{xt} & 0 & 0 & g_x & g_{yt} & 0 & 0 & g_y \\
0 & g_{xx} & 0 & 0 & 0 & 0 & 0 & 0 \\
0 & 0 & g_{xx} & 0 & 0 & g_{yy} & 0 & 0 \\
0 & g_{xt} & 0 & g_{xx} & 0 & g_{yt} & 0 & 0 \\
0 & 0 & g_{xt} & 0 & 0 & 0 & g_{yt} & g_{yy} \\
0 & 0 & 0 & 0 & 0 & 0 & g_{yy} & 0 \\
0 & 0 & 0 & g_{xt} & 0 & 0 & 0 & g_{yt}
\end{pmatrix}
\begin{pmatrix}
u_1 \\
u_{1_x} \\
u_{1_y} \\
u_{1_t} \\
u_2 \\
u_{2_x} \\
u_{2_y} \\
u_{2_t}
\end{pmatrix}
= -
\begin{pmatrix}
g_t \\
g_{tx} \\
g_{ty} \\
g_{tt} \\
0 \\
0 \\
0 \\
0 \\
0 \\
0
\end{pmatrix}
\tag{4}$$

Since this represents an overdetermined system of equations, we prepare for the standard pseudoinverse approach by multiplying this system of equations from the left with the transpose of the coefficient matrix on the left hand side of equ. (4). The result can be written in the following form:

$$\left(G_{xx} + \underline{g_x}\,\underline{g_x}^T \right) \underline{u_1} + \left(G_{xy} + \underline{g_y}\,\underline{g_x}^T \right) \underline{u_2} = -G_{tx}$$

$$\left(G_{xy} + \underline{g_x}\,\underline{g_y}^T \right) \underline{u_1} + \left(G_{yy} + \underline{g_y}\,\underline{g_y}^T \right) \underline{u_2} = -G_{ty}
\tag{5}$$

with the 4×1 matrices G_{tx} as well as G_{ty} given by

$$G_{tx} = \left(\underline{g_x}^T\underline{g_t} \,,\, g_x g_{tx} \,,\, g_x g_{ty} \,,\, g_x g_{tt} \right)^T
\tag{6a}$$

and

$$G_{ty} = \left(\underline{g_y}^T\underline{g_t} \,,\, g_y g_{tx} \,,\, g_y g_{ty} \,,\, g_y g_{tt} \right)^T
\tag{6b}$$

The 4×4 matrices G_{xx}, G_{xy}, and G_{yy} are given by

$$G_{xx} = \begin{pmatrix}
g_{xx}^2 + g_{xt}^2 & 0 & 0 & 0 \\
0 & g_x^2 + g_{xt}^2 & 0 & 0 \\
0 & 0 & g_x^2 + g_{xx}^2 + g_{xt}^2 & 0 \\
0 & 0 & 0 & g_x^2 + g_{xx}^2
\end{pmatrix}
\tag{7a}$$

$$G_{xy} = \begin{pmatrix}
g_{xt} g_{yt} & 0 & 0 & 0 \\
0 & g_x g_y + g_{xt} g_{yt} & 0 & 0 \\
0 & 0 & g_x g_y + g_{xt} g_{yt} & 0 \\
0 & 0 & 0 & g_x g_y
\end{pmatrix}
\tag{7b}$$

$$G_{yy} = \begin{pmatrix} g_{yy}^2 + g_{yt}^2 & 0 & 0 & 0 \\ 0 & g_y^2 + g_{yy}^2 + g_{yt}^2 & 0 & 0 \\ 0 & 0 & g_y^2 + g_{yt}^2 & 0 \\ 0 & 0 & 0 & g_y^2 + g_{yy}^2 \end{pmatrix} \tag{7c}$$

3.2 Solution Approach

A straightforward check will show that $(\underline{g_y}^T, -\underline{g_x}^T)^T$ represents an eigenvector associated with the eigenvalue 0 of the coefficient matrix in equ. (5). The coefficient matrix thus does not have full rank. Similarly, one can show that the scalar product of this eigenvector with the right hand side of the system of equations (5) vanishes, too:

$$\underline{g_y}^T G_{tx} - \underline{g_x}^T G_{ty} = 0$$

As a consequence, the entire system of equations (5) is linearly dependent upon each other, i.e., one can omit (at least) one equation from this system. This implies, of course, that not all unknowns can be determined. One can choose one among them as a free parameter, upon which the solutions for the other unknowns depend, and attempt to invert the resulting system of equations symbolically. A symbolic algebra program like MAPLE gives a result from which one can learn two things: (i) the resulting 7×7 coefficient matrix does indeed have full rank and (ii) the symbolic solutions for the unknowns run across many pages and do not provide any immediate insight.

Therefore, a different approach is chosen here. Closer inspection of the eigenvector corresponding to the eigenvalue 0 of the coefficient matrix from equ. (5) shows that it is practical to select, for example, u_{2_x} as a free parameter. This parameter can be arbitrarily set to zero (a nonzero value will later be reintroduced once the results have become a bit more perspicuous!) which has the consequence that one can omit the corresponding column from the coefficient matrix as well as the corresponding equation from the system given by equ. (5). The variables are then reordered according to $(u_1, u_2, u_{1_x}, u_{1_y}, u_{2_y}, u_{1_t}, u_{2_t})^T$ and the coefficient matrix as well as the system of equations are reordered correspondingly. The idea is to first concentrate on an investigation of the first five unknowns, suppressing for the moment the unknowns u_{1_t} and u_{2_t} and the related equations.

The system of equations for the remaining five unknowns $u_1, u_2, u_{1_x}, u_{1_y}, u_{2_y}$ can then be written in the following form:

$$\begin{pmatrix} g_x^2 + g_{xx}^2 & g_x g_y & g_x g_{xx} & 0 & 0 \\ g_x g_y & g_y^2 + g_{yy}^2 & 0 & g_x g_{yy} & g_y g_{yy} \\ g_x g_{xx} & 0 & g_x^2 + g_{xx}^2 & 0 & 0 \\ 0 & g_x g_{yy} & 0 & g_y^2 + g_{xx}^2 & g_x g_y \\ 0 & g_y g_{yy} & 0 & g_x g_y & g_y^2 + g_{yy}^2 \end{pmatrix} \begin{pmatrix} u_1 \\ u_2 \\ u_{1_x} \\ u_{1_y} \\ u_{2_y} \end{pmatrix} = - \begin{pmatrix} g_x g_t + g_{xx} g_{tx} \\ g_y g_t + g_{yy} g_{ty} \\ g_x g_{tx} \\ g_x g_{ty} \\ g_y g_{ty} \end{pmatrix} \tag{8}$$

In order to expose the structure of the solution, the coefficient matrix will be subdivided and some abbreviating terms will be introduced for the right hand side:

$$\begin{pmatrix} A & B \\ C & D \end{pmatrix} \begin{pmatrix} u_1 \\ u_2 \\ u_{1_x} \\ u_{1_y} \\ u_{2_y} \end{pmatrix} = - \begin{pmatrix} r_1 \\ r_2 \\ r_3 \\ r_4 \\ r_5 \end{pmatrix} \tag{9}$$

This system of equations can now be solved by using an identity for the inverse of a matrix which has been divided into submatrices:

$$\begin{pmatrix} A & B \\ C & D \end{pmatrix}^{-1} = \begin{pmatrix} A^{-1} + A^{-1}B(D - CA^{-1}B)^{-1}CA^{-1} & -A^{-1}B(D - CA^{-1}B)^{-1} \\ -(D - CA^{-1}B)^{-1}CA^{-1} & (D - CA^{-1}B)^{-1} \end{pmatrix}$$

(10)

If we equate

$$A = \begin{pmatrix} g_x^2 + g_{xx}^2 & g_x g_y \\ g_x g_y & g_y^2 + g_{yy}^2 \end{pmatrix}$$

(11a)

$$B = C^T = \begin{pmatrix} g_x g_{xx} & 0 & 0 \\ 0 & g_x g_{yy} & g_y g_{yy} \end{pmatrix}$$

(11b)

and

$$D = \begin{pmatrix} g_x^2 + g_{xx}^2 & 0 & 0 \\ 0 & g_x^2 + g_{xx}^2 & g_x g_y \\ 0 & g_x g_y & g_y^2 + g_{yy}^2 \end{pmatrix}$$

(11c)

we obtain the formal solutions

$$\begin{pmatrix} \hat{u}_{1_x} \\ \hat{u}_{1_y} \\ \hat{u}_{2_y} \end{pmatrix} = (D - CA^{-1}B)^{-1} \left[CA^{-1} \begin{pmatrix} r_1 \\ r_2 \end{pmatrix} - \begin{pmatrix} r_3 \\ r_4 \\ r_5 \end{pmatrix} \right]$$

(12a)

and

$$\begin{pmatrix} \hat{u}_1 \\ \hat{u}_2 \end{pmatrix} = -A^{-1} \begin{pmatrix} r_1 \\ r_2 \end{pmatrix} - A^{-1}B \begin{pmatrix} \hat{u}_{1_x} \\ \hat{u}_{1_y} \\ \hat{u}_{2_y} \end{pmatrix}$$

(12b)

3.3 Discussion of the Solution

This way of writing the solutions makes it immediately obvious that neglecting the partial derivatives of the optical flow given by equ. (12a) introduces a bias into the estimate for u as given by equ. (12b) – unless the partial derivatives given by equ. (12a) all turn out to be zero. In order to study the implications of such a condition, we write the solutions of equ. (12a) explicitly:

$$\begin{pmatrix} \hat{u}_{1_x} \\ \hat{u}_{1_y} \\ \hat{u}_{2_y} \end{pmatrix} = \frac{\begin{pmatrix} g_{xx}g_{yy} \\ -g_x g_{yy} \\ -g_y g_{xx} \end{pmatrix}^T \begin{pmatrix} g_t \\ g_{tx} \\ g_{ty} \end{pmatrix}}{g_x^4 g_{yy}^4 + g_y^4 g_{xx}^4 + g_{xx}^2 g_{yy}^2 (g_x^2 + g_{xx}^2)(g_y^2 + g_{yy}^2)} \begin{pmatrix} g_x^2 g_{yy}^3 \\ g_x g_y g_{xx} g_{yy}^2 \\ g_y^2 g_{xx}^3 \end{pmatrix}$$

(13)

One can immediately recognize two conditions where the right hand side vanishes without either the implication that $\hat{u}_1 = \hat{u}_2 = 0$ or the implication that the denominator vanishes, which would be equivalent to the coefficient matrix being singular:

(i) $g_x = g_y = 0$ and $g_{xx} \neq 0$ as well as $g_{yy} \neq 0$,

which is the condition for a local extremum of the gray value pattern at the point under scrutiny, i.e. the center of our local window;

(ii) $g_{xx}g_{yy}g_t - g_xg_{yy}g_{tx} - g_yg_{xx}g_{ty} = 0.$

In order to see the consequences of condition (ii), we write the solution given by equ. (12b) explicitly :

$$\begin{pmatrix} \hat{u}_1 \\ \hat{u}_2 \end{pmatrix} = -\frac{1}{g_x^2 g_{yy}^2 + g_y^2 g_{xx}^2 + g_{xx}^2 g_{yy}^2}\begin{pmatrix} g_y^2 + g_{yy}^2 & -g_x g_y \\ -g_x g_y & g_x^2 + g_{xx}^2 \end{pmatrix}\begin{pmatrix} g_x g_t + g_{xx}g_{tx} \\ g_y g_t + g_{yy}g_{ty} \end{pmatrix} - A^{-1}B\begin{pmatrix} \hat{u}_{1_x} \\ \hat{u}_{1_y} \\ \hat{u}_{2_y} \end{pmatrix} \quad (14)$$

If we concentrate on the first term on the right hand side of equ. (14) and disregard for the moment the second term – the bias term – with the partial derivatives of the optical flow, which is assumed to be zero according to the starting point of this discussion, we see that the denominator comprises terms which differ from zero only in the two well known cases (Nagel 1983, 1987) of a gray value corner – at least one of the two terms $g_x^2 g_{yy}^2$ or $g_y^2 g_{xx}^2$ is different from zero although $g_{xx}^2 g_{yy}^2$ may vanish – or of a gray value extremum with $g_x = g_y = 0$ and $g_{xx} \neq 0$ as well as $g_{yy} \neq 0$. If we evaluate the product $A^{-1}(r_1, r_2)^T$ on the right hand side of equ. (12b) under the condition (ii) above, we obtain:

$$\begin{pmatrix} \hat{u}_1 \\ \hat{u}_2 \end{pmatrix} = -\begin{pmatrix} \frac{g_{tx}}{g_{xx}} \\ \frac{g_{ty}}{g_{yy}} \end{pmatrix} \qquad \text{for the case} \quad \hat{u}_{1_x} = \hat{u}_{1_y} = \hat{u}_{2_y} = 0 \qquad (15)$$

which is exactly the solution obtained in Nagel (1987) for a gray value extremum. With other words, the solutions obtained here with a generalized approach comprising the partial derivatives of optical flow specialize to those obtained earlier if the appropriate conditions are chosen. This makes sense since these earlier results have been obtained by postulating that the optical flow is locally constant.

The observation that the coefficient matrix of equ. (5) has only rank 7 rather than the desired rank 8 can be rephrased by the statement that a random multiple of the eigenvector associated with the eigenvalue 0 of this matrix can always be added to any solution. This aspect can be used to rewrite the subset of solutions describing the spatial derivatives of u in a more symmetrical fashion :

$$\begin{pmatrix} \hat{u}_{1_x} \\ \hat{u}_{1_y} \\ \hat{u}_{2_x} \\ \hat{u}_{2_y} \end{pmatrix} = \frac{\begin{pmatrix} g_{xx}g_{yy} \\ -g_xg_{yy} \\ -g_yg_{xx} \end{pmatrix}^T \begin{pmatrix} g_t \\ g_{tx} \\ g_{ty} \end{pmatrix}}{g_x^4 g_{yy}^4 + g_y^4 g_{xx}^4 + g_{xx}^2 g_{yy}^2 (g_x^2 + g_{xx}^2)(g_y^2 + g_{yy}^2)}\begin{pmatrix} g_x^2 g_{yy}^3 \\ g_{yy}(\frac{1}{2}g_xg_yg_{xx}g_{yy} - w) \\ g_{xx}(\frac{1}{2}g_xg_yg_{xx}g_{yy} + w) \\ g_y^2 g_{xx}^3 \end{pmatrix} \quad (16)$$

where w represents the free parameter due to the rank deficiency of the coefficient matrix in equ. (5). This explicit expression facilitates two observations. First, the divergence of $\boldsymbol{u}$ does not depend on the parameter w and, since it is an invariant with respect to rotations of the coordinate system, always has the value :

$$div\ \hat{\boldsymbol{u}} = \hat{u}_{1_x} + \hat{u}_{2_y} = \frac{\begin{pmatrix} g_{xx}g_{yy} \\ -g_x g_{yy} \\ -g_y g_{xx} \end{pmatrix}^{\mathrm{T}} \begin{pmatrix} g_t \\ g_{tx} \\ g_{ty} \end{pmatrix}}{g_x^4 g_{yy}^4 + g_y^4 g_{xx}^4 + g_{xx}^2 g_{yy}^2 (g_x^2 + g_{xx}^2)(g_y^2 + g_{yy}^2)} \left(g_x^2 g_{yy}^3 + g_y^2 g_{xx}^3 \right) \tag{17}$$

calculated in a coordinate system with $g_{xy} = 0$. The difference compared to the result reported in Nagel (1990) is due to the fact that there a slightly different approach has been used, namely to obtain the equations by minimizing the squared analogy to equ. (3), integrated over the entire image plane with Gaussian weights in order to enforce integrability of the Taylor series approximations. The exposition presented here is easier to follow. The essential arguments concerning the rank of the coefficient matrix carry over.

Similarly, one obtains an expression for $curl\ \boldsymbol{u} = u_{2_x} - u_{1_y}$ which, however, will transform if the coordinate system is rotated away from the alignment with the principal axes of the gray value pattern in the image plane since the terms containing g_{xy} have been set to zero during the derivation of equ. (18) :

$$curl\ \hat{\boldsymbol{u}} = \frac{\begin{pmatrix} g_{xx}g_{yy} \\ -g_x g_{yy} \\ -g_y g_{xx} \end{pmatrix}^{\mathrm{T}} \begin{pmatrix} g_t \\ g_{tx} \\ g_{ty} \end{pmatrix}}{g_x^4 g_{yy}^4 + g_y^4 g_{xx}^4 + g_{xx}^2 g_{yy}^2 (g_x^2 + g_{xx}^2)(g_y^2 + g_{yy}^2)} \left(\frac{1}{2} g_x g_y g_{xx} g_{yy} (g_{xx} - g_{yy}) + w(g_{xx} + g_{yy}) \right)$$
$$\tag{18}$$

The free parameter w influences $curl\ \boldsymbol{u}$, unless the point under scrutiny is located on a zerocrossing contour of the Laplacian, i.e. unless it is a saddle point with $g_{xx} = -g_{yy}$. Note that the case $g_{xx} = g_{yy} = 0$ – in conjunction with the assumption that we have aligned the coordinate system with the principal axes of $g(\boldsymbol{x}, \mathrm{t})$ and thus $g_{xy} = 0$ – does not allow to determine both components of the optical flow and thus needs not be discussed separately. Obviously, the free parameter w will also influence the solutions for $\hat{u}_1$ and $\hat{u}_2$ – see equ. (14).

3.4   Comparison with Related Approaches

Relations between various gradient-based approaches towards the estimation of optical flow have been studied in Nagel (1987). The subsequent discussion will thus be restricted to later publications. Variants of the observation by Nagel (1983) that both components u_1 and u_2 of optical flow can be estimated at gray value extrema, gray value corners, and their environment have been reported subsequently. Reichardt et al. (1988) (for a more detailed derivation see Reichardt and Schlögl 1988) derive an equation – in particular their equ. (13) – which can be written in the notation used here as

$$\frac{\mathrm{d}}{\mathrm{d}t}\nabla g\ =\ \begin{pmatrix} g_{xx}u_1 + g_{xy}u_2 + g_{xt} \\ g_{yx}u_1 + g_{yy}u_2 + g_{yt} \end{pmatrix} = 0 \tag{19}$$

Uras et al. (1988) (see, too, Verri et al. 1989a and 1992; Girosi et al. 1989) introduce this same equation, based on the postulate that the spatial gradient of the local gray value pattern is temporally stationary. Compared with the OFCE, it comprises two constraints and thus allows to estimate both components of optical flow – whenever the Hessian of $g(\boldsymbol{x})$ at time t is nonsingular – whereas the OFCE provides only one constraint. It should be clear that equ. (19) defines 'another' optical flow which is in general not identical with the one introduced by the OFCE, by the variant of optical flow introduced implicitly by the approach of Nagel (1983) or – to be precise, still another one – by equ. (3). The question then turns to studying which approach comprises most or all of the alternatives as special cases so that one can concentrate on the analysis of the most general one.

Since the approach by Uras *et al.* (1988) does not comprise the first derivatives of the gray value pattern, these authors do not obtain optical flow estimates at gray value corners as in the case of Nagel (1983). If one chooses the special case of a gray value extremum, the more general approach of Nagel (1983) gives exactly the same solution – see the discussion related with equ. (15) – which becomes obvious if the coordinate system in equ. (19) is aligned with the principal axes of the gray value pattern at the origin, i.e. $g_{xy} = 0$. Another way to note the inherent relation of the approach by Reichardt *et al.* (1988) and Uras *et al.* (1988) with earlier approaches of Nagel consists in applying the spatial gradient operator to the OFCE :

$$\nabla\left(\frac{dg}{dt}\right) = \nabla(g_x u_1 + g_y u_2 + g_t) = \left(\begin{array}{c} g_{xx}u_1 + g_{yx}u_2 + g_{tx} + g_x u_{1_x} + g_y u_{2_x} \\ g_{xy}u_1 + g_{yy}u_2 + g_{ty} + g_x u_{1_y} + g_y u_{2_y} \end{array}\right) = 0 \quad (20)$$

One sees immediately that the spatial gradient of the OFCE simplifies to the equation studied by Reichardt *et al.* (1988) and Uras *et al.* (1988) if the spatial derivatives of the optical flow can be assumed to be zero, i.e. that the optical flow does not change locally around $\boldsymbol{x}_0$.

Verri *et al.* (1990) start from the observation that a-priori knowledge about the temporal stationarity of some linear combination of spatial derivatives of optical flow provides an additional constraint equation for the components of optical flow itself. They consider in particular linear combinations of spatial optical flow derivatives which correspond to pure divergence, pure rotation, or pure shear of the optical flow field. Combination of the scalar equation for u_1 and u_2, which results for each case, with the OFCE equ. (1) then allows to estimate both components u_1 and u_2, provided the spatiotemporal changes of $g(\boldsymbol{x}, t)$ are compatible with the postulated flow field properties.

Optical flow and its first order spatial derivatives have been estimated by Campani and Verri (1990) – see, too, Verri *et al.* (1992) in this volume – based on the assumption that the spatial variation of optical flow can be locally well approximated by a Taylor expansion up to first order around $\boldsymbol{x}_0$, i.e.

$$\boldsymbol{u}(\boldsymbol{x}) = \boldsymbol{u}(\boldsymbol{x}_0) + \left(\begin{array}{cc} u_{1_x} & u_{1_y} \\ u_{2_x} & u_{2_y} \end{array}\right) \left(\begin{array}{c} x - x_0 \\ y - y_0 \end{array}\right) = \boldsymbol{u}(\boldsymbol{x}_0) + \mathrm{M}(\boldsymbol{x} - \boldsymbol{x}_0) \quad (21)$$

Insertion of this expression for $\boldsymbol{u}(\boldsymbol{x})$ into the OFCE then yields a constraint equation for each location $\boldsymbol{x}$ within a square N×N pixel patch around $\boldsymbol{x}_0$:

$$\nabla g^{\mathrm{T}}(\boldsymbol{x})\,\boldsymbol{u}(\boldsymbol{x}) + g_t(\boldsymbol{x}) = \nabla g^{\mathrm{T}}(\boldsymbol{x})\,\boldsymbol{u}(\boldsymbol{x}_0) + \nabla g^{\mathrm{T}}(\boldsymbol{x})\,\mathrm{M}(\boldsymbol{x} - \boldsymbol{x}_0) + g_t(\boldsymbol{x}) = 0 \quad (22)$$

Since even for a small patch of 3×3 pixels the number of equations exceeds the number of unknowns, namely the two components of $\boldsymbol{u}(\boldsymbol{x}_0)$ and of its four first spatial derivatives at $\boldsymbol{x}_0$, a pseudoinverse approach allows to determine these unknowns. In case the spatial variation of the gray value pattern could be described by a first order Taylor expansion of

$$\nabla g(\boldsymbol{x} + \delta\boldsymbol{x}, t) = \left(\, g_x(\boldsymbol{x}, t) + g_{xx}(\boldsymbol{x}, t)\delta x + g_{xy}(\boldsymbol{x}, t)\delta y,\; g_y(\boldsymbol{x}, t) + g_{yx}(\boldsymbol{x}, t)\delta x + g_{yy}(\boldsymbol{x}, t)\delta y\,\right)^{\mathrm{T}},$$

one encounters the same problem as with equ. (3), namely that the coefficient matrix of the pseudoinverse becomes singular. Therefore, the patch has to be large enough to comprise spatial variations of the gray value gradient beyond the linear ones. Then one has the advantage to be able to estimate all six unknowns, but the disadvantage that the relation between the spatial gray value variation and optical flow estimates can not be studied in more detail. Partial derivatives of $\boldsymbol{u}$ with respect to time have not been taken into account by these authors. Experiences with experimental investigations of this approach will be discussed in section 4.

A more general approach than the one discussed in subsection 3.2 has been published by Werkhoven and Koenderink (1990). They use the local jet of the gray value function $g(\boldsymbol{x}, t)$, which they denote as $L(\boldsymbol{p}; t, \sigma)$ at image position $\boldsymbol{p} = (x, y)^{\mathrm{T}}$ at time t and resolution level σ. The local jet at image location $\boldsymbol{p}_0$ is represented by the truncated Taylor series with a local coordinate system $\boldsymbol{r} = (\zeta, \eta)^{\mathrm{T}}$:

$$L(\boldsymbol{p}_0 + \boldsymbol{r}; t, \sigma) = \sum_{n=0}^{N} \frac{1}{n!} \left(\zeta \frac{\partial}{\partial x} + \eta \frac{\partial}{\partial y} \right)^n L(\boldsymbol{p}; t, \sigma)\Big]_{\boldsymbol{p}=\boldsymbol{p}_0} \tag{23}$$

which is written as

$$L(\boldsymbol{p}_0 + \boldsymbol{r}; t, \sigma) = \sum_{n=0}^{N} \sum_{m=0}^{n} l_{n-m,m} \frac{\sigma^{-n}}{n!} \binom{n}{m} \zeta^{n-m} \eta^m \tag{24}$$

The components $l_{n-m,m}$ of the local jet around image position $\boldsymbol{p}_0$ at resolution parameter σ are found as the output of receptive fields $\Psi_{n-m,m}(\boldsymbol{r}; \sigma)$ centered at $\boldsymbol{p}_0$:

$$\Psi_{n-m,m}(\boldsymbol{r}; \sigma) = \sigma^n \frac{\partial^n}{\partial \zeta^{n-m} \partial \eta^m} \frac{1}{2\pi\sigma^2} e^{-\frac{\zeta^2 + \eta^2}{2\sigma^2}} \tag{25}$$

This approach yields

$$l_{n-m,m} = \sigma^n \frac{\partial^n}{\partial x^{n-m} \partial y^m} L(\boldsymbol{p}; t, \sigma)\Big]_{\boldsymbol{p}=\boldsymbol{p}_0} = \int L(\boldsymbol{p}_0 + \boldsymbol{r}; t, \sigma) \Psi_{n-m,m}(\boldsymbol{r}; \sigma) d\boldsymbol{r} \tag{26}$$

It is assumed that the temporal change of a locally restricted gray value pattern between two images recorded with a small time difference Δt can be described by an affine coordinate transformation of the local environment around the image point under scrutiny ('convected invariance of intensity'). This affine transformation is represented by Δt times an expression for the optical flow as in equ. (21). The authors exploit the possibility to shift this time dependence from the unknown gray value pattern to the known receptive field profile functions $\Psi_{n-m,m}(\boldsymbol{r}; \sigma)$. A Taylor expansion of these receptive field functions up to first order then allows them to formulate a system of linear equations in the unknown optical flow components and their first spatial derivatives. Temporal derivatives of the optical flow are not taken into account (see section 3.5). Eventually, the temporal change of the local jet components $l_{n-m,m}$ from t to $t + \Delta t$ are replaced by their partial derivative with respect to time. As a result, they obtain (using u_1 and u_2 in place of u_0 and v_0, respectively, the latter denotation employed by Werkhoven and Koenderink 1990 for the optical flow components)

$$\left(\frac{u_1}{\sigma}, u_{1_x}, u_{1_y}, \frac{u_2}{\sigma}, u_{2_x}, u_{2_y} \right)^{\mathrm{T}} = \mathbf{C}_I^{-1} \left(\frac{\partial l_{00}}{\partial t}, \frac{\partial l_{10}}{\partial t}, \frac{\partial l_{01}}{\partial t}, \frac{\partial l_{20}}{\partial t}, \frac{\partial l_{11}}{\partial t}, \frac{\partial l_{02}}{\partial t} \right)^{\mathrm{T}} \tag{27}$$

where the matrix C_I is given by (C_I is written here instead of the matrix C of Werkhoven and Koenderink 1990 in order to distinguish this matrix from the one introduced in equ. (11c))

$$
C_I = \begin{pmatrix}
-l_{10} & l_{20} & l_{11} & -l_{01} & l_{11} & l_{02} \\
-l_{20} & l_{10}+l_{30} & l_{21} & -l_{11} & l_{01}+l_{21} & l_{12} \\
-l_{11} & l_{21} & l_{10}+l_{12} & -l_{02} & l_{12} & l_{01}+l_{03} \\
-l_{30} & 2l_{20}+l_{40} & l_{31} & -l_{21} & 2l_{11}+l_{31} & l_{22} \\
-l_{21} & l_{11}+l_{31} & l_{20}+l_{22} & -l_{12} & l_{02}+l_{22} & l_{11}+l_{13} \\
-l_{12} & l_{22} & 2l_{11}+l_{13} & -l_{03} & l_{13} & 2l_{02}+l_{04}
\end{pmatrix}
\tag{28}
$$

This system of six equations, which depend on up to fourth order spatial and on up to third order spatiotemporal derivatives of the gray value pattern, is a case investigated by these authors in more detail. In general, this system of equations has full rank and can be inverted to obtain the unknown optical flow components and their first spatial derivatives.

In order to explicate the relation between this formulation and the one represented by equ. (4), we neglect all spatiotemporal derivatives of order three or higher in equ. (27), replace $l_{n-m,m}$ by the corresponding spatial derivatives of $g(x,t)$ and absorb the powers of σ into the definition of these spatial derivatives of g. One has to note that due to the way in which Werkhoven and Koenderink (1990) define the partial derivatives of g, all partial spatial derivatives of overall odd order contain a minus sign. These manipulations yield, for example, $l_{11} \Rightarrow g_{xy}$, $(\partial l_{00}/\partial t) \Rightarrow g_t$ and $(\partial l_{01}/\partial t) \Rightarrow -g_{yt}$. Let the coordinate system be aligned with the principal axes of the gray value function at p_0, i.e. $g_{xy} = 0$. As a result, one obtains instead of equ. (27) the following expression:

$$
\begin{pmatrix}
g_x & g_{xx} & 0 & g_y & 0 & g_{yy} \\
-g_{xx} & -g_x & 0 & 0 & -g_y & 0 \\
0 & 0 & -g_x & -g_{yy} & 0 & -g_y \\
0 & 2g_{xx} & 0 & 0 & 0 & 0 \\
0 & 0 & g_{xx} & 0 & g_{yy} & 0 \\
0 & 0 & 0 & 0 & 0 & 2g_{yy}
\end{pmatrix}
\begin{pmatrix}
u_1 \\
u_{1_x} \\
u_{1_y} \\
u_2 \\
u_{2_x} \\
u_{2_y}
\end{pmatrix}
=
\begin{pmatrix}
g_t \\
-g_{tx} \\
-g_{ty} \\
0 \\
0 \\
0
\end{pmatrix}
\tag{29}
$$

This is a linear combination of the system of equations given by equ. (4), provided we omit from equ. (4) the temporal derivatives u_{1_t} and u_{2_t} as well as all equations derived from non-zero powers of δt in equ. (3), i.e. rows four, seven, eight and ten in equ. (4). The second and third equation in equ. (29) has to be multiplied by -1, the fourth and sixth by 1/2. Subtracting the fourth and sixth row from the first one will then yield the equivalence, up to a missing minus sign on the right hand side of equ. (29). In order to prepare the background for additional aspects, a further discussion of the relations between these two approaches is postponed to the concluding section 6.

Extensive experiments with synthetic image data have been performed by Werkhoven and Koenderink (1990), based on a numerical inversion of the coefficient matrix by singular-value-decomposition. Closed form solutions are given only for the spatial derivatives of optical flow in the special case of tracking, i.e. for $u_1 = u_2 = 0$.

3.5   Estimating Temporal Derivatives of Optical Flow

The advantage of being able to discuss explicit expressions for area-based estimates of the spatial derivatives of optical flow hopefully has prepared the reader to accept some steps by analogy for the case in which the temporal derivatives are taken into account, too. We go back to equ. (5) and eliminate the column corresponding to u_{2_x} as well as the corresponding equation determined by the symmetry of the coefficient matrix. After having reordered the variables such that u_{1_t} and u_{2_t} are the last ones, we apply again the technique to subdivide the coefficient matrix into a 5×5 upper left quadratic matrix A^*, corresponding to the one given by equ. (8), into a 2×5 matrix B^*, into a 5×2 matrix C^* and a 2×2 matrix D^* which is given by

$$
D^* = \begin{pmatrix} g_x^2 + g_{xx}^2 + g_{xt}^2 & g_x g_y + g_{xt} g_{yt} \\ g_x g_y + g_{xt} g_{yt} & g_y^2 + g_{yy}^2 + g_{yt}^2 \end{pmatrix} \tag{30}
$$

Using A^*, B^*, C^*, and D^* instead of A, B, C, and D in an analogy to the approach described in connection with equs. (10) through (12) yields immediately the result that the solutions for u_1, u_2, u_{1_x}, u_{1_y}, u_{2_y} – obtained above under the assumption that the temporal derivatives of the optical flow can be neglected – depend on the estimates for u_{1_t} and u_{2_t} in a manner formally analogous to the dependance of $\hat{u}_1$ and $\hat{u}_2$ on $\hat{u}_{1_x}$, $\hat{u}_{1_y}$, and $\hat{u}_{2_y}$ as given by equ. (12b). The solutions for u_1, u_2, u_{1_x}, u_{1_y}, and u_{2_y} will thus be biased. Special gray value patterns may, however, result in configurations for which the estimates $\hat{u}_{1_t}$ and $\hat{u}_{2_t}$ and, therefore, the bias do indeed vanish.

4   On Comparisons Between Optical Flow Estimates and Ground Truth

As a preliminary consequence of the exposition presented so far, it appears necessary to check experimentally whether the various interactions between estimates for the optical flow and its spatiotemporal derivatives can be corroborated by the evaluation of digitized image sequences from real world scenes. Given the multitude of experimental difficulties to be expected for such an endeavour, one should find out first how much is actually known about quantitative comparisons between estimates of optical flow based on image sequences and the corresponding ground truth. The subsequent discussion will be restricted, due to space limitations, to publications from 1990 and 1991. Publications which contain only qualitative indications about the accuracy of optical flow estimates from recorded image sequences or only data from synthetic image sequences – these two categories together represent the great majority of contributions with experimental results – are not discussed systematically in this context. The presentation is organized into a subsection on estimates based on an average over some image region, into a subsection on estimates of time-to-collision based on optical flow, into one on depth estimates derived from optical flow, and into one with comments.

4.1   Error Rates for Estimates Based on some Average over an Image Region

Koch and coworkers (see Battiti *et al.* 1991, Koch *et al.* *1991*) studied a variant of a multigrid estimation technique based on the approach of Horn and Schunck (1981) in which they investigated the dependence of the estimate on choosing the proper scale for the determination of the required spatial and temporal derivatives of the image intensity. The thrust of this investigation aimed at reducing the computational expense of a multigrid iterative approach for solving the Euler-Lagrange equations by spacevariant decisions regarding when to stop refinement steps. The authors present an error analysis for this approach and a comparison of results obtained by their approach with known correct values. Apart from experiments with synthetic data, they evaluated an image sequence produced by translating a pine cone parallel to the image plane by accurately measured amounts between successive video-frames. The authors report a flow estimate – averaged over a significant fraction of a digitized S-VHS frame – which deviated from the veridical value by about 10% under favourable conditions. No details are given about the precision of individual estimates. Moreover, the use of a smoothness term in their iterative approach may prevent their results from being used to assess a direct area-based approach.

Root mean square error estimates for individual optical flow vectors, obtained under precisely calibrated laboratory conditions, have been reported by Singh (1991). In this setup, a video camera translates in a plane perpendicular to its optical axis, past a textured poster with interframe displacements of 0.6 inches. The optical axis is not perpendicular to the plane of the poster which is about 12 inches away from the camera (not clear whether initially or at the end). The author uses a Kalman Filter approach to propagate local optical flow estimates into their spatial as well as temporal environment. The local estimates are obtained by a correlation search based on a sum-of-squared-differences (SSD) scoring function (see Anandan 1989) with window sizes of 3×3 and 5×5 pixels. The resulting score as a function of search displacements during the interframe match is converted into a confidence measure and exploited to average optical flow estimates for the propagation step. The RMS error, averaged over the entire field of view, for u_1 and u_2 amounts to 2.7% after twelve frames. Initially, the displacement vectors between accepted matches amount to 4 pixels where the poster is closest to the camera, and to 2 pixels where the poster is farthest away from the camera. For displacements between the last two frames, the corresponding figures are 6 and 3 pixels, respectively. Other experiments with an image sequence from a road scene yield qualitatively correct results. For this sequence, no ground truth was available. More detailed error statistics are given for this experiment in Singh (1990) from which it can be learned that – presumably prior to the averaging process across twelve frames – 83-87% of all optical flow estimates (depending on the particular selection criteria) deviated less than 25% from the correct value. Since the experimental setup excluded the appearance of discontinuities of the optical flow field, gross mismatches seem to have slipped through occasionally.

In the course of experiments to detect obstacles in front of an autonomously driving vehicle by the evaluation of optical flow fields, Enkelmann (1991) obtained an indirect quantitative average error estimate for the magnitude of optical flow. He compared the estimate for the 3D velocity component along the optical axis obtained from his estimates of the optical flow field with the corresponding value obtained by interactively matching the images of prominent scene points and using the camera calibration. Both values agreed to within 10%. It should be noted, however, that in order to obtain the values to be compared, optical flow estimates had to be integrated over a significant image area corresponding to the road immediately in front of the vehicle. Moreover, since the optical flow estimation approach was based on a smoothness requirement, the error margin reported by Enkelmann (1991) is not necessarily applicable to optical flow estimates based on smaller local image patches.

The so-called 'velocity functional method' (see Waxman and Wohn 1988) assumes that an (up to) second order Taylor expansion represents an appropriate description for the local spatial variation of the displacement rate field. This approximation actually represents the correct spatial variation for the displacement rate associated with the scene space motion of a planar surface. Exploiting such an approach requires an additional hypothesis, the so-called 'convected invariance of contours' which states that a contour element is locally carried along with the displacement rate. This hypothesis couples the displacement rate to the extraction of contour elements from images and thus converts the estimate into one of a variant of what is here called optical flow. It has already been mentioned earlier that it has recently been shown by this same group that – at least for zero-crossing contours of the Laplacian of the Gaussian-filtered image – the hypothesis of convected invariance of contour can result in incompatibilities with the assumption of 'convected invariance of intensity' (Wu and Wohn 1991). Assuming that the 'convected invariance of contour'-hypothesis represents a good approximation, Wohn *et al.* (1991) thoroughly analyze the errors for optical flow estimation based on an approach which matches each contour point to its closest neighbor on a contour extracted from the subsequent image. The resulting estimates of contour point displacements are then fitted – in this particular example – to a first order Taylor expansion of the optical flow field in a small neighborhood in order to obtain a representation for the full optical flow field in this neighborhood. The authors show under which conditions this approach provably converges. Various careful experiments with recorded image sequences from different laboratory scenes compare their estimates with the true values of the displacement rate, giving less than 5% discrepancy for the optical flow and about 10% or less discrepancy for the estimated spatial derivatives of optical flow. It should be noted, however, that these error estimates appear to have been obtained by comparing the optical flow parameters computed from the robot manipulator motion of the camera with those estimated by integrating over the field of view containing a more or less planar object arrangement. Error margins may be significantly greater if the estimates for optical flow and its derivatives are derived from only a small subsection of the entire image frame.

Negahdaripour and Lee (1991) estimate egomotion and gross scene structure from image sequences. They assume that the scene is composed mostly from planar panels and that the optical flow variation can be well approximated as a linear function of image plane coordinates within the image of a panel. Each panel is assumed to be of significant size and to be covered with well-textured surface markings. Starting from slightly overlapping square image regions of 19×19 pixel size in an 128×128 digitized image, they fit a version of the OFCE containing linear expressions for the optical flow field in the form $g_{t,i} + g_{x,i}(a_1 + a_3 x_i + a_5 y_i) + g_{y,i}(a_2 + a_4 x_i + a_6 y_i) = 0$ to the gray value pattern for all pixels i within each region. A hierarchical clustering algorithm is then employed to merge neighboring regions compatible with a fit to this linear expression until an acceptance threshold for the residuum of the fit within a tentative pairing of regions is exceeded. The resulting coefficients are used to determine the translation and rotation vector describing the egomotion as well as the normals of the panels in the scene which are assumed to correspond to the image regions surviving the clustering step. At least two regions are required corresponding to panels with significantly different normals in the scene. The authors report errors of 4-6% for the translation vectors obtained from image sequences recorded from two different real world scenes. The estimation error for the (non existent) rotation appears to be negligible. The estimation errors for the panel normals are somewhat larger, in particular in areas far from the recording video camera. Although the authors do not give error estimates for optical flow distributions within an image region, their good quantitative results support further attempts to estimate first order optical flow derivatives in addition to optical flow components themselves.

4.2   Indirect Error Margins Related to Time-to-Collision and Rotation Estimates

Campani and Verri (1990) report only indirectly about a quantitative error estimate for optical flow and its derivatives, based on a real image sequence recorded while a camera translates obliquely towards a planar poster. Optical flow estimates around the focus of expansion obtained for each of about 20 consecutive frames are converted to the time-to-collision which deviates usually with errors smaller than 10% from the veridical value, although larger deviations of up to 20% occur occasionally. It is not obvious from the publication of Campani and Verri (1990), whether the image area used to determine the time-to-collision had been restricted to a single 41×41 pixel patch employed for the pseudoinverse solution or was larger, possibly comprising several such patches around the focus of expansion[1]. Uras *et al.* (1988) report analogous experiments where they estimated time-to-collision from a neighborhood of the flow field singularity for the case of a purely translational motion between camera and scene. Based on estimates for 10×10 smoothed optical flow vectors around the singularity, they found a time-to-collision of 28.9±2.8 units, to be compared with the veridical value of 31.1 units. In a similar experiment with a purely rotational motion, they found agreement to within 10% for the angular velocity. Based on a different image sequence with 50 frames, this group confirmed the attainable estimation precision for the angular velocity of a purely rotational optical flow field to be within ±5%. Again, this figure results from averaging over an essential fraction of the field of view in each image frame (Baraldi *et al.* 1989, De Micheli *et al.* 1990). The latter authors point out that their experimental setup in this case, with a highly textured plane parallel to the image plane, is optimal from a theoretical point of view and also proved during the experiments to be the most favourable one.

As discussed in section 3.4, Verri *et al.* (1990) exploited a-priori knowledge about the character of the optical flow field to estimate both components u_1 and u_2 from a short sequence of real images. In the case of a purely diverging optical flow field, their estimate for time-to-collision derived from a pointwise average of optical flow estimates over a small neighborhood of the singular point of this optical flow field agreed with the veridical value to within 10-15%. An analogous experiment for the case of pure rotation yielded similar agreement, again after some averaging over a small neighborhood of the singular point.

4.3   Indirect Error Margins Related to Depth Estimates Based on Optical Flow

Jiang and Weymouth (1990) report about an approach to estimate depth based on the relative normal components of optical flow. For a short image sequence recorded from three keys on a table, these authors report about 5% error for the estimated normal optical flow component. In a more difficult short sequence showing a pyramid with low contrast between some of the adjacent faces, their estimates were reported to have an error mostly within 10%. This publication is a bit ambiguous about the specific method used by the

---

[1]According to a private communication by A. Verri on February 6, 1992, the area evaluated in the quoted experiment comprised essentially the entire image

authors, although it appears that the OFCE has been applied to estimate the normal flow components.

Based on the approach described by Uras *et al.* (1988) (see section 3.4) to estimate both components of optical flow, Baraldi *et al.* (1989) reported depth estimates for three different planar surfaces recorded by a video camera which translated obliquely towards the scene. The depth estimates derived from optical flow yielded 35±5 cm to be compared with a true value of 37 cm for one surface, 61 ± 9 cm and 62 ± 13 cm for the second and third surface to be compared with 56 cm and 72 cm, respectively, as the true values. The relative errors thus vary between 5% and 15%, depending on the distance.

Sandini and Tistarelli (1990) employ a gradient-based approach to estimate the optical flow component perpendicular to zero-crossing contours of the Laplacian of the Gaussian-filtered image in order to estimate a depth map for objects in a scene recorded by a moving video-camera. This represents an alternative to the contour matching employed by Wohn *et al.* (1991). Sandini and Tistarelli (1990) emphasize the importance of control over the motion of the recording camera such that the optical flow estimates obtained by their approach remains reliable enough to form the basis for an acceptable depth map estimation. Although their publication comprises an error analysis of the influence of the camera motion (see also Tistarelli and Sandini 1990), it does not give quantitative comparisons with respect to their optical flow estimates. Errors in the individual estimates of optical flow vectors are compensated to some extent in their approach by tracking contour points through an extended subsequence of images. An essentially similar approach by Vernon and Tistarelli (1990) reported range errors of about 2%, employing the estimation of normal components for optical flow based on the OFCE along zero-crossing contours, but does not give error margins for the estimated optical flow components themselves.

A similar setup has been described and evaluated by Hayes and Fisher (1990) who estimated optical flow using the OFCE and reported depth errors which – upon averaging over seven images – ranged between less than 10% and up to 20% in extreme cases. Again, no errors were given for the optical flow estimates used to arrive at the reported depth values.

4.4   Comments and Additional Remarks Regarding Error Estimates

Some investigations will be mentioned here despite the fact that they do not fall squarely into one of the selected categories. The observation of outliers in the optical flow field recommends an approach of Schnörr (1991) who applied an analysis based on a singular value decomposition of the coefficient matrix for the determination of the optical flow components u_1 and u_2. He modified the estimation procedure in order to avoid unreasonable results in case of an ill-conditioned coefficient matrix.

As an example for another approach which provides some figure of confidence for the optical flow estimate, Bigün *et al.* (1991) should be mentioned. They estimate optical flow locally by analysing the eigenvalues and associated eigenvectors of a matrix constructed from the scatter matrix of spatiotemporal gradient components which is averaged by convolution with an 11×11×11 pixel wide Gaussian kernel. Combinations of the eigenvalues can be converted into a confidence measure of the estimates. These authors have tested their approach using – in addition to synthetic image sequences – a video-subsequence from a street scene. Although the results are qualitatively satisfactory for those estimates considered reliable, no detailed comparisons with the unknown veridical values were possible for the road scene sequence.

Jähne (1990) discusses the bias introduced into optical flow estimates by inappropriate transfer functions of operators applied to estimate the spatiotemporal derivatives of the gray value pattern. His analysis – see, too, related work by Fleet and Jepson (1991) – is corroborated by investigations on one-dimensional synthetic signals without and with added Gaussian noise. An iterative improvement loop appears to counterbalance this effect well enough to represent a practical solution, although at higher computational costs.

Fleet and Jepson (1990) subject an image sequence to a series of linear, shift-invariant filters with only local spatiotemporal support, each filter being tuned to a narrow range of orientation, speed, and scale. They argue that the evolution of phase contours in the output of these filters yields a better approximation to the motion or displacement rate field than constructs which rely more on the signal amplitude. Similar arguments have been discussed, for example, by Weng (1990). Situations where problems may occur are analyzed by Fleet *et al.* (1991) (see, too, Jenkin *et al.* 1991 and Fleet and Jepson 1991). In a careful investigation, Fleet and Jepson (1990) determine error margins for their results obtained from a variety of input image sequences. Based on their conversion of these errors to relative optical flow errors, they report a majority of results with such errors in the range of 6% to 10%. This compares favourably with deviations between estimated and expected optical flow reported by these authors from earlier references of other approaches. An independent evaluation of this approach and a quantitative comparison with results from other approaches appears worthwhile.

Murray and Buxton (1990) report error frequency histograms for some of their experiments to estimate optical flow based on interframe matching of edge elements, i.e. a token-based approach [2]. The errors of optical flow estimates depend on the structure of the image, the edge element size, and the veridical motion field. Murray and Buxton (1990) obtained under favorable conditions a symmetrical unimodal distribution centered at zero with a half width at half maximum (HWHM) of 0.1 pixel per frame for an interframe shift of 2 pixels, i.e. about 5%. Under less favorable conditions, the error histogram became asymmetrical and had a larger HWHM, with a non-negligible tail exceeding 10% for greater interframe shifts of gray value structures.

In another token-based approach, Huber and Graefe (1991) track the images of two spots which are known a-priori to correspond to scene features with the same distance from the recording camera, the latter being mounted on a translating robot. The authors determine the distance in the image plane between these features to subpixel precision by polynomial interpolation of cross-correlation results obtained from template matching to the gray value pattern. The temporal change of this image distance between the segments can be converted, based on the known camera velocity in the scene, to a distance estimate. Under these conditions, the authors report a decrease in the distance error over travelled distance from 10% after 0.2 m down to less than 1% after the camera had travelled about 4 m. The particular experimental setup and the possibility to continuously improve the current distance estimate, based on the model-based accumulation of measurements over time, represents a special case which can not be carried over to assess error margins for optical flow estimates in a more general situation.

The sensitivity of depth determination based on single displacement estimates from token matching has been analyzed by Snyder (1989) and extended by Dutta and Snyder (1990 and 1991), in particular for situations where already small errors in the image

---

[2]Note the distinction between an optical flow vector obtained by matching, e. g., a corner token – which has been extracted by a more or less heuristic procedure – on the one hand and, on the other hand, that particular characteristics of a continuous gray value variation, which are necessary to estimate both components of the optical flow, can be considered to represent a corner

displacement estimate can result in gross depth errors. Although Dutta and Snyder (1990) report depth estimates with errors of 7-13% by a correspondence-based approach, provided some refinement steps are applied to their raw estimates, they point out the considerable fraction of outliers due to mismatches. Given a currently attainable 'regular' displacement estimation precision of the order of half a pixel, they draw attention to the plausible observation that depth estimates can only be expected with reasonably small errors if the depth range is restricted to about ten times the camera translation relative to the scene, i.e. about ten times the 'base line length'.

The results reported in this section 4 are those found during an attempt to thoroughly search available regular publications – about 40000 pages, see the Appendix – for the years 1990 and 1991 (no preprints). Although this survey should be extended to publications which have appeared in years prior to 1990, the available evidence accumulated so far indicates that optical flow has been estimated under favourable circumstances with errors between 5% and 10%. Only very few estimates for spatial derivatives of optical flow have been reported. Error margins for these estimates, representing an average over a sizable image area, are of the order of 10% in case of Wohn *et al.* (1991). Good results for entities derived from the estimated spatial variation of optical flow bound errors for these derivatives as in the experiments by Torre, Verri and coworkers or by Negahdaripour and Lee. No cues have been found in the literature about measuring the temporal derivative of optical flow, not to speak about any quantitative error estimates for these temporal derivatives.

5   Experimental Setup for Checking Optical Flow Estimates

Given the situation regarding experimental corroboration for approaches to estimate optical flow and its derivatives from real world image sequences, one could argue that optical flow is a qualitative concept anyway (Verri and Poggio 1987) and all suggested estimation approaches which deliver qualitatively acceptable results should be considered to be equivalent. On the other hand, it appears unsatisfactory to have no generally accepted experimental methodology to assess quantitatively under which conditions and by how much the estimates obtained by particular approaches deviate from the veridical motion or displacement rate field.

It thus appears desirable to prepare an experimental environment which should facilitate quantitative comparisons between optical flow estimates and true values for the displacement rate field. In order to have full control over all aspects of image recording, it has been decided to record image sequences by a solid state video-camera mounted on the hand of a 6-axes robot. This robot is controlled by the same computer which also evaluates the optical flow estimation approaches.

The objects in the field of view of this camera will initially consist of close approximations to a polyhedral scene. A system has recently been completed which implements the theory of Sugihara (1986) for quasipolyhedral scenes (Müller 1991, see also Hirsch *et al.* 1991). Unavoidable effects such as a finite curvature of the intersection between different faces, imperfect object vertices, specular reflections on visible edges of holes stamped into otherwise planar sheet metal faces etc. have been coped with by various heuristics – see Figure 1. A stationary trinocular video-camera-system in addition to the video-camera on the robot hand facilitates an independent quantitative measurement of the object configuration in space.

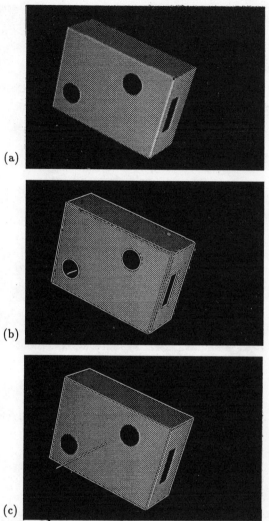

Figure 1: *(a) Digitized image of a quasipolyhedral object stamped and bent from sheet metal. The effects of various deviations from an ideal polyhedral object such as intersections with finite radius of curvature between adjacent faces, holes not bound by polygons, imperfect vertices due to the finite thickness of the material, specular reflections etc. are clearly discernible. (b) Edge elements extracted from this digitized image. (c) Edge segments finally selected for the interpretation as a line drawing, together with the normal vector to the facet which is used for picking up this workpiece by a suction gripper. (By permission reproduced from Müller 1991)*

Adaptation of this setup should enable to record various configurations of quasi-polyhedral objects by a video-camera moved under computer control by a robot. The fact that each single image frame recorded from such a quasipolyhedral scene can be interpreted based on Sugihara's theory helps to explore quantitative estimates of straight line segment arrangements surrounding a planar facet (see, e.g., Murray and Buxton 1990).

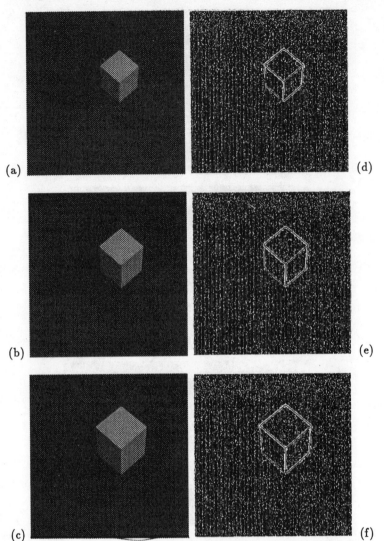

Figure 2: *(a) Digitized image frame 1 from an image sequence of a cube recorded by a video-camera mounted on the hand of a robot; (b) Frame 11; (c) Frame 21; (d) Location in frame 1 where the gradient magnitude attains a local maximum in the gradient direction – without any threshold applied; (e) Analogous for frame 11; (f) Analogous for frame 21 (results of an approach described in Otte and Nagel 1992).*

Moreover, the hypotheses obtained about the polyhedral scene based on the video-input from the camera on the robot can be compared to those obtained by the trinocular camera arrangement. This is expected to provide data which can be compared quantitatively with those derived from direct, area-based optical flow estimates in the vicinity of images of edges and vertices. In particular, the knowledge that faces are essentially planar can be exploited in at least two ways:

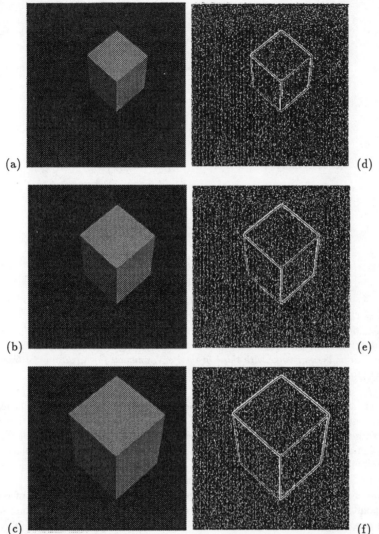

Figure 3: *Analogous to Figure 2, but for frame 31 (a + d), frame 41 (b + e) and frame 50 (c + f) from the same image sequence.*

(i) It can be tested whether line or curve segments shift from frame to frame in compliance with an hypothesis that their counterparts in the scene are situated on the same planar facet.

(ii) Given recognized boundary segments of a planar facet for which correspondence has been established between frames, it becomes possible to compute the image displacement of any point on such a facet between image frames.

This should facilitate a quantitative check of optical flow estimates for small surface elements selected regarding a variety of gray value patterns close to edges and junctions or

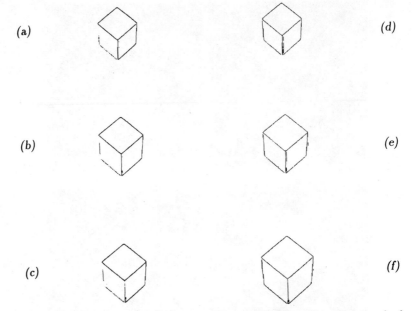

Figure 4: *(a) Locations of maxima of the gradient magnitude in gradient direction for frame 1, with a gradient magnitude maximum $\geq$ 6; (b) Analogous for frame 11; (c) Analogous for frame 21; (d) Edge element chains for frame 1 if subthreshold edge elements are retained provided their orientation is compatible with those in their environment; (e) Analogous for frame 11; (f) Analogous for frame 21.*

well separated from image areas where discontinuities of optical flow have to be expected. Effects of occlusion and shadowing should become accessible to investigation without undue additional experimental efforts. Workpieces with a combination of polyhedral as well as non-polyhedral bounding surfaces and combinations of such objects should enable a controlled extension of experiments for the estimation of optical flow to more general conditions.

In order to avoid time consuming and error-fraud interactive selections of particular image points, a reliable detection and localisation of edge element and junction positions even in image areas with weak contrast are important. An approach to exploit systematically various spatial configurations of edge elements in up to 5×5 neighborhoods around a pixel under scrutiny (Otte and Nagel 1992) provides the background for the study of optical flow estimation procedures. Figures 2a-c and 3a-c show six frames from an image sequence of a cube recorded by a video-camera moving on the hand of a robot. Figures 2d-f and 3d-f show the location of maxima of the gradient magnitude in the gradient direction without any threshold. If a (rather small!) threshold of 6 is applied, results as given by Figures 4a-c and 5a-c are obtained. Although all of the background 'noise edge elements' have disappeared, so have quite some edge elements on the images of cube edges. Figures 4d-f and 5d-f show edge chains obtained by the approach described in Otte and Nagel (1992) where even subthreshold edge elements are accepted provided their orientation is compatible with the one derived by a detailed analysis of edge elements in their environment. Note that these are chains of edge elements extracted from the digitized

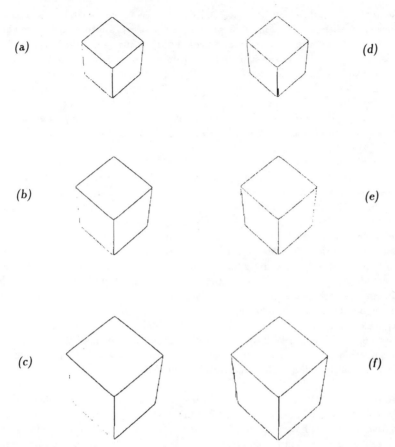

Figure 5: *Analogous to Figure 4, but for frame 31 (a + d), frame 41 (b + e) and frame 50 (c + f) from the same image sequence.*

image, not any straight line segments fitted to remaining edge elements such as those shown in Figure 4a-c and 5a-c.

6   Concluding Discussion

The considerations presented in preceding sections illustrate the problems associated with attempts to define and estimate optical flow. Nevertheless, it appears advantageous to explore how well such estimation approaches can be applied in image areas with significant gray value variations which, however, need not to be edge elements, gray value corners, or junctions. Understanding optical flow estimates in such areas can contribute to hypothesize potential discontinuities of optical flow or to link optical flow estimates obtained at different edge curves in analogy to the way edge-based and region-based segmentation approaches are thought to complement each other – see, e.g., Verri *et al.* (1989b) for very illustrative experimental evidence.

It has been postulated that the 'Optical Flow Constraint Equation (OFCE)' should remain exactly valid for all values of $(\boldsymbol{x}, t + \Delta t)$ in some local environment around $(\boldsymbol{x}_0, t)$

provided the dependency on $(x - x_0, \Delta t)$ can be accounted for by a first order Taylor expansion of all entities. This is a generalisation of an approach presented in Nagel (1987) by including also temporal derivatives of optical flow (see, too, Nagel 1990). Based on this postulate, a system of eight linear equations for the optical flow components as well as for their first partial derivatives with respect to x and t has been derived. The resulting system of linear equations comprises only up to second order spatiotemporal derivatives of the gray value pattern. Although it turned out that one of these eight equations depends linearly on the others, the introduction of a free parameter related to the rotational component of optical flow allows to solve the remaining set of equations in symbolic form.

An expression has been derived which shows explicitly how the omission of partial spatial derivatives of optical flow can introduce a bias into the estimate of the optical flow components themselves. Similarly, the omission of partial temporal derivatives of optical flow can introduce a bias both into the estimate of the optical flow components themselves as well as into the estimate of their spatial derivatives.

In order to simplify the discussion of the resulting symbolic expressions, the temporal derivatives of optical flow are set to zero. Without making additional assumptions, a closed form solution for the divergence of the optical flow field can then be derived which does not depend on the free parameter. Moreover, the connection between the local spatiotemporal gray value structure and the local spatial structure of the optical flow field can be clarified. Consequences of – often implicit – assumptions such as those incorporated into various other approaches to estimate optical flow can be discussed on the basis of a still manageable symbolic solution obtained in this case.

In particular, the approach by Werkhoven and Koenderink (1990) has been recast in the notation used in section 3 to facilitate a more detailed comparison with the approach studied in sections 3.1-3.3 and 3.5. Neither approach is a straightforward specialization of the other one, but both can be specialized in such a manner that one arrives at an equivalent system of linear equations. For this purpose, one has to suppress the partial derivatives of optical flow with respect to time in the approach investigated in section 3.1. In the approach of Werkhoven and Koenderink (1990), one has to suppress spatiotemporal derivatives of the gray value pattern of order three and higher. It is illustrative to inspect for this special case under which conditions both components of optical flow and their spatial derivatives can be estimated. The denominator of equ. (14), i.e. $det(A)$ with A given by equ. (11a), has to differ from zero in order to obtain a solution for both u_1 and u_2. One can consider the inverse condition number of A as a quantitative indication for the degree with which the gray value pattern at (x_0, t) is sufficient to estimate both components u_1 and u_2. Analogously, one can consider the inverse condition number of the matrix $(D - CA^{-1}B)$ – see equation (12a) – as a quantitative indication for the existence of sufficient gray value structure to estimate three spatial derivatives of optical flow.

This idea will now be applied to the approach of Werkhoven and Koenderink (1990). If the higher order spatiotemporal derivatives of $g(x, t)$ are retained, the system of linear equations presented by Werkhoven and Koenderink (1990) will allow in principle to estimate both components of optical flow and all its four spatial derivatives. Assume that the unknowns are reordered as indicated in equ. (31) below. In order to obtain a more symmetrical coefficient matrix, the first line of equ. (24) from Werkhoven and Koenderink (1990) is transferred to a line between the fourth and the fifth from the original system of linear equations. Moreover, the first and second equation of the resulting system of equations have been multiplied on both sides by -1. Using the notation of section 3, one obtains

$$
C_I \begin{pmatrix} u_1 \\ u_2 \\ u_{1_x} \\ u_{1_y} \\ u_{2_x} \\ u_{2_y} \end{pmatrix} = \begin{pmatrix} C_{IA} & C_{IB} \\ C_{IC} & C_{ID} \end{pmatrix} \begin{pmatrix} u_1 \\ u_2 \\ u_{1_x} \\ u_{1_y} \\ u_{2_x} \\ u_{2_y} \end{pmatrix} = - \begin{pmatrix} r_1 \\ r_2 \\ r_3 \\ r_4 \\ r_5 \\ r_6 \end{pmatrix} = - \begin{pmatrix} g_{xt} \\ g_{yt} \\ g_{xxt} \\ g_t \\ g_{xyt} \\ g_{yyt} \end{pmatrix}
\tag{31}
$$

Note that the right hand side of equ. (31) has been multiplied by -1 compared to equ. (24) of Werkhoven and Koenderink (1990) in order to obtain compatibility with the OFCE for the case of vanishing second and higher order derivatives of $g(\boldsymbol{x}, t)$. The submatrices of the coefficient matrix C_I are given by the following expressions :

$$
C_{IA} = \begin{pmatrix} g_{xx} & g_{xy} \\ g_{xy} & g_{yy} \end{pmatrix}
\tag{32a}
$$

$$
C_{IB} = \begin{pmatrix} g_x + g_{xxx} & g_{xxy} & g_y + g_{xxy} & g_{xyy} \\ g_{xxy} & g_x + g_{xyy} & g_{xyy} & g_y + g_{yyy} \end{pmatrix}
\tag{32b}
$$

$$
C_{IC} = \begin{pmatrix} g_{xxx} & g_{xxy} \\ g_x & g_y \\ g_{xxy} & g_{xyy} \\ g_{xyy} & g_{yyy} \end{pmatrix}
\tag{32c}
$$

and

$$
C_{ID} = \begin{pmatrix} 2g_{xx} + g_{xxxx} & g_{xxxy} & 2g_{xy} + g_{xxxy} & g_{xxyy} \\ g_{xx} & g_{xy} & g_{xy} & g_{yy} \\ g_{xy} + g_{xxxy} & g_{xx} + g_{xxyy} & g_{yy} + g_{xxyy} & g_{xy} + g_{xyyy} \\ g_{xxyy} & 2g_{xy} + g_{xyyy} & g_{xyyy} & 2g_{yy} + g_{yyyy} \end{pmatrix}
\tag{32d}
$$

In analogy to the steps discussed in connection with equ. (12), the inverse of the condition number of $(C_{ID} - C_{IC} C_{IA}^{-1} C_{IB})$ can be considered as a scoring function to indicate the reliability with which the spatial variation of $g(\boldsymbol{x}, t)$ around $(\boldsymbol{x}_0, t)$ allows to determine all four spatial derivatives of optical flow – provided $\hat{u}_1$ and $\hat{u}_2$ are determined as given above. Of course, these estimates will be biased since we suppressed the temporal derivatives of optical flow. Nevertheless, we arrive at an explicit quantitative connection between the spatial gray value structure given by derivatives of up to order four *and the resulting first order structure of optical flow*. If we suppress, for example, all derivatives of $g(\boldsymbol{x}, t)$ of order three or higher and set $g_{xy} = 0$, C_{ID} will become singular and we reproduce the degenerate situation discussed in section 3.4 in connection with equ. (29). If, on the other hand, the inverse condition number of $(C_{ID} - C_{IC} C_{IA}^{-1} C_{IB})$ approaches 1, then all four spatial derivatives ∇u_1 and ∇u_2 can be determined. If they turn out to be zero, i.e. if the optical flow does not change locally around this image position in the image plane, the analogy to equ. (12b) will specialize to the case investigated by Reichardt *et al.* (1988) and Uras *et al.* (1988):

$$\begin{pmatrix} \hat{u}_1 \\ \hat{u}_2 \end{pmatrix} = -C_{1A}^{-1} \begin{pmatrix} g_{xt} \\ g_{yt} \end{pmatrix} = -C_{1A}^{-1} \nabla g_t \tag{33}$$

Another point worthy to be investigated is the trade-of between products of up to second order spatial derivatives of $g(x,t)$ in the coefficient matrix of equ. (8) and the linear appearance of up to fourth order spatial derivatives in the coefficient matrix of equ. (31) where the latter matrix has full rank in the general case. Obviously, the approach by Werkhoven and Koenderink (1990) should be extended to incorporate temporal derivatives of optical flow. This should allow to study the contribution of all first-order spatiotemporal derivatives of the optical flow on an equal footing, without the necessity to employ products of spatiotemporal derivatives of $g(x,t)$ in the coefficient matrix as it is required in section 3.5.

The detailed discussion presented so far should contribute to consolidate various approaches in a joint framework. It naturally raises the question under which conditions the estimates deviate – and by how much – from the veridical motion or displacement rate vector field. A survey of recent literature shows that only few quantitative error estimates have been reported for results obtained from recorded image sequences. In favourable cases, the agreement between estimated and veridical values for optical flow components u_1 and u_2 is better than 10% although frequently the error margin applies only to an entity derived from optical flow estimates after some averaging processes. Only very few attempts have been reported to compare estimates for spatial derivatives of optical flow, again implying an averaging process prior to the comparison with veridical values.

The majority of investigations regarding estimation errors for optical flow is concerned either with synthetic image sequences or yields only qualitative results. It thus appears desirable to investigate the determination of optical flow under controlled conditions in order to study the dependency of estimation errors on the spatiotemporal gray value variations. Image sequences recorded from quasipolyhedral scenes by a camera mounted on the moving hand of a robot are expected to facilitate such investigations. An environment to perform such experiments has been realized and first attempts to use it have been started.

It should have become clear that detailed comparisons, both between various approaches to define and estimate optical flow and its derivatives as well as between theoretical estimation approaches and experimental results, are now feasible, but need further attention. This should strengthen the fundament for investigations about how living creatures evaluate the spatiotemporal variation of irradiance patterns impinging on their retina.

7   Acknowledgements

Stimulating discussions as well as comments on a draft version of this contribution by K. Daniilidis, W. Enkelmann, M. Otte, and in particular by G. Orban are gratefully acknowledged. Special thanks go to M. Otte for his great help in converting this text to the final text system. The image sequence referred to in Figures 2-5 has been recorded by M. Otte and K. Daniilidis in cooperation with V. Gengenbach and C. Müller. The solutions for optical flow estimates have been cross-checked by exercising the symbolic algebra program MAPLE to which I have been introduced by J. Rieger. K. Daniilidis provided help in setting up some of these calculations.

Part of the research reported in this contribution has been supported by the ESPRIT Basic Research Action 3001, project INSIGHT, of the European Community.

Appendix

Set of Literature Searched for the Survey on Optical Flow Error Estimates

A thorough attempt has been made to search the available publications for the years 1990 and 1991. In order to indicate the scope of the survey on which the statements concerning the sparseness of quantitative error assessments for optical flow estimates have been based, the set of literature taken into account is enumerated here. Just for fun, the number of pages scanned during the search are given in parentheses. They add up to about *40000 (!) pages of refereed publications* which have appeared in essentially two years. Books and literature prior to 1990 are not included in this page count.

* Artificial Intelligence Journal (4670);

* Biological Cybernetics (2290);

* Computer Vision, Graphics, and Image Processing (3000);

* DAGM-Symposium Mustererkennung 1990+1991 (1260)

* International Journal of Computer Vision (830);

* International Journal of Visual Communication and Image Representation (600);

* International Journal of Robotics Research (1420);

* IEEE Transactions on Pattern Analysis and Machine Intelligence (2490);

* IEEE Transactions on Robotics and Automation (1640);

* IEEE Transactions on Systems, Man, and Cybernetics (2410);

* Image and Vision Computing (760);

* Journal of the Optical Society of America A (4130)

* Machine Vision and Applications (560);

* Pattern Recognition (2670);

* Pattern Recognition Letters (1580);

* Proc. British Machine Vision Conference 1990 as well as 1991 (830)

* Proc. DARPA Image Understanding Workshop 1990 (920)

* Proc. First European Conference on Computer Vision 1990 (620)

* Proc. IEEE (3510)

* Proc. IEEE Conf. on Computer Vision and Pattern Recognition CVPR '91 (760)

* Proc. IEEE Workshop on Visual Motion, Princeton/NJ 1990 (340)

* Proc. International Conference on Computer Vision ICCV '90 (750)

* Proc. International Conference on Pattern Recognition ICPR '90 (1610).

References

Anandan, P. (1989) A Computational Framework and an Algorithm for the Measurement of Visual Motion. International Journal of Computer Vision 2, 283-310

Baraldi, P., De Micheli, E., Uras, S. (1989) Motion and Depth from Optical Flow. Proc. Fifth Alvey Vision Conference, University of Reading, Reading/UK, 25-28 September 1989, pp. 205-208

Battiti, R., Amaldi, E., Koch, C. (1991) Computing Optical Flow Across Multiple Scales: An Adaptive Coarse-to-Fine Strategy. International Journal of Computer Vision 6, 133-145

Bigün, J., Granlund, G.H., Wiklund, J. (1991) Multidimensional Orientation Estimation with Applications to Texture Analysis and Optical Flow. IEEE Trans. Pattern Analysis and Machine Intelligence PAMI-13, 775-790

Campani, M., Verri, A. (1990) Computing Optical Flow from an Overconstrained System of Linear Algebraic Equations. Proc. Third International Conference on Computer Vision ICCV '90, 4-7 December 1990, Osaka/Japan, pp. 22-26

De Micheli, E., Uras, S., Torre, V. (1990) The Analysis of Time Varying Image Sequences. Proc. First European Conference on Computer Vision - ECCV '90, Antibes/France, 23-27 April 1990. O. Faugeras (ed.), Lecture Notes in Computer Science 427, Springer-Verlag Berlin Heidelberg New York/NY etc., pp. 595-597

Deriche, R., Faugeras, O., Giraudon, G., Papadopoulo, T., Vaillant, R., Viéville, T. (1992) Four Applications of Differential Geometry to Computer Vision. This volume, 93-141

Dutta, R., Snyder, M.A. (1990) Robustness of Correspondence-Based Structure from Motion. Third International Conference on Computer Vision ICCV '90, 4-7 December 1990, Osaka/Japan, pp. 106-110

Dutta, R., Snyder, M.A. (1991) Robustness of Structure from Binocular Known Motion. Proc. IEEE Work-shop on Visual Motion, Princeton/NJ, 7-9 October 1991, pp. 81-86

Enkelmann, W. (1991) Obstacle Detection by Evaluation of Optical Flow Fields from Image Sequences. Image and Vision Computing 9, 160-168

Faugeras, O. (1990) On the Motion of 3D Curves and Its Relationship to Optical Flow. Proc. First European Conference on Computer Vision - ECCV '90, Antibes/France, 23-27 April 1990. O. Faugeras (ed.), Lecture Notes in Computer Science 427, Springer-Verlag Berlin Heidelberg New York/NY etc., pp. 107-117

Fleet, D.J., Jepson, A.D. (1990) Computation of Component Image Velocities from Local Phase Information. International Journal of Computer Vision 5, 77-104

Fleet, D.J., Jepson, A.D. (1991) Stability of Phase Information. Proc. IEEE Workshop on Visual Motion, Princeton/NJ, 7-9 October 1991, pp. 52-60

Fleet, D.J., Jepson, A.D., Jenkin, M.R.M. (1991) Phase-Based Disparity Measurement. CVGIP: Image Understanding 53, 198-210

Girosi, F., Verri, A., Torre, V. (1989) Constraints for the Estimation of Optical Flow. Proc. IEEE Workshop on Visual Motion, 20-22 March 1989, Irvine/CA, pp. 116-124

Gong, S. (1989) Curve Motion Constraint Equation and Its Applications. Proc. IEEE Workshop on Visual Motion, 20-22 March 1989, Irvine/CA, pp. 73-80

Gong, S., Brady, M (1990) Parallel Computation of Optical Flow. Proc. First European Conference on Computer Vision - ECCV '90, Antibes/France, 23-27 April 1990. O. Faugeras (ed.), Lecture Notes in Computer Science 427, Springer-Verlag Berlin Heidelberg New York/NY etc., pp. 124-133

Hayes, G.M., Fisher, R.B. (1990) Evaluation of a Real-Time Kinetic Depth System. Proc. British Machine Vision Conference, 24-27 September 1990, Oxford/UK, pp. 315-318

Hildreth, E.C. (1984) Computations Underlying the Measurement of Visual Motion. Artificial Intelligence 23, 309-354

Hirsch, E., Paillou, Ph., Müller, C., Gengenbach, V. (1991) A Versatile Parallel Computer Architecture for Machine Vision. Application to the Comparison of Real Images and CAD-Based Representations. Proc. Computer Architecture for Machine Perception CAMP '91, B. Zavidovique and P.-L. Wendel (eds.), 16-18 December 1991, Paris/France, pp. 251-261

Horn, B.K.P. (1986) Robot Vision. The MIT Press, Cambridge/MA and London/UK

Horn, B.K.P., Schunck, B.G. (1981) Determining Optical Flow. Artificial Intelligence 17, 185-203

Huber, J., Graefe, V. (1991) Quantitative Interpretation of Image Velocities in Real Time. Proc. IEEE Workshop on Visual Motion, Princeton/NJ, 7-9 October 1991, pp. 211-216

Jähne, B. (1990) Motion Determination in Space-Time Images. Proc. First European Conference on Computer Vision - ECCV '90, Antibes/France, 23-27 April 1990. O. Faugeras (ed.), Lecture Notes in Computer Science 427, Springer-Verlag, Berlin, Heidelberg, New York/NY etc., pp. 161-173

Jenkin, M.R.M., Jepson, A.D., Tsotsos, J.K. (1991) Techniques for Disparity Measurement. CVGIP: Image Understanding 53, 14-30

Jiang, F., Weymouth, T.E (1990) Depth from Relative Normal Flows. Pattern Recognition 23, 1011-1022

Koch, C., Wang, H.T., Battiti, R., Mathur, B., Ziomkowski, C. (1991) An Adaptive Multi-Scale Approach for Estimating Optical Flow: Computational Theory and Physiological Implementation. Proc. IEEE Workshop on Visual Motion, Princeton / NJ, 7-9 October 1991, pp. 111-117

Müller, C. (1991) Verwendung von Bildauswertungsmethoden zur Erkennung und Lage-bestimmung von generischen polyedrischen Objekten im Raum. Dissertation, Fakultät für Informatik der Universität Karlsruhe (TH), Karlsruhe/Deutschland (November 1991)

Murray, D.W., Buxton, B.F. (1990) Experiments in the Machine Interpretation of Visual Motion. The MIT Press, Cambridge/MA and London/UK

Nagel, H.H. (1983) Displacement Vectors Derived from Second Order Intensity Variations in Image Sequences. Computer Vision, Graphics, and Image Processing 21, 85-117

Nagel, H.H. (1984) Recent Advances in Image Sequence Analysis. Proc. Premier Colloque Image - Traitement, Synthèse, Technologie et Applications, 21-25 May 1984, Biarritz/France, pp. 545-558

Nagel, H.H. (1987) On the Estimation of Optical Flow: Relations between Different Approaches and Some New Results. Artificial Intelligence 33, 299-324

Nagel, H.H. (1989) On a Constraint Equation for the Estimation of Displacement Rates in Image Sequences. IEEE Trans. Pattern Analysis and Machine Intelligence PAMI-11, 13-30

Nagel, H.H. (1990) Extending the 'Oriented Smoothness Constraint' into the Temporal Domain and the Estimation of Derivatives of Optical Flow. Proc. First European Conference on Computer Vision - ECCV '90, O. Faugeras (ed.), Lecture Notes in Computer Science 427, Springer-Verlag, Berlin, Heidelberg, New York/NY etc., pp. 139-148

Negahdaripour, S., Lee, S. (1991) Motion Recovery from Image Sequences Using First-Order Optical Flow Information. Proc. IEEE Workshop on Visual Motion, Princeton/NJ, 7-9 October 1991, pp. 132-139

Okutomi, M., Kanade, T. (1990) A Locally Adaptive Window for Signal Processing. Proc. Third International Conference on Computer Vision ICCV '90, 4-7 December 1990, Osaka/Japan, pp. 190-199

Orban, G.A. (1992) The Analysis of Motion Signals and the Question of the Nature of Processing in the Primate Visual System. This volume, 24-56

Otte, M., Nagel, H.H. (1992) Extraction of Line Drawings from Gray Value Images by Non-Local Analysis of Edge Element Structures. To appear in Proc. Second European Conference on Computer Vision, Genoa/Italy, May 1992

Reichardt, W.E., Schlögl, R.W. (1988) A Two Dimensional Field Theory for Motion Computation. Biological Cybernetics 60, 23-35

Reichardt, W.E., Schlögl, R.W., Egelhaaf, M. (1988) Movement Detectors Provide Sufficient Information for Local Computation of 2-D Velocity Field. Naturwissenschaften 75, 313-315

Sandini, G., Tistarelli, M. (1990) Active Tracking Strategy for Monocular Depth Inference over Multiple Frames. IEEE Trans. Pattern Analysis and Machine Intelligence PAMI-12, 13-27

Schnörr, C. (1991) Funktionalanalytische Methoden zur Gewinnung von Bewegungsinformation aus TV-Bildfolgen. Dissertation, Fakultät für Informatik der Universität Karlsruhe (TH), Karlsruhe/Deutschland, Juni 1991

Singh, A. (1990) An Estimation-Theoretic Framework for Image-Flow Computation. Proc. Third International Conference on Computer Vision ICCV '90, 4-7 December 1990, Osaka/Japan, pp. 168-177

Singh, A. (1991) Incremental Estimation of Image Flow Using a Kalman Filter. Proc. IEEE Workshop on Visual Motion, Princeton/NJ, 7-9 October 1991, pp. 36-43

Snyder, M.A. (1989) The Precision of 3-D Parameters in Correspondence-Based Techniques: The Case of Uniform Translational Motion in a Rigid Environment. IEEE Trans. Pattern Analysis and Machine Intelligence PAMI-11, 523-528

Sugihara, K. (1986) Machine Interpretation of Line Drawings. The MIT Press, Cambridge/MA and London/UK

Tistarelli, M., Sandini, G. (1990) Estimation of Depth from Motion Using an Anthropomorphic Visual Sensor. Image and Vision Computing 8, 271-278

Uras, S., Girosi, F., Verri, A., Torre, V. (1988) A Computational Approach to Motion Perception. Biological Cybernetics 60, 79-87

Vernon, D., Tistarelli, M. (1990) Using Camera Motion to Estimate Range for Robotic Parts Manipulation. IEEE Trans. Robotics and Automation RA-6, 509-521

Verri, A., Poggio, T. (1987) Against Quantitative Optical Flow. Proc. First International Conference on Computer Vision, London/UK, 8-11 June 1987, pp. 171-180

Verri, A., Girosi, F., Torre, V. (1989a) Mathematical Properties of the Two-Dimensional Motion Field: From Singular Points to Motion Parameters. Journal of the Optical Society of America A 6, 698-712

Verri, A., Uras, S., De Micheli, E. (1989b) Motion Segmentation from Optical Flow. Proc. Fifth Alvey Vision Conference, University of Reading, Reading/UK, 25-28 September 1989, pp. 209-214

Verri, A., Girosi, F., Torre, T. (1990) Differential Techniques for Optical Flow. Journal of the Optical Society of America A/7, 912-922

Verri, A., Straforini, M., Torre, V. (1992) Computational Aspects of Motion Perception in Natural and Artificial Vision Systems. This volume, 71-92

Waxman, A.M., Wohn, K. (1988) Image Flow Theory: A Framework for 3-D Inference from Time-Varying Imagery. In C. Brown (ed.), Advances in Computer Vision, Vol. 1, Lawrence Erlbaum Associates, Hillsdale/NJ, pp. 165-224

Weng, J. (1990) A Theory of Image Matching. Proc. Third International Conference on Computer Vision ICCV '90, 4-7 December 1990, Osaka/Japan, pp. 200-209

Werkhoven, P., Koenderink, J.J. (1990) Extraction of Motion Parallax Structure in the Visual System. Biological Cybernetics 63, 185-191 (Part I) and 193-199 (Part II)

Wohn, K.Y., Wu, J., Brockett, R.W. (1991) A Contour-Based Recovery of Image Flow: Iterative Transformation Method. IEEE Trans. Pattern Analysis and Machine Intelligence PAMI-13, 746-760

Wu, J., Wohn, K. (1991) On the Deformation of Image Intensity and Zero-Crossing Contours under Motion. CVGIP: Image Understanding 53, 66-75

The Cracking Plate and Its Parallel Implementation

Marc Proesmans and André Oosterlinck

ESAT MI2, Katholieke Universiteit te Leuven

1   Introduction

This chapter introduces a scheme for the simultaneous suppression of noise and sharpening of discontinuities in visual data such as intensity or depth. Like other regularization type of approaches it can be described as based on a physical analogue: plates prone to crack initiation and propagation.

Traditional edge detection methods detect discontinuities at a very local scale. In general the local change in amplitude of the original data is estimated by the outcome of some gradient operator, and if somewhere a threshold is exceeded, the site u.c. can be considered to be a discontinuity. A sequential tracking algorithm can be used to link the different sites.

These methods are in fact very noise sensitive. Several researchers (e.g. Blake and Zisserman 1986; Terzopoulos 1983) have been investigating more global approaches such as plate or membrane models to reconstruct the noisy original data. In that case the solution is to be found by a regularizing functional. Regularization techniques have been proven to be quite suitable for solving ill-posed problems such as optical flow, stereo matching, shape from shading, surface reconstruction and others. For such problems the analytic formulation is not enough to find a reliable solution on the basis of the data alone. The solution might be not unique or unstable or may not even exist. Regularizing the solution, it can be rendered well-posed in the Hadamard sense (Tikhonov and Arsenin 1977). Therefore regularization forms a unifying and sound mathematical basis for the treatment of ill-posed problems. The underlying idea is to make additional assumptions, i.e. to impose some physical plausible constraints on the solution(s). Often one assumes the eventual solution to be smooth (the so-called "smoothness constraint"). One then obtains a variational problem which can be solved using e.g. finite element techniques. In section 2 we will discuss regularization more in detail.

Although adding smoothness constraints offer a suitable framework for the solution of ill-posed problems, a number of important problems remain unaccounted for. It was evident from the start that simple smoothing conditions do not yield satisfactory results, especially with jump discontinuities since these are blurred. Some refinements to the

original membrane and plate models have therefore been put forward e.g. weak continuity constraints (Blake and Zisserman 1986; Terzopoulos 1986). Others (Gamble and
Poggio 1987; Geman and Geman 1984) introduced line processes which are guided by intensity data. Intensity also played a crucial role in the discontinuity preserving algorithms
proposed by Cornelius and Kanade (Cornelius and Kanade 1983) and Nagel and Enckelmann (Nagel and Enckelmann 1986; Nagel 1987). One attractive feature of line processes
is that they include facilitation effects. Once a discontinuity is detected, it will more
easily be extended ("propagated") towards neighbouring regions, especially if the trajectory is smooth. A strong drawback, however, is the fact that the different discontinuity
configurations have to be listed explicitly, i.e. one has to explicitly specify the structure
and "cost" of different discontinuity parts such as corners, straight linear pieces, etc.

We introduce a scheme for discontinuity propagation that is akin to the weak continuity
approach. The idea is to fit the original data with a cracking plate, based on theories of
fracture mechanics. In other words, the plate model which regularizes the reconstruction
problem, will be provided with a mechanism by which the discontinuities propagate very
similar to cracks in solid materials. In the theory of elasticity, if a local discontinuity is
present as an initial tear or crack, it will influence the local stress field much stronger than
in current regularization models. In fracture mechanics, these processes are governed by
large stress concentrations nearby the crack tip. Of course, from our point of view these
are mere analogies. The crucial element is that crack growth shows a facilitation effect
in the longitudinal direction and an inhibitory effect in the transversal direction. In that
sense, the behaviour of a plate equipped with such a crack growth would come close to
e.g. Grossberg's filling-in and boundary processes (Grossberg and Mingolla 1985) or other
models for neurophysiological phenomena (Orban and Gulyas 1988). The implementation
and characteristics of this scheme will be discussed in section 3.

The system governing equations which result from the finite element discretization
are highly suitable for iterative implementations on parallel structures. Some important
neurophysiological issues fit into this framework, such as cooperative-competitive mechanisms and multi-resolution interactions. Section 4 will go deeper into the parallel nature
of our plate bending problem.

2   The Plate Bending Problem

2.1   Variational Formulation

The representation of visible surfaces has attracted considerable interest as an intermediate goal of computer vision, since Gibson made the conjecture that human visual
perception amounts to perception of visible surfaces and this research gained even more
momentum since the introduction of ideas about $2\frac{1}{2}D$ sketch by Marr. This section will
develop a particular computational approach to intermediate-level vision. The problem
itself is a non-trivial inverse problem. The visual field is only known in a limited number
of samples. As such the surface will not be determined uniquely everywhere, there remain infinitely many feasible surfaces. Furthermore, the estimates are subject to errors,
and high spatial frequency noise can locally perturb the surface radically. From these
considerations it is clear that our problem is ill-posed, since they do not guarantee that a
solution exists, or that it will be unique or stable. Regularization provides a systematic
approach to reformulating this ill-posed problem as a well-posed and effectively solvable
variational principle.

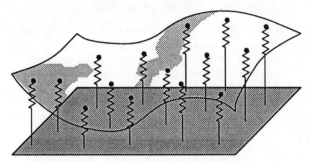

Figure 1: *Physical interpretation of the plate model.*

This method is also applicable on the visual problem of optical flow, as clearly explained by the work of Prof. H.H. Nagel in this book.

The regularized plate bending problem tries to find a surface $w = f(x, y)$ which approximates as good as possible a given set of points (i.e. the image data), while the surface itself should possess some degree of "smoothness". The variational formulation of the plate problem has frequently been used to reconstruct some function which is only known in a number of discrete points. Let K be a linear space of admissible functions. The regularized problem is formulated mathematically finding $w \in K$ according to the variational principle

$$\Pi(w) = \min_{v \in K} \Pi(v), \quad \Pi(v) = \mathbf{E}(v) + \mathbf{D}(v).$$

where the term $\mathbf{D}(v)$ must insure that the solution w does not deviate too much from the original measurements c_i (e.g. measurements such as intensity values, depth information...), and $\mathbf{E}(v)$ should guarantee some degree of continuity of the function. A physical model is depicted in figure 1. The visual data apply forces in the z-direction which deflect the surface from its nominally planar state. The penalty function $\mathbf{D}(v)$ is the total deformation energy of a set of ideal springs attached to the data.

Experiments carried out by Grimson (Grimson 1981) seem to indicate that second order plate surfaces (C^2) exhibit an appropriate degree of smoothness. From a mathematical point of view, the energy density of a C^2-surface depends upon the principal curvatures of the surface (e.g. Blake and Zisserman 1986; Terzopoulos 1986).

2.2 The Energy Density E of the Plate

The potential energy of a plate under deformation is given by an integral of a quadratic form in the principal curvatures of the surface κ_1, κ_2 (Courant and Hilbert 1953). The potential energy density is given by an expression of the form

$$\mathbf{e} = \frac{A}{2}(\kappa_1^2 + \kappa_2^2) + B\kappa_1\kappa_2$$

Usually one defines two additional quantities , the *mean curvature* $\mathbf{H}$ and the *Gaussian curvature* $\mathbf{K}$.

$$\mathbf{H}(p) = \frac{1}{2}(\kappa_1 + \kappa_2) \quad \mathbf{K}(p) = \kappa_1\kappa_2$$

The potential energy can easily be rewritten as a function of these quantities

$$\mathbf{e} = \frac{A}{2}(\kappa_1 + \kappa_2)^2 + (B - A)\kappa_1\kappa_2$$

$$\mathbf{e} = 2A\mathbf{H}^2 + (B - A)\mathbf{K} \tag{1}$$

What we need is the expression of the energy in terms of the unknown solution w. However, the general substitution of $\mathbf{H}$ and $\mathbf{K}$ as a function of w, introduces second order derivatives which complicate the discretization of the governing equations and the incorporation of the crack growth mechanism. Therefore the order of the functional should be reduced. This can be achieved by introducing two additional functions β_x and β_y which are the tangents on the surface in the x and y direction. An additional advantage is that we can achieve a greater viewpoint independence. The mean and gaussian curvature can then be given by

$$\mathbf{H} = \frac{1}{2}(\frac{\partial \beta_x}{\partial x} + \frac{\partial \beta_y}{\partial y})$$

$$\mathbf{K} = \frac{\partial \beta_x}{\partial x}\frac{\partial \beta_y}{\partial y} - \left(\frac{1}{2}(\frac{\partial \beta_x}{\partial y} + \frac{\partial \beta_y}{\partial x})\right)^2$$

If we divide equation (1) by $\frac{A}{2}$ and define $\nu = \frac{B}{A}$, then the potential energy takes the form

$$\mathbf{e} = (\frac{\partial \beta_x}{\partial x})^2 + (\frac{\partial \beta_y}{\partial y})^2 + 2\nu\frac{\partial \beta_x}{\partial x}\frac{\partial \beta_y}{\partial y} + \frac{1-\nu}{2}(\frac{\partial \beta_x}{\partial y} + \frac{\partial \beta_y}{\partial x})^2 \tag{2}$$

and ν turns to be the well-known Poisson coefficient in elasticity theory. For real plates, this quantity accounts for the fact that, even if the stresses act only in x and y direction, there might be some constriction along the z-axis. More important for us is to know that the latter substitution of course assumes $A \neq 0$. It means that the square of the principal curvatures themselves are always present in the energy density equation. In fact the Gaussian curvature is inappropriate as a measure of surface consistency, since a difference in bending forces does not necessarily result in a difference in Gaussian curvature. In other words it cannot distinguish sinusoidal and planar surfaces since both have zero Gaussian curvature. On the other hand the mean curvature does not remain invariant.

2.3   The Plate's Functional

At this stage we did not yet account for the external forces which are responsible for the bending of the plate. We consider our plate to be connected to the original data by a number of springs. Let $c(i)$ be the value of the i-th sample of the original data and $w(i)$ the corresponding value of the plate to be found. The energy involved in stretching those springs can be written as $D = \sum_{i=1}^{N} \frac{\xi_i}{2}(w(i) - c_i)^2$ in which ξ_i denotes the stiffness of a spring with linear force-displacement characteristic at location i , $F = \xi_i(w(i) - c_i)$ (Hooke's law). The surface will then reach equilibrium as the energy of the whole system is minimum. So we have to minimize a functional of the form

$$\int_\Omega \mathbf{e}(x,y)dxdy + \sum_{i=1}^{N} \frac{\xi_i}{2}(w(i) - c_i)^2$$

The three unknown functions, w, β_x, β_y, however, are not independent. In fact the functional should be subjected to some constraints $\beta_x = \frac{\partial w}{\partial x}$, $\beta_y = \frac{\partial w}{\partial y}$ which can be introduced in our functional by the Lagrange multiplier method

$$\int_\Omega \mathbf{e}(x,y) + k\mathbf{c}(x,y)$$

where $k\mathbf{c}$ ensures that the constraints are nearly satisfied. The term $k\mathbf{c}$ is the *penalty function*. Thus the overall surface response to the given loads can be found by minimizing the functional

$$\int_{\Omega} \left(\frac{\partial \beta_x}{\partial x}\right)^2 + \left(\frac{\partial \beta_y}{\partial y}\right)^2 + 2\nu \frac{\partial \beta_x}{\partial x} \frac{\partial \beta_y}{\partial y} + \frac{1-\nu}{2} \left(\frac{\partial \beta_x}{\partial y} + \frac{\partial \beta_y}{\partial x}\right)^2$$

$$+ k \left((\beta_x - \frac{\partial w}{\partial x})^2 + (\beta_y - \frac{\partial w}{\partial y})^2 \right) \, dxdy + \sum_{i=1}^{N} \frac{\xi_i}{2}(w(i) - c_i)^2 \tag{3}$$

2.4   Finite Element Discretization

In the finite element method representation a function is approximated as a linear combination of local support basis functions. In visual applications, a natural tessellation would follow the image sampling pattern, i.e. the domain is tessellate into square subdomains with sides of length h. Within the subdomain or "*element*", the function values depend upon the node values (pixel values) by an interpolation function.

$$w = \sum_{i=1}^{N} h_i w_i \qquad \beta_x = \sum_{i=1}^{N} h_i \beta_{x,i} \qquad \beta_y = \sum_{i=1}^{N} h_i \beta_{y,i}$$

In order for the finite element approach to converge, the element interpolation functions should be appropriately chosen. In finite element terms, they have to be complete and compatible (Reddy 1986). For our plate problem we can restrict ourselves to a very simple bilinear interpolation function:

$$w^e(x,y) = \mathbf{a}xy + \mathbf{b}x + \mathbf{c}y + \mathbf{d}$$

(w^e denotes the function values w within an element e). Using these approximations for w, β_x, β_y, we can write the functional as

$$\Pi = \sum_{e=1}^{n_e} \int_{\Omega_e} \left(\frac{\partial \beta_x^e}{\partial x}\right)^2 + \left(\frac{\partial \beta_y^e}{\partial y}\right)^2 + \frac{1}{2}\left(\frac{\partial \beta_x^e}{\partial y} + \frac{\partial \beta_y^e}{\partial x}\right)^2 + k \left((\beta_x^e - \frac{\partial w^e}{\partial x})^2 + (\beta_y^e - \frac{\partial w^e}{\partial y})^2 \right) \, dxdy$$

$$+ \sum_{i=1}^{N} \frac{\xi_i}{2}(w(i) - c_i)^2$$

in which each of the terms can be evaluated separately. n_e is the number of elements: $n_e = (n-1)^2$, n being the size of the image. This discrete functional can be minimized by setting the partial derivatives with respect to each of the node values $w_{i,j}, \beta_{x,i,j}, \beta_{y,i,j}$, to zero.

$$\frac{\partial \Pi}{\partial w_{i,j}} = 0$$

$$\frac{\partial \Pi}{\partial \beta_{x,i,j}} = 0 \quad i,j = 1..N$$

$$\frac{\partial \Pi}{\partial \beta_{y,i,j}} = 0$$

We omit the details of the calculation of these derivatives. The final result is a linear system of $3n^2$ equations given below by means of computational molecules (Terzopoulos 1984). This $3n^2$ system of equations can be interpreted as a 3×3 system in which the unknowns are the *images* w, β_x, β_y, and the coefficients convolution masks.

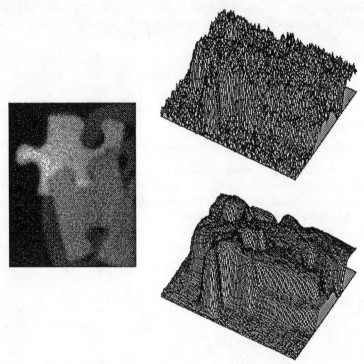

Figure 2: *Left: original noisy intensity image. Upper right: corresponding intensity surface. Lower right: solution of the plate bending problem.*

$$\left(\begin{bmatrix} -1 & 0 & -1 \\ -2 & 8 & -2 \\ -1 & 0 & -1 \end{bmatrix} + \frac{1}{9}kh^2 \begin{bmatrix} 1 & 4 & 1 \\ 4 & 16 & 4 \\ 1 & 4 & 1 \end{bmatrix} \right) * \beta_x + \frac{1}{2} \begin{bmatrix} 1 & 0 & -1 \\ 0 & 0 & 0 \\ -1 & 0 & 1 \end{bmatrix} * \beta_y - \frac{1}{3}kh \begin{bmatrix} -1 & 0 & 1 \\ -4 & 0 & 4 \\ -1 & 0 & 1 \end{bmatrix} * w = 0$$

$$\frac{1}{2} \begin{bmatrix} 1 & 0 & -1 \\ 0 & 0 & 0 \\ -1 & 0 & 1 \end{bmatrix} * \beta_x + \left(\begin{bmatrix} -1 & -2 & -1 \\ 0 & 8 & 0 \\ -1 & -2 & -1 \end{bmatrix} + \frac{1}{9}kh^2 \begin{bmatrix} 1 & 4 & 1 \\ 4 & 16 & 4 \\ 1 & 4 & 1 \end{bmatrix} \right) * \beta_y - \frac{1}{3}kh \begin{bmatrix} 1 & 4 & 1 \\ 0 & 0 & 0 \\ -1 & -4 & -1 \end{bmatrix} * w = 0$$

$$kh \begin{bmatrix} -1 & 0 & 1 \\ -4 & 0 & 4 \\ -1 & 0 & 1 \end{bmatrix} * \beta_x + kh \begin{bmatrix} 1 & 4 & 1 \\ 0 & 0 & 0 \\ -1 & -4 & -1 \end{bmatrix} * \beta_y + \left(4k \begin{bmatrix} -1 & -1 & -1 \\ -1 & 8 & -1 \\ -1 & -1 & -1 \end{bmatrix} + 6\xi \right) * w = 6\xi c$$

As an example

$$\begin{bmatrix} -1 & 0 & -1 \\ -2 & 8 & -2 \\ -1 & 0 & -1 \end{bmatrix} * \beta_x \Bigg|_{i,j} = \begin{array}{l} -1.\beta_{x,i-1,j-1} + 0.\beta_{x,i-1,j} - 1.\beta_{x,i-1,j+1} \\ -2.\beta_{x,i,j-1} + 8.\beta_{x,i,j} - 2.\beta_{x,i,j+1} \\ -1.\beta_{x,i+1,j-1} + 0.\beta_{x,i+1,j} - 1.\beta_{x,i+1,j+1} \end{array}$$

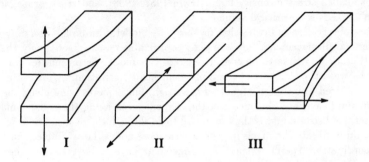

Figure 3: *3 modes of crack growth for a plate in vertical direction: I. opening mode, II. sliding mode, III. tearing mode.*

This system of equations features computationally desirable properties. Its matrix is positive definite, sparse, banded and symmetric, due to the local support of the finite element representation. The size, on the other hand, may become extremely large, since the number of pixels in typical images can range from 10^4 to 10^5 or more. This combination of properties suggests the application of iterative techniques such as relaxation methods. We used a coupled Gauss-Seidel scheme in which for each point (i, j) three equations are solved simultaneously, and the results are immediately exploited in subsequent iterations. Figure 2 shows the finite element representation of the plate for the intensity image on the left. The figure clearly shows that the noise is dramatically reduced, at the expense of edges that are blurred.

3   The Cracking Plate

As mentioned, the straightforward application of smoothness conditions doesn't allow for satisfactory result in the neighbourhood of discontinuities. Our model uses a scheme of facilitated discontinuity growth, integrated in the solution of a plate bending problem.

3.1   Cracks

The presence of cracks in solids has been the subject of intensive research which has led to a new branch in mechanics, namely fracture mechanics (Broek 1986; Sneddon and Lowengrub 1969; Lawn and Willshaw 1975)

A first observation is that any stress situation can be described by three independent modes of crack growth: the opening, tearing and sliding mode (figure 3). Secondly, for each mode it has been found that in the neighbourhood of the crack tip, there is a very large stress concentration. Applied mathematics shows us that there exists a singularity[1] in the stress function

$$\sigma_{ij} = \frac{K}{\sqrt{2\pi r}} f_{ij}(\theta) \tag{4}$$

---

[1] An infinite stress is of course not very realistic. In a real world situation, one encounters a plastic zone with radius r_p^* which is related to the maximum stress for the solid.

where $f_{ij}(\theta)$ is a function of the geometry of the solid in the neighbourhood of the crack, K is the so-called stress intensity factor(Irwin 1957, 1958), and crack propagation will occur if K exceeds some critical value k_c.

The first observation indicates that for our plate model, assuming it can only undergo (small) vertical displacements (i.e. the pixel coordinates remain fixed), only the tearing mode (III) crack growth is to be considered. For this mode, discontinuities can be positioned in between the elements of the finite element discretization. A discontinuity may thus be considered to be a concatenation of straight linear pieces lying along the sides of the surrounding elements. In our case, the element mesh being the square image sampling pattern, a discontinuity is thus supposed to go through the pixels. Consequently the model should duplicate the pixels in order to use more nodes (i.e. variables) for the same spatial coordinates. Therefore special arrangements have to be taken for the eventual equations, since the element contributions have now been split apart. Our plate model is capable of handling this kind of discontinuities since the equations only depend on the element contributions of the node under consideration.

For instance, the first computational molecule we met in the equations described above, can be written as a summation of the contributions of the 4 surrounding elements (the center pixel is depicted in bold).

$$
\begin{bmatrix} -1 & 0 & -1 \\ -2 & \mathbf{8} & -2 \\ -1 & 0 & -1 \end{bmatrix} = \begin{bmatrix} -1 & 0 \\ -1 & \mathbf{2} \\ -1 & 0 \end{bmatrix} + \begin{bmatrix} -1 & 0 \\ -1 & \mathbf{2} \\ -1 & 0 \end{bmatrix} + \begin{bmatrix} 0 & -1 \\ \mathbf{2} & -1 \\ 0 & -1 \end{bmatrix} + \begin{bmatrix} \mathbf{2} & -1 \\ 0 & -1 \end{bmatrix}
$$

We only need an additional data structure to keep up with the relation between the elements and the corresponding nodes, a kind of connectivity matrix (Proesmans 1989).

The second observation depicts how applied mechanics offers us a way to describe the stress concentrations in the vicinity of the crack. It is assumed that crack extension will occur when the stress intensity factor, and consequently the stresses at the tip, exceed some critical value. It turns out that the whole process depends upon the geometry of the crack as well as upon the external applied loads (Irwin 1957, 1958; Sneddon and Lowengrub 1969).

The stress intensity factor is closely related to the energy release rate originally introduced by Griffith (Griffith 1921). This criterion states that a crack can only exist when the surface energy necessary for crack extension can be delivered by the work of the external applied forces. This criterion is the starting point for the further development of our scheme.

3.2   Propagation

Thus far the crack system is considered to be static. However, if an unbalanced force acts on any volume element within the cracked body, the element will be accelerated, and thereby acquire kinetic energy. The system is then dynamic and the normal equilibrium conditions no longer apply. However, for practical reasons, the general crack-path problem can readily be treated in terms of the above thermodynamical fracture criteria (Griffith). By the way, these criteria do hold in case crack extension proceeds sufficiently slow.

If we consider an incremental extension δc of some crack, it may logically - not only on a physical, but also on a mathematical basis - be proposed that the favoured direction θ will be the one which maximizes the decrease in total energy. In other words, crack growth will take place in the direction for which the energy release rate is maximum. If we define the energy release rate as

$$R = \frac{\partial E}{\partial c}$$

then this criterion of maximum release rate would be

$$\frac{\partial R}{\partial \theta} = 0 \qquad \frac{\partial^2 R}{\partial \theta^2} < 0$$

R has to be evaluated as a function of the crack growth angle. First one calculates the energy stored in the solid for some crack length c, then one considers a crack extension δc in some direction θ and calculates the energy again. The latter will differ with an amount δE from the first. Since only a finite crack propagation can be evaluated, the energy release rate should be approximated by

$$R = \frac{\delta E}{\delta c}$$

This can be realized for different angles θ. The direction in which R is maximum is supposed to belong to the crack path. The question that arises is what forces are responsible for the crack propagation in our plate model, or what is the relationship between the deformation of the plate and the energy release during propagation. The criterion of maximum energy release rate has been compared to the principle of maximum normal stress. The latter states that the plate will fail in the direction of maximum transversal stress, which is perpendicular to the orientation of the principal stresses. In our plate model, the forces responsible for mode III crack propagation are mainly the transversal stresses. Theory of elasticity learns us that these are given by the penalty term, for the $x-$ and $y-$direction:

$$\tau_x = \frac{\partial w}{\partial x} - \beta_x$$

$$\tau_y = \frac{\partial w}{\partial y} - \beta_y$$

So, in terms of energy, we would have to evaluate (the details will be omitted)

$$E_1 = \beta_x^{1001^2} + \frac{3-\nu}{6}\beta_x^{1010^2} + \beta_y^{1100^2} + \frac{3-\nu}{6}\beta_y^{1010^2}$$

$$+2\nu\beta_x^{1001}\beta_y^{1100} + \frac{1-\nu}{2}(\beta_x^{1100} + \beta_y^{1001})^2$$

$$E_2 = (\frac{h}{2}\beta_x^{1111} - w^{1001})^2 + (\frac{h}{2}\beta_y^{1111} - w^{1001})^2 + (\frac{h^2}{12})^2(\beta_x^{1001^2} + \beta_y^{1100^2})$$

$$+(\frac{h}{2}\beta_x^{1100} + w^{1010})^2 + (\frac{h}{2}\beta_y^{1001} - w^{1010})^2 + (\frac{h^2}{36})^2(\beta_x^{1010^2} + \beta_y^{1010^2})$$

is minimal. ($f^{xxxx} = +/ - f_{i+1,j+1} + / - f_{i,j+1} + / - f_{i,j} + / - f_{i+1,j}$, the operator $+/-$ is determined by $x = 1/0$).

Experiments show that the energy distribution of each of the terms is in general concentrated in the neighbourhood of large stress regions. Using these terms, the penalty term $(\beta_x - \frac{\partial w}{\partial x})^2 + (\beta_y - \frac{\partial w}{\partial y})^2$ shows the appropriate behaviour for discontinuity detection which corresponds to the characteristics of the transversal forces in elasticity theory.

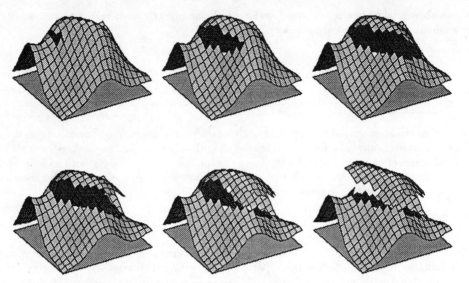

Figure 4: *Subsequent step of the crack propagation process.*

Furthermore, the energy distribution in the neighbourhood of the crack increases hyperbolically towards the crack tip in correspondence with fracture mechanics. Figure 4 shows some intermediate steps of a propagation process based on these energy functions.

The solution method is analogous to the original algorithm in which we use an iterative coupled Gauss-Seidel scheme which covers the whole domain Ω During the discontinuity propagation on assumes that, if the discontinuity has grown over a small distance, the spatial influence of this growth will be limited compared with the domain Ω. The pixels will be updated starting from the crack tip on outwards along some closed contour, by systematically increasing its radius or size. One can easily verify that, if convergence is reached by these iteration steps, it will also be reached by an ordinary iteration step over the whole domain Ω.

Because of the rather time consuming nature of the algorithm, the model has been implemented on a transputer network. This parallel implementation helped a great deal in understanding the influence of the different energy functions and discontinuities on the behaviour of the model.

4   Parallel Implementation

Computer vision tasks require an enormous amount of computation, especially when the data is in image form, demanding high performance computers for practical real-time applications. In general, however, parallelism appears to be the only economical way to achieve the performance required for vision tasks.

There have been several methods of classifying computer architectures. The most common classification, based on control (Flynn) considers the presence or absence of potential multiplicity in the instruction and data stream of the computer. Given that, four classes of computers result, SISD (single instruction single data), MISD (multiple instruction, single data), SIMD (single instruction multiple data) and MIMD (multiple instruction multiple data).

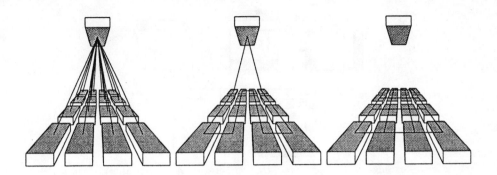

Figure 5: *Schematic representation of Master-Slave, Ring and Neighbour communication channels. On top: host computer. Below: transputer platform.*

A SIMD computer is implemented as a set of identical synchronized processing elements capable of simultaneously performing the same operation on different data. Although the processor elements execute in parallel, processor elements may be programmed to ignore any particular instruction. SIMD computers are also called processor arrays. MIMD computers consist of a number of fully programmable processors each capable of executing its own program.

The applicability of an architecture to image processing problems depends upon their nature, ranging from low level image processing to more complex image analysis. Historically, a fundamental bottleneck in processing capability has been perceived with the low level image processing task. Many SIMD and pipeline architectures have been developed for low level image processing applications, whereas special architectures for high level image analysis have received less attention. Among the lowlevel architectures, algorithms have been suggested for binary array processors (Reeves 1980), pipeline processors (Sternberg 1979), special function processors such as systolic arrays (Kung and Song 1982, Yen and Kulkarni 1981), or pyramid arrays (Dyer 1982). Lin and Kumar (Lin and Kumar 1991) for example, present $2D$ parallel algorithm for discrete relaxation techniques. Derin and Won (Derin and Won 1987) introduce a segmentation algorithm based on deterministic relaxation with varying neighbourhood structures. Carver Mead (Carver Mead 1989) investigated the possibility of implementing relaxation type algorithms directly in VLSI systems.

In general, low-level vision tasks require computations corresponding to each pixel of the image. This is indicative to the amount of computation required. Fortunately, these computations are usually highly regular in that sense that the same computations are performed for all portions of the image. These low level tasks comprise for example smoothing operations, convolutions, histogram generation, hough transforms, clustering algorithms etc. The key issue in designing parallel algorithms is thus to distribute the execution of the various parts of the algorithm over a number of communicating processors. We have chosen for a transputer network since these processors can be linked together to give high performance systems with arbitrary topology and thus can be considered as "building blocks". In order to be efficient, the parallel network should support the realization of an arbitrary communication network. For the plate model, the communication channels can be subdivided into 3 basic configurations.

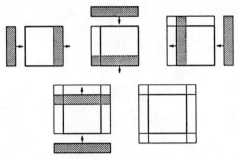

Figure 6: *Subsequent steps in data transfer between transputers at each iteration.*

- The *Master-Slave* configuration supports the communication between the main processor (the host computer) and the transputers. It is used for the division of the original image from the main processor to the available transputers. Each transputer will receive a subimage for which it calculates the (partial) solution.

- The *Ring* structure is a pipeline through the transputer network with the advantage that it represents the shortest communication channel passing each of the transputers in their user-defined configuration. The ring is therefore very suitable for passing commands or global status information between the main processor and the underlying transputers, without loosing too much communication time.

- The *Neighbour* channels support the communication between the transputers without interference of the main processor. This way the transputers can interchange information about the plate's behaviour or the existence of discontinuities at their boundaries. The timing constraints of these channels will have a large influence on the global performance of the transputer implementation, since at each iteration each transputer will gather the necessary information from the updated pixels in the neighbouring regions. It is therefore of crucial importance that this communication would be realized as efficient as possible.

The network has been proven successful not only for the plate model but also for similar relaxation schemes. Note that due to the intrinsic sequential nature of the algorithm, i.e. iterations have to be carried out one after the other, parallelization only concerned the space domain.

The implementation on the parallel network, however, is not straightforward. First, the Gauss-Seidel scheme as such is not suited for parallel implementation since new values are immediately exploited in the subsequent computations. If such a scheme would be used by each transputer separately, the global system would not be equivalent to the one processor system and its performance would approximate that of the Jacobi relaxation scheme which has proven to be not as effective as the Gauss-Seidel scheme. However, by an appropriate choice of the order in which the nodes are updated according to the plates equations, the relaxation scheme can be made independent of the number of transputers and/or their configuration. An obvious approach is to divide the collection of nodes in a number of groups, each of which are treated separately. We then have a "coloured" relaxation scheme if we associate a colour to each of the groups [2]. The number of colours depends upon the topology of the nodes neighbouring points in order to carry out an iter-

---

[2]E.g. the parallel implementation of the classical Poisson equation would require a red-black scheme.

Figure 7: *The segmentation of a toothed wheel. Upper left : original image, upper right : plate model, below : discontinuity line ($\nu = 0$, $k = 0.1$, $h = 1$, $\xi = 1.0$).*

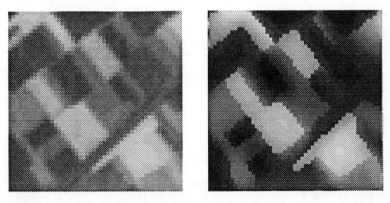

Figure 8: *landscape image of a SPOT-satellite : original (left) and the cracked plate (right) ($\nu = 0$, $k = 0.1$, $h = 1$, $\xi = 1.0$).*

ation. For our plate model the necessary points for a node lie in its 3×3 neighbourhood. One can easily verify that in this case a four-coloured scheme would be appropriate. Each transputer will be assigned a code that represents the colour in which it has to operate. Second, special care has to be taken for the communication between neighbouring transputers, in order to allow for the necessary interchange of information at each iteration step. The data flow has been organized into four subsequent synchronous bidirectional data transfers in each direction to gain maximum efficiency (figure 6).

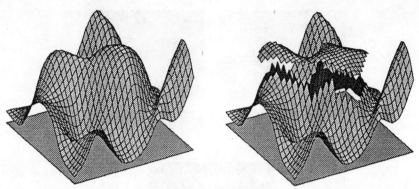

Figure 9: *Two intermediate steps in the propagation process on the SPOT-image : the blob corresponds to the most central white patch in the original greyvalue image.*

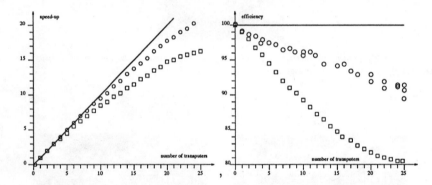

Figure 10: *Speed-up and efficiency : the squares denote a worst case simulation, the circles correspond to realistic timing measurements.*

5 Results

Figure 7 and 8 show some results of the reconstruction problem for. The plate model shown in these figures features a number of interesting properties compared with other plate models in the literature. By the introduction of the angular functions β_x and β_y, we achieved that the finite element representation of the plate could be realized with smaller support interpolation functions, which are moreover complete and compatible. This simplified the introduction of the integration of discontinuities a great deal. Furthermore, we gain more viewpoint independence and the angular functions allow direct incorporation of surface normals. The plate's functional shows the nice characteristic that it explicitly shows a shear deformation term which is responsible for the mode III propagation, which enabled us to implement the scheme for facilitated discontinuity propagation.

The cracking plate approach has also been investigated within the framework of texture segmentation related to the work of Prof. D.H. Foster. In this case the visual texture elements can be characterized by a curvature measure, which can be used directly as external forces to a similar plate bending problem.

The quality of a parallel algorithm is determined by a number of quantities, the most important being the speed-up (figure 10). The *speed-up* is defined as the running time of the sequential algorithm executed on one processor, divided by the running time of the parallel algorithm executed on a number of processors. The *processor utilization* or *efficiency* is defined as the speed-up divided by the number of processors used to execute the algorithm. Clearly, the best one can do with a parallel algorithm is to attain a speed-up equal to the number of processors and an efficiency equal to one. If the speed-up increases linearly with increasing number of processors, we say that the speed-up is linear. A linear speed-up implies that there is no saturation with an increasing number of processors. For our plate model, the speed-up can be approximated by the formula

$$S \sim \frac{1}{(\frac{1}{\sqrt{x}} + 2f\frac{d}{n})^2}$$

with n the size of the image, d the width of the boundary information, and x the number of transputers. f reflects the importance of the communication between the transputers towards the calculation time. It is clear that saturation does show up with increasing number of transputers, since the time necessary for communication becomes relatively larger compared with the computations involved in solving the equations. Analogously, the efficiency can be approximated by

$$\eta = \frac{1}{(1 + 2f\frac{d}{n}\sqrt{x})^2}$$

Clearly, if no communication were necessary ($d = 0$), the efficiency would reach 100%.

References

Bathe, K.J. (1982) Finite element procedures in engineering analysis. Prentice Hall

Blake, A., Zisserman, A. (1986) Invariant surface reconstruction using weak continuity constraints. IEEE conf. on Computer vision and Pattern recognition

Broek, D. (1986) Elementary engineering fracture mechanics. 4th revised edition, Martinus Nijhoff Publishers

Carver Mead (1989) Analog VLSI and Neural systems. Addsion-Wesley publishing Co.

Cornelius, N., Kanade, T. (1983) Adapting optical flow to measure object motion in reflectance and X-ray sequences. Proc. ACM Siggraph/Sigart Interdisciplinary workshop on motion: representation and perception, pp. 50-58

Courant, R., Hilbert, D. (1953) Methods of mathematical physics. vol. 1, Interscience, N.Y.

Derin, H., Won, C. (1987) A parallel image segmentation algorithm using relaxation with varying neighbourhoods and its mapping to array processors. Comp. Vision Graphics and Image Proc., Vol 40

Dyer, C.R. (1982) A VLSI Pyramid Machine for hierarchical parallel image processing. Proceedings of the IEEE Conf. on Patt. Recog. and Image Proc., pp. 381-386

Gamble, E., Poggio, T. (1987) Visual integration and detection of discontinuities: the key role of intensity edges. MIT, A.I. Memo 970, oct. 1987

Geman, S., Geman, D. (1984) Stochastic relaxation, Gibbs distributions and the Bayesian restoration of images. IEEE Trans. on pattern analysis and machine intelligence, Vol. PAMI-6, No.6, Nov. 1984

Griffith, A.A. (1921) The phenomena of rupture and flow in solids. Phil. Trans. Roy. Soc. London A/221, pp. 163-197

Griffith, A.A. (1925) The theory of rupture. Proc. 1st Int. Congress Applied Mechanics, pp. 55-63, Biezeno, Burgers Ed. Waltman

Grimson, W.E.L. (1981) From images to surfaces: a computational study of the human early visual system. MIT press, Cambridge, MA

Grossberg, S., Mingolla, E. (1985) Neural dynamics of form perception: boundary completion, illusory figures, and neon color spreading. Psychological Review 92, pp. 173-211

Irwin, G.R. (1957) Analysis of stresses and strains near the end of a crack traversing a plate. Trans. ASME J. Appl. Mech.

Irwin, G.R. (1958) Fracture. In: Handbuch der Physik, Vol.VI Springer, Berlin

Kung, H.T., Song, S.W. (1982) A systolic 2-D convolution chip. In Multicomputers for Image Processing: Algorithms and Programs, Academic Press, N.Y.

Lin, W.M., Kumar, V.K. (1991) Parallel algorithms and architectures for discrete relaxation techniques. Proc. Comp. Vision and Pattern Recognition, pp. 514

Lipschutz, M.M. (1969) Theory and problems of differential geometry. McGraw-Hill

Lawn, B.R., Wilshaw, T.R. (1975) Fracture of brittle solids. Cambridge Solid states Science Series, Cambridge University Press

Nagel, H.-H., Enckelmann, W. (1986) An investigation of smoothness constraints for the estimation of displacement vector fields from image sequences. IEEE Trans. pattern analysis machine intelligence, Vol. PAMI-8, No.5, pp. 565-593, sept. 1986

Nagel, H.-H. (1987) On the estimation of optical flow: relations between different approaches and some new results. Artificial intelligence, pp. 299-324

Orban, G., Gulyas, B. (1988) Image segragation by motion: cortical mechanisms and implementation in neural networks. In: Eckmiller and von der Malsburg (eds.) Neural Computers, NATO ASI series, Vol. F41, Springer Verlag, Berlin

Proesmans, M. (1989) Calculation and use of intrinsic characteristics. Internal report, ESAT-MI2

Reddy, J.N. (1986) Applied functional analysis and variational methods in engineering.

Reeves, A.P. (1980) A systematically designed binary array processor. IEEE Trans. Comput. C-29, pp. 278-287

Sternberg, R. (1979) Parallel architectures for Image processing. Proceedings on the 3rd International IEEE COMPSAC, Chicago, pp. 712-717

Sneddon, I.N., Lowengrub, M. (1969) Crack problems in classical theory of elasticity. Siam series of applied mathematics

Terzopoulos, D. (1983) The Role of Constraints and Discontinuities in Visible Surface Reconstruction. Proceedings of the 8th International Joint Conference A.I., Karlsruhe, Germany, pp. 1073-1077

Terzopoulos, D. (1984) Multilevel reconstruction of visual surfaces: variational principles and finite element representations, M.I.T. In: A. Rosenfeld (ed.) Multiresolution Image Proc. and Anal., Springer Verlag, N.Y., pp. 237-310

Terzopoulos, D. (1986) Image analysis using multigrid relaxation methods. IEEE Trans. on Pattern Analysis and Machine Intelligence, Vol. PAMI-8, No.2, pp. 129-139, march 1986

Yen, D.W.L., Kulkarni, A.V. (1981) The ESL systolic processor for signal and image processing. IEEE Comp. Soc. Workshop on Comp. Arch. for Patt. Anal. and Image Database Management, Hot Springs, pp. 273-277

Tikhonov, A., Arsenin, V. (1977) Solutions of ill-posed problems. Winston, Washington DC

The Perception and Representation of Depth and Slant in Stereoscopic Surfaces

Brian J. Rogers

Department of Experimental Psychology, University of Oxford

1 Surface Slant and Deformation

1.1 Introduction

The accurate perception of the shapes of three-dimensional surfaces and their layout is an essential capacity of any visual system, biological or man-made, for it to function effectively in a natural human environment. Two of the most important and precise sources of information which provide this ability in human vision are (i) binocular disparities and (ii) the optic flow produced when the observer moves with respect to the environment. It has been shown empirically that human observers are able to exploit both sources of information effectively even in the situation where each is presented in isolation (e.g. Julesz 1960; Rogers and Graham 1979). In addition, it has been shown that the way we perceive 3-D structure and layout from these two cues is very similar (Rogers and Graham 1982, 83). For example, these authors have shown that our sensitivity for detecting the 3-D structure of sinusoidal corrugations specified by binocular disparities or motion parallax is very comparable. In both cases, thresholds for detecting the 3-D structure are lowest when the corrugation frequency is around 0.3-0.5 cycles/deg and the shapes of the sensitivity functions show similar bandpass characteristics. Further evidence of the similarity between the two systems comes from Graham and Rogers (1982a) who found comparable simultaneous contrast effects (spatially-induced modifications of perceived depth) and successive contrast effects (after-effects resulting from prolonged viewing of 3-D surfaces) for disparity- and parallax-specified surfaces. Moreover they reported that the 3-D aftereffects produced by prolonged viewing of surfaces specified by one source of information could be directly nulled or cancelled with depth from the other source (Graham and Rogers 1982b; Rogers and Graham 1984). Finally, Rogers and Graham (1983) showed that the appearance of 3-D surfaces specified by these two cues could be markedly anisotropic – i.e. the appearance and the perceived depth can depend on the orientation of the surface. Low spatial frequency sinusoidal corrugations appeared to have more depth and had lower absolute thresholds when the spatial frequency corrugations were oriented horizontally compared to when the same corrugations were oriented vertically.

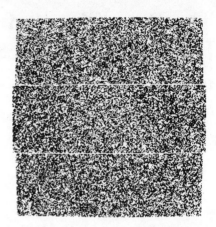

Figure 1: *A simultaneous contrast effect in the disparity domain. The two halves of the random dot stereogram may be fused by crossing the eyes. The flanking regions at the top and bottom of the stereogram contain an identical disparity gradient between the images and each should be seen as slanting from left to right, out of the plane of the page. The central region has no disparity gradient and should be seen as lying in the plane of the page, but simultaneous contrast results in the region being seen as slanting in the opposite direction to that of the flanking regions (see also Graham and Rogers 1982).*

The same authors also showed that the depth analogue of the Craik-O'Brien-Cornsweet illusion, first reported by Anstis *et al.* (1978) for disparity-specified surfaces, was also seen with surfaces specified by monocular motion parallax. In both cases, the illusion was of greater magnitude when the discontinuity of the depth profile was oriented vertically rather than horizontally.

These results lend support to the idea that the mechanisms used to extract information about the shapes of 3-D surfaces from binocular disparities and optic flow share much in common. Within the machine vision community, it is generally accepted that the initial stage of representing 3-D surfaces from disparity information should be to construct a depth or "range map" which specifies the disparity values of each different direction in space (e.g. Marr and Poggio 1976, 1979; Mayhew and Frisby 1980). Indeed, some of the binocular cells in the cat and monkey visual cortex, first described by Bishop *et al.* (1971) and Barlow, Blakemore and Pettigrew (1967), which responded best when a spot or bar of light fell on slightly different (disparate) positions of the two retinae, could be candidate mechanisms for computing the horizontal disparity of a point or line in 3-D space. However, the fact that the human perception of 3-D surfaces can often be significantly non-veridical, as demonstrated in the simultaneous contrast effects reported by Graham and Rogers (1982a) and Brookes and Stevens (1989), is necessarily inconsistent with the idea of a simple "range map", unless we assume that there are subsequent processes (perhaps involving lateral inhibition) which act to modify the initially derived depth values (figure 1). The first part of this chapter will be concerned with idea that the human visual system might use the amount of **deformation** (over time, in the case of optic flow transformations and between the eyes, in the case of binocular vision) in order to extract information about the local slant of 3-D surfaces.

1.2 Motion and Disparity Gradients

In the optic flow domain, no one has ever suggested that the motion analogue of a point disparity – the speed of motion of a single point over the retina – could ever provide information about 3-D structure. Instead, Gibson (1950) and others have stressed the potential importance of the spatial gradients of retinal motion that would be generated, for example, when an observer translates with respect to a ground plane surface. The significance of gradients in the disparity domain was first brought out by Tyler (1973, 1974) who showed that the upper threshold for perceiving disparity modulations was subject to a disparity gradient limitation. Burt and Julesz (1980) subsequently showed that the fusion of individual pairs of disparate points was also constrained by a disparity gradient limit of 1 for surfaces which slanted around either a horizontal or a vertical axis. More recently, Pollard *et al.* (1985) successfully incorporated a disparity gradient limit into their stereo algorithm in order to limit the number of incorrect binocular matches. These studies provide good evidence that disparity gradients can play a role in limiting possible binocular pairings, but it is also possible that disparity gradients may play an important role as primitives for representing 3-D surfaces (Marr 1982). In the optic flow domain, Koenderink and van Doorn (1975, 1976a) and Longuet-Higgins and Prazdny (1980) have shown that information about local surface slant (up to a scaling factor) may be obtained by computing the amount of deformation needed to map successive images in an optic flow sequence. Computing the amount of deformation in an optical flow sequence involves measuring the local spatial changes of velocity and is therefore related to the first spatial derivative or gradient of the flow field. In the disparity domain, Koenderink and van Doorn (1976b) have demonstrated that the amount of deformation needed to map the left eye's image on to the right eye's image contains information about the local surface slant of stereoscopic surfaces (again, up to a scaling factor). This observation is important because it highlights the formal similarity of the information provided by binocular disparities and optic flow and, at the same time, it suggests an alternative way of representing information about 3-D slant.

Is there any evidence to suggest that the human visual system might compute the amount of deformation needed to map one eye's image on to the other in order to determine local surface slant? In 1986, Rogers and Koenderink showed that there was a motion parallax analogue of the "induced effect" first reported by Ogle (1938, 1950, 1955) for stereoscopic surfaces. In Ogle's induced effect, a vertical magnification of just one eye's image (leaving the horizontal sizes the same) produces the impression of a surface which slants away from the fronto-parallel in a left-right direction. Because the transformation needed to map the vertically-magnified image presented to one eye on to the other can be decomposed into both a deformation and an dilatation (figure 2), Rogers and Koenderink argued that the appearance of surface slant in the induced effect situation was consistent with the idea that the human visual system computed the amount of deformation between the two eyes' images. In their motion parallax analogue of the induced effect, a single image was presented to just one eye which expanded and contracted along a vertical axis with to side-to-side movements of the observer's head. The resulting flow field can be decomposed similarly into a deformation and a dilatation component. The fact that the surface was seen as slanting from left to right with respect to the fronto-parallel was interpreted as evidence that the human visual system also computed the amount of deformation in a sequence of images over time in order to recover information about local surface slant.

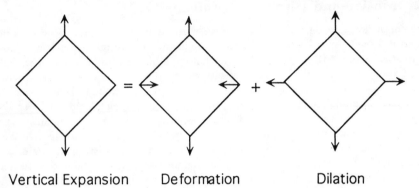

Vertical Expansion   Deformation   Dilation

Figure 2: *A vertical expansion of one image with respect to another (as is present in Ogle's induced effect) can be decomposed into a component of deformation plus a component of dilation. The fact that slant is seen in Ogle's induced effect is therefore consistent with the idea that the visual system uses the amount of deformation needed to map one eye's image onto the other as a source of information about surface slant.*

1.3   The Deformation Hypothesis

What other evidence is there that the human visual system extracts the amount of deformation needed to map *either* one eye's image on to the other (in the binocular case) *or* a sequence of images over time (in the optic flow case)? If we consider the stereoscopic images projected by a surface which is slanted (inclined) around a horizontal axis, those images are related (to a first approximation) by a **horizontal shear** (figure 3a). Decomposing the transformation into the differential invariants suggested by Koenderink and van Doorn (1976b) reveals that the horizontal shear is made up of both a deformation and a rotation component (figure 3b). The fact that we perceive slant in such a stereo pair is therefore consistent with the deformation hypothesis, but it is also consistent with the hypothesis that the human visual system extracts conventional horizontal point disparities. If the two stereoscopic images are related by a **vertical shear** (figure 3c), the decomposition consists of both a deformation and a rotation (figure 3d), as before, but in this situation the horizontal disparities between all corresponding points are zero. Hence, if slant is perceived in the situation in which the stereoscopic images are related by a vertical shear, this ought to be strong evidence in favour of the deformation hypothesis.

1.4   Vertically-Sheared Images

In 1990, Cagenello and Rogers studied the appearance of stereoscopic surfaces in which the images were related by a vertical shear (figure 3c). The images were 50% random dot patterns subtending 20 by 20 deg visual angle and they were displayed on two large screen display oscilloscopes arranged in a modified Wheatstone configuration (Rogers and Graham 1982). The vertically-sheared "test" images were presented to the observer for either 1, 2.5 or 6.5 seconds, after which they were replaced by a pair of images related by a horizontally-shear for a further 2 seconds. These horizontally-sheared "matching" images had an adjustable disparity gradient and the observer's task was to adjust the disparity gradient until the perceived slant matched that produced by the vertically-sheared "test" images. Although there is no real world situation which could produce a pair of stereo images related by a vertical shear, Cagenello and Rogers reported that observers reliably

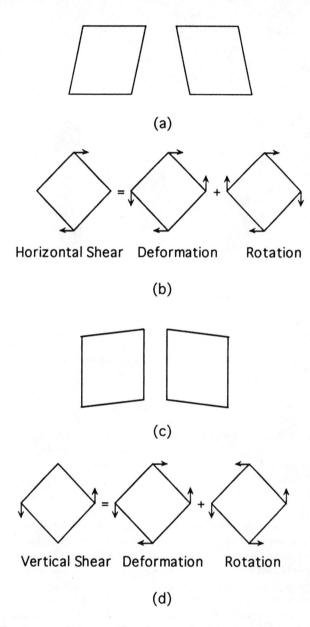

Figure 3: *The stereoscopic images of a planar surface slanting about a horizontal axis are related (to a first approximation) by a horizontal shear (3a), which can be decomposed into a component of deformation and a component of rotation (3b). A pair of stereoscopic images related by a vertical shear (3c) can also be decomposed into a component of deformation and a component of rotation (3d). If the visual system uses the amount of deformation needed to map one eye's image onto the other in order to provide information about surface slant, we would predict that images related by a vertical shear should also be seen as slanting in depth, despite the fact that there are no horizontal disparities between corresponding points.*

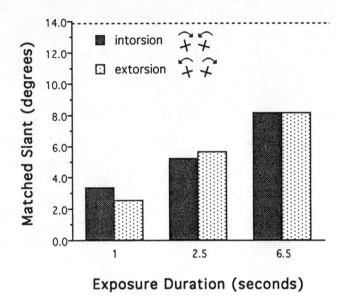

Figure 4: *The perceived slant seen with binocular images related by a vertical shear as a function of the exposure duration. Perceived slant was measured using a matching procedure (see text). Longer viewing (6.5 seconds) of these images resulted in perceived slant which was up to two-thirds of that predicted from the amount of deformation present (indicated by the dashed line), but shorter exposures produced only small amounts of perceived slant.*

perceived the fused images to be a surface slanting in depth about a horizontal axis. At first sight, this would seem to be good evidence in favour of the deformation hypothesis that the human visual system computes the amount of local deformation needed to map one eye's image on to the other in order to determine surface slant. The authors noted, however, that (i) the degree of perceived slant was very small when the vertically-sheared images were only presented for a short time and (ii) the amount of perceived slant steadily increased as observers viewed the images over longer intervals (figure 4).

A possible explanation of the results obtained in this experiment suggested by Ca-genello and Rogers was that the observers' eyes made equal and opposite cyclotorsional eye movements (**cyclovergence**) when presented with the vertically-sheared images. Such a pattern of eye movements would "convert" the vertically-sheared images on the screens into horizontally-sheared images on the retinae and all theories would predict the appearance of surface slant in this case (figure 5). Moreover, the fact that the amount of apparent surface slant increased with the longer viewing times is consistent with the known temporal characteristics of the cyclovergence system (Kertesz and Sullivan 1978; Howard and Zacher 1991).

1.5   The "Double Stimulus"

The question of whether the perceived slant seen in vertically-sheared stereoscopic images was an artefact of cyclovergent eye movements was answered in two ways. First, torsional eye movements were monitored directly (see 1.8 below) and, second, a stereoscopic "double stimulus" was created which had the effect of minimising the possibility of cyclotorsional

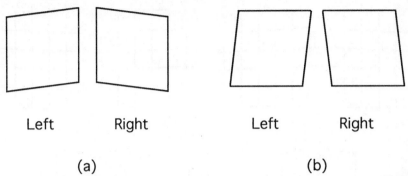

Left   Right   Left   Right

(a)   (b)

Figure 5: *If the eyes made cyclovergent eye movements (torsional movements in opposite directions), binocular stimuli related by a vertical shear (5a) will be converted into images related by horizontal shear on the retinae (5b). As a consequence, horizontal disparities would be generated between the two eye's images and all theories of binocular stereopsis would predict the appearance of slant.*

eye movements occurring. The pattern in each half of the stereopair was contained within a dumbbell-shaped aperture and an equal and opposite vertical shear was applied on the left and right sides. Thus the left sides of the stereopair were related by a vertical shear in one direction and the right sides of the stereopair by a vertical shear in the opposite direction (figure 6). The presence of equal and opposite vertical shears within each of the stereo images ought to have kept any cyclovergent eye movements to a minimum but, even if such movements did occur, they would introduce horizontal disparities in the *same* direction in the left and right halves of the dumbbell pattern and thus no differential slant should be seen between the two halves. On the other hand, if the human visual system does indeed compute the amount of deformation between the images reaching the two eyes, the presence of equal and opposite vertical shear transformation in the left and right halves of the dumbbell pattern should produce the impression of differential slant in the two halves. The patterns themselves were composed of *either* 50% density random dots *or* grids of ± 45 or 0/90 deg lines. The two halves of the dumbbell pattern each subtended 12 deg visual angle in diameter.

To further test the deformation hypothesis, Cagenello and Rogers also created "double stimuli" in which the transformation was *either* (i) an equal and opposite rotation *or* (ii) an equal and opposite deformation. If local surface slant is computed from the amount of deformation needed to map one eye's image on to the other, stereoscopic images related by a rotation should *not* produce an impression of differential slant, because there is no deformation present. On the other hand, stereoscopic images related by a deformation should produce an enhanced impression of differential slant between the left and right sides because the amount of deformation is double that produced by *either* the vertically-sheared *or* the horizontally-sheared image pairs. The predicted results derived from the deformation hypothesis are therefore: (i) maximal differential slant from the images related by a deformation, (ii) intermediate differential slant from images related by a vertical shear and (iii) no differential slant from images related by a rotation.

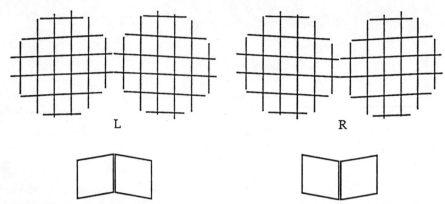

Figure 6: *A stereoscopic "double stimulus" was created by subjecting the left halves of the stereopair to vertical shear in one direction and the right halves of the stereopair to vertical shear in the opposite direction. The presence of equal and opposite vertical shear in the two halves of the "double stimulus" should not induce any cyclovergent eye movements. The actual stereograms were composed of either (i) a grid of 0/90 deg lines (as shown here), or (ii) a grid of ± 45 deg lines or (iii) random dots.*

1.6   Results

The obtained pattern of results was quite different (figure 7). The maximum perceived differential slant was obtained when observers viewed the "double stimulus" consisting of patterns related by a rotation, even though the mapping between these images contained no deformation. The deformation "double stimulus" yielded an impression of differential slant which was slightly less, even though it contained the most deformation of all three "double stimuli". The vertical shear "double stimulus" was perceived to have a small amount of differential slant between the left and right halves of the dumbbell, but it was in the *wrong* direction to that predicted by the deformation hypothesis.

1.7   A Modified Deformation Hypothesis

What can be concluded from these results? The most obvious conclusion would seem to be that the stereoscopic system does *not* extract the amount of deformation needed to map one eye's image on to the other, as was originally suggested by Koenderink and van Doorn. There is, however, an alternative interpretation. In order to extract the amount of deformation needed to map one image on to another, either between the eyes or over time in an optic flow sequence, it is necessary to monitor the position or orientation changes that occur with respect to two or more differently oriented axes (Koenderink 1985, 1986). Intuitively, this can be understood as being necessary in order to separate out the consequences of a rotation between the images (which generates *equal* changes in the orientation of elements at *all* absolute orientations) from a deformation (which generates changes of orientation which are *different* at *different* absolute orientations).

In optic flow fields, there is no inherent restriction on the sort of image deformation that may be produced when the observer moves with respect to the 3-D world (or a 3-D surface moves with respect to an observer) but in binocular stereopsis, real 3-D surfaces never produce differences in the orientation or position of elements close to the horizontal meridian, apart from when the eyes are misaligned torsionally. The differences in position

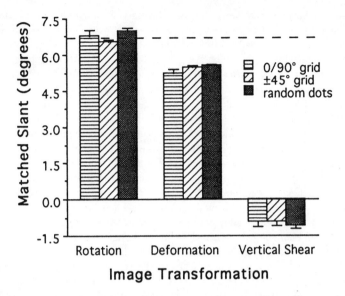

Figure 7: *The perceived differential slant between the left and right halves of the "double stimulus" as a function of the transformation between the binocular images. Perceived slant was measured using a matching procedure (see text). Perceived slant was largest when the binocular images were related by a rotation (left-hand bars), despite the fact that there was no deformation between the images. In contrast, only a very small amount of differential slant was seen in the binocular images related by a vertical shear and it was in the wrong direction to that predicted by the deformation hypothesis (right-hand bars).*

and orientation of corresponding elements in a surface slanting around the horizontal axis are maximal close to the vertical meridian, whilst for a surface slanting around the vertical axis, they are maximal along the ± 45 deg diagonals (Cagenello and Rogers 1988). Therefore, if the eyes are torsionally aligned, there is no need for the visual system to separately monitor the orientation and positional changes around both the horizontal and vertical axes. A simpler solution would be to use any positional or orientational misalignments close to the horizontal meridian as a signal to drive the cyclovergence system and thereby correct those torsional misalignments. The stereo system could then use any remaining differences in position or orientation as information about the shape of the 3-D surface under inspection.

1.8 The Role of Cyclovergent Eye Movements

This alternative hypothesis makes an important prediction about the stimuli capable of driving the cyclovergence system. It suggests that positional or orientational misalignments close to the horizontal meridian should provide a very effective signal to drive the cyclovergence system. Conversely, because orientation and positional differences between corresponding elements close to the vertical meridian in the two eyes may be a consequence of *either* (i) cyclotorsional misalignment *or* (ii) the slant of surfaces inclined around a horizontal axis, it is unlikely that positional or orientational misalignments close to the vertical meridian will be a very effective signal to drive the cyclovergence system, because their cause is ambiguous (Rogers and Howard 1991).

These two predictions were tested by monitoring the torsional state of the two eyes directly whilst presenting the observer with various dichoptic patterns which were transformed over time. In particular, the modified deformation hypothesis predicts that dichoptically-presented patterns which are subject to an equal and opposite *vertical* shearing transformation in the two eyes should be a very effective stimulus for driving cyclovergence. This is because a vertically-shearing pattern generates maximal positional and orientational disparities between elements which are close to the horizontal meridian and, conversely, it generates no orientational or positional disparities between elements close to the vertical meridian. The second set of stimuli used to test the hypothesis consisted of dichoptically-presented patterns which were subjected to an equal and opposite *horizontal* shearing transformation in the two eyes. If the modified deformation hypothesis is correct, we should predict that an equal and opposite horizontal shearing between the two eyes would be a very ineffective stimulus to drive the cyclovergence system because this transformation generates maximal positional and orientational disparities between elements close to the vertical meridian and, conversely, generates no orientational or positional disparities between elements which are close to the horizontal meridian.

Cyclovergent eye movements were also monitored in response to patterns which were subjected to an equal and opposite *rotation* in the two eyes. It is, after all, a rotation transformation which is created by accidental torsional misalignment between the two eyes in normal viewing and, therefore, should be the best stimulus to drive the cyclovergence system. In a fourth condition, observers were presented with patterns which were subjected to an equal and opposite deformation between the two eyes. In this case, the equal and opposite deforming patterns generated positional and orientational disparities around both the horizontal and vertical meridians, but in *opposite* directions. The deformation transformation consisted of a vertical shear in one direction, which created positional and orientational disparities around the horizontal meridian, and a horizontal shear in the opposite direction, which created positional and orientational disparities around the vertical meridian. The significance of this condition is that it directly pits the contribution of the vertical disparities of corresponding elements close to the horizontal meridian with the horizontal disparities of corresponding elements close to the vertical meridian.

The fact that the best response of the cyclovergence system is to low temporal frequencies suggests that cyclovergence is a position rather than a movement driven system (Howard and Zacher 1991). In addition, Howard and Zacher reported that the maximum gain of the cyclovergent response could be as high as 0.85 to patterns which rotated in equal and opposite directions in the two eyes at a temporal frequency of 0.05 Hz and a peak-to-peak amplitude of 2 deg of cyclodisparity. This result shows that the cyclovergence system generates equal and opposite torsional movements of the two eyes which follow the counter-rotating patterns almost perfectly. Howard and Zacher have also shown that a high gain of the cyclovergence system is only elicited when the counter-rotating patterns occupy a large part of the visual field.

1.9   Methods

In the experiments reported here, subjects were presented with 80 deg diameter patterns which were subjected to *either* (i) an equal and opposite vertical shear between the two eyes, *or* (ii) an equal and opposite horizontal shear, *or* (iii) an equal and opposite rotation *or* (iv) an equal and opposite deformation. The patterns were rear-projected on to Mylar screens which were positioned on the side faces of a cube frame on either side of subject's head (figure 8). Two mirrors at ± 45 deg were positioned at the centre of the cube frame,

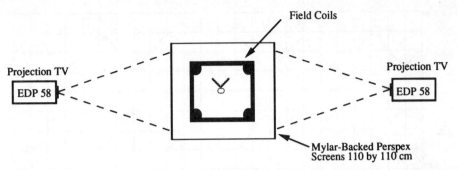

Figure 8: *Plan view of the experimental apparatus. The observer viewed the two rear-projection screens via a pair of mirrors at ± 45 deg to the line of sight. The images, which were created on two video boards in a Macintosh II fx computer, were rear-projected on to the screens from two Electrohome EDP 58 video projectors. Torsional eye movements were recorded using the system of field coils, arranged in a frame surrounding the observer, and close-fitting contact lenses with embedded coils.*

just in front of the subject's eyes, so that the left eye saw only the left screen and the right eye only the right screen. The patterns themselves were generated on a Macintosh IIfx computer and projected on to the Mylar screens using two Electrohome EDP 58 video projectors.

Three different stimulus patterns were used – (i) a video captured image of a piece of coarsely woven burlap cloth containing multiple grey levels, (ii) a 50% radial random dot pattern in which the size of the dot segments increased linearly from the centre of the pattern and (iii) a video-captured image of a piece of wrapping paper containing randomly-shaped black blotches on a white background. A sequence of 71 images was created from each of the original images by subjecting it one of the four different transformations. In each case, the new position of the every element was computed and grey-level interpolation was used to simulate subpixel displacements. The sequence of images represents the 71 equally-spaced temporal steps of the pattern undergoing a sinusoidal oscillation between the maximum displacement values of ± 1.9 deg. The peak-to-peak amplitude of cyclodisparity between the two eyes was thus 7.6 deg. A complete oscillation of 140 images took 10 seconds to present, corresponding to a temporal frequency of 0.1 Hz. The horizontal and vertical resolutions of the patterns were 480 by 480 pixels and the images were refreshed at a rate of 67 Hz. Although only 14 new images were presented every second, the small size of the displacement steps, together with the grey level interpolation, produced the impression of a smoothly transforming pattern.

1.10   Eye Movement Recording

Torsional eye movements were recorded using the system of field coils and close-fitting scleral contact lenses with embedded coils manufactured by Skalar of Delft. The lenses were individually calibrated using an artificial eye and were found to have gains of around 0.25 volts/degree. The noise level of the signals from the Skalar amplifiers was such that torsional eye movements of less than 1 *arc min* could be resolved. Six channels of the Skalar amplifiers (horizontal, vertical and torsional for both left and right eyes) were each monitored at a 500 Hz sampling rate by a second Macintosh computer using a National Instruments A to D interface board. The digitised signals were stored on

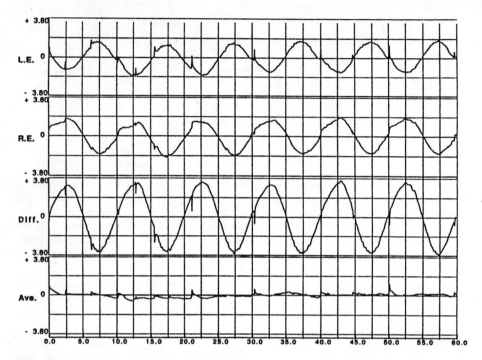

Figure 9: *A set of eye movement traces for one observer (JW) recorded over the sixty second duration (x-axis) of a trial in which the images presented to the two eyes counter-rotated in opposite directions at a frequency of 0.1 Hz. The upper two traces show the torsional eye movements generated in the left and right eyes respectively. The torsional eye movements are in opposite directions (cyclovergence). The third trace shows the difference signal between the upper two traces and indicates the extent of cyclovergence. The amplitude of cyclorotation of the two patterns was ± 3.8 deg and is indicated by the outer horizontal lines surrounding the eye movement record. For this observer, the amplitude of cyclovergence created by the counter-rotating patterns is nearly as large as the stimulus amplitude itself. The lowest trace shows the average torsional response in the same direction in the two eyes (cycloversion) which was minimal in this situation.*

disc for subsequent analysis and displayed using hard copy output. The subjects were asked to keep their eyes directed towards the centre of the transforming pattern which was marked by a small cross. Keeping the subject's gaze directed towards the centre of the pattern minimised the possibility of artefacts arising from crosstalk between the horizontal, vertical and torsional signals.

1.11   Results

Figure 9 shows a typical record of the torsional eye movements of subject JW using the burlap dot pattern undergoing an equal and opposite rotation to the two eyes, over the sixty seconds of the trial. The upper two traces show the left and right torsional movements respectively. The third trace shows the difference signal between the left and right movements and thereby indicates the extent of cyclovergent eye movements in *opposite* directions. The lowest trace shows the average of the left and right movements and thereby

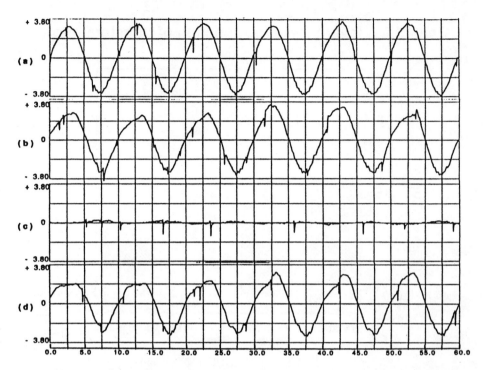

Figure 10: *A comparison of the amount of cyclovergence (the difference between the left and right eyes' torsional signals) as a function of the equal and opposite transformation presented to the two eyes. The upper trace (10a) shows the extent of cyclovergence in response to patterns which rotated in opposite directions in the two eyes (cf. third trace in figure 9). The second trace down (10b) shows the extent of cyclovergence, which is also high, when the patterns were subjected to an equal and opposite vertical shear in the two eyes. By comparison, very little cyclovergence was generated when the patterns to the two eyes were subjected to an equal and opposite horizontal shear (10c). The lowest trace (10d) shows the extent of cyclovergence in response to patterns which were subjected to an equal and opposite deformation in the two eyes.*

indicates the extent of cycloversional eye movements in the *same* direction. The peak-to-peak amplitude of the eye movements was calculated by measuring the distance between successive peaks in the eye movement trace and averaging. The gain of the cyclovergence was determined by dividing the average peak-to-peak amplitude of the cyclovergent response by the peak-to-peak amplitude of the cyclodisparity in the stimulus patterns (i.e. 7.6 deg).

The general pattern of responses can be seen by comparing the eye movement records for subject JW under the four transformation conditions. In (10a) it can be seen that the amplitude of the cyclovergent eye movements was nearly as large as the peak-to-peak amplitude of the transforming stimulus when the patterns *counter-rotated* in opposite directions in the two eyes. In (10b), the amplitude of cyclovergence to patterns which *sheared vertically* in opposite directions in the two eyes is smaller, but still very high. This is in marked contrast to the results obtained when the patterns *sheared horizontally* in opposite directions in the two eyes which are shown in (10c). In this case, the amount

Gain of Cyclovergence as a function of transformation

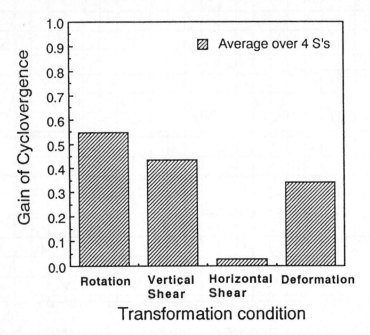

Figure 11: *The gain of cyclovergence, averaged over the four observers and the three different pattern types, plotted as a function of the transformation type. The gain of cyclovergence to vertically-shearing patterns was found to be nearly 80% of that to counter-rotating patterns, even though the visual system is never subjected to such a transformation in real life. In contrast, the gain of cyclovergence to horizontally-shearing patterns averaged only 0.03 – a factor of 13 lower than that produced by vertically-shearing patterns.*

of cyclovergence produced by the patterns is negligible. Finally in (10d), it can be seen that cyclovergent eye movements are elicited by patterns which were subjected to an equal and opposite deformation in the two eyes despite the fact that the cyclodisparities are in opposite directions along the horizontal and vertical meridians. The direction of the cyclovergence indicates that it is the orientation or positional disparities close to the horizontal meridian which are responsible for the driving of the torsional eye movements.

The quantitative results are shown in the histograms of figure 11 as a function of the four transformation types (rotation, vertical shear, horizontal shear and deformation), averaged over the three pattern types (burlap, radial random dot pattern and wrapping paper) which all produced similar amounts of cyclovergence. The gains of the cyclovergent response are also averaged over the four observers. The results show that the maximum gain of the cyclovergence system was obtained when the patterns presented to the two eyes rotate in equal and opposite directions in the two eyes. The average gain over four subjects was 0.55. The effectiveness of the counter-rotating patterns in driving cyclovergence is not surprising given that the purpose of the system is to eliminate any torsional misalignment between the two eyes. Moreover the gain found in these experiments is very comparable to that obtained by Howard and Zacher (0.58) using 80 deg fields, an oscillation frequency of 0.1 Hz and a slightly smaller peak-to-peak amplitude of 6.0 (compared to the 7.6 deg used here).

More surprising are the results for vertically- and horizontally-shearing patterns. The cyclodisparity in the stimuli averaged over all orientations is the same for both shearing conditions and is only one half of that produced by the counter-rotating patterns. The gain of cyclovergence was, however, very different in the two cases. The average gain for vertically-shearing patterns is 0.43 (roughly 78% of that for rotating patterns) but for horizontally-shearing patterns was only 0.03. The average ratio of the gains for vertically- and horizontally-shearing patterns over the four subjects was 13:1. The difference is very clear cut and was present in the results of each subject. Pattern type (burlap, radial random dot pattern or wrapping paper) had no significant effect on the amplitudes of the cyclovergent eye movements in any of the transformation conditions.

The average gain of cyclovergence for the patterns subjected to an equal and opposite deformation between the two eyes was 0.34. In this condition, there were cyclodisparities between corresponding elements close to both the horizontal and vertical meridian, but in opposite directions. The direction of the cyclovergence response was found to be appropriate to the direction of vertical shear and thus was consistent with the cyclodisparities of elements close to the horizontal meridian. The extent of the gain is particularly impressive because the cyclodisparities for elements between $\pm$ 45 deg of the vertical meridian (i.e. half the total) were all in the *wrong* direction (if the eyes made no cyclovergent response) and over an even greater range given that the eyes did make cyclovergent movements with an average amplitude of over one-third of the stimulus amplitude.

1.12 Stereoscopic Deformation and Cyclovergence – Conclusions

The fact that (i) cyclovergent eye movements are driven almost as well by equal and opposite vertically-shearing patterns as by cyclorotating patterns presented to the two eyes and (ii) equal and opposite horizontally-shearing patterns are relatively ineffective at driving cyclovergence strongly suggests a *dissociation* between the roles of disparate elements close to the horizontal and vertical meridia. In particular, it suggests that the vertical disparities of corresponding elements close to the horizontal meridian are used to drive cyclovergent eye movements in order to eliminate those disparities, whilst the horizontal disparities of corresponding elements close to the vertical meridian are used as a source of information about the 3-D shape of surfaces. Such a dissociation is possible within the binocular stereoscopic system because (i) gradients of vertical disparities across the horizontal meridian do *not* contain any information about 3-D shape and (ii) the elimination of cyclovergence errors is essential in order for horizontal disparities to be interpreted correctly.

This pattern of results is not incompatible with the original deformation hypothesis proposed by Koenderink and van Doorn (1976a, 1976b). In order to recover information about surface slant, it is necessary to detect the orientation or position differences between corresponding elements at two or more different orientations. This could be done explicitly by making a direct comparison between orientation or position differences along differently oriented axes. In the case of the stereoscopic system however, the possible differences between the images are constrained by the fact that the eyes are separated horizontally (strictly speaking, along the axis joining the two eyes). This means that no 3-D surfaces can ever generate orientation or positional differences along the horizontal meridian and, conversely, if there are orientation or positional differences along the horizontal meridian, they must be due to torsional misalignment. Consequently, it is possible for the human visual system to adopt the simple strategy of using orientation or positional differences of corresponding elements close to the horizontal meridian to drive the cyclovergence system (and thereby eliminate those differences) and then use the positional or orientation disparities elsewhere in the field to provide information about surface shape (Rogers and Howard 1991). Recent work by De Bruyn *et al.* (1992) suggests that it is the *position* differences of elements close to the horizontal meridian, rather than the *orientation* differences, which are used to drive cyclovergence.

1.13   Motion Parallax and Deformation

The situation within the motion parallax system is significantly different. It cannot be assumed that changes in the vertical position of elements close to the horizontal meridian over time are due to changes in the torsional state of the eye. Therefore, in order to compute 3-D shape from motion parallax information, it is necessary to monitor changes in the positions of corresponding elements around both horizontal and vertical axes. In other words, the visual system would have to compute the amount of local deformation in the way described by Koenderink and van Doorn (1975, 1976a).

There is some evidence to suggest that there are differences in the way surface slant is computed from disparity- and motion parallax-specified surfaces which are consistent with the above. In the stereoscopic experiments using the dumbbell-shaped "double stimulus" described earlier, it was found that differential slant was not perceived when the transformation between the two eyes' images was a vertical shear (even though there was deformation between the two eyes' images) but was perceived when the transformation between the two eyes' images was a rotation (even though there was no deformation between the two eyes' images). The analogous experiment was set up using motion parallax as the source of 3-D information. A dumbbell-shaped double stimulus was again constructed but, in the motion parallax analogue, the left and right halves of the single dumbbell pattern were dynamically transformed in equal and opposite directions with each side-to-side movement of the subject's head. In the first case, the two halves were transformed with an equal and opposite vertical shear (coupled to the head movement) and in the second case, the two halves were transformed with an equal and opposite rotation (again coupled to the head movement).

The results for the equivalent experiment in the motion parallax domain were quite different. Whilst the presence of equal and opposite vertical shear yielded no impression of differential slant in the disparity experiment, clear differential slant was seen in the motion parallax case. Conversely, whilst the presence of equal and opposite rotation yielded a clear impression of differential slant in the disparity experiment, no differential slant was seen in the motion parallax case.

This pattern of results suggests that in order to recover information about surface slant, the human visual system extracts some characteristic of the optic flow field which is essentially equivalent to the monitoring the amount of local deformation over time. In the stereoscopic system, on the other hand, the human visual system has been able to exploit (and any machine vision system potentially could exploit) the fact that the range of possible transformations between the two eyes is constrained by their physical locations. Differences in the orientation or position of corresponding elements close to the horizontal meridian in the two eyes are necessarily the result of torsional misalignment rather than the 3-D shape of the fixated surface and therefore can be used as a stimulus to drive the cyclovergence system, in order to eliminate that misalignment. If the eyes are prevented from making the appropriate torsional corrections (as they were with our "double stimulus") surface slant will be misperceived.

2 Vertical Disparities in Binocular Stereopsis

2.1 Introduction

The fact that the eyes are separated horizontally means that most of the small differences in the relative position (disparities) between the images on the two retinae are in a horizontal direction, parallel to the axis joining the two eyes. Simple geometric considerations show that, to a first approximation, the magnitude of the horizontal disparity between two points on an 3-D object or surface will vary inversely with the square of the viewing distance. Consequently, it is necessary to know the *absolute distance* from the observer to the surface in order to correctly scale the magnitude of the disparity and thereby recover the depth separation of the two points. Traditionally, it has been assumed that information about absolute distance is derived from non-visual sources such as the vergence angle between the eyes. Evidence in support of the role of vergence in disparity scaling comes from the observation that the magnitude of the perceived separation of two disparate surfaces is affected by manipulations of the vergence state of the two eyes (Foley 1980; Foley and Richards 1972). The second part of this chapter will be concerned with the computational theory of vertical disparities as a source of information about absolute distance and the empirical evidence for their use by the human visual system.

2.2 The Vertical Disparity Hypothesis

Ten years ago, Mayhew and Longuet-Higgins (1982) described an elegant new computational theory of binocular stereopsis which showed that there was information in the pattern of vertical differences between the images on the two retinae (vertical disparities) to specify the absolute distance to the fixation point. The existence of vertical disparities in the positions of corresponding images had been recognised previously by both Helmholtz (1925) and Ogle (1950), but neither author had suggested that the vertical disparities could be used to derive an estimate of viewing distance (but see Howard 1970). The fact that there is information in the pattern of vertical disparities can be appreciated intuitively by considering the images generated by a rectangular surface viewed at different distances (figure 12). When the surface is close to the observer, the projected images will be slightly trapezoidal in shape because the left side of the rectangle is slightly closer to the left eye than the right side of the rectangle and therefore subtends a slightly larger visual angle (figure 12a). The converse is true for the images projected on to the right retina. However, when the surface is at a large distance from the observer, the distances from the two sides of the rectangle are similar and so the projected images will be approximately rectangular in shape (figure 12b). If we consider the differences between the images in the two eyes, it is clear that there will be a gradient of vertical disparity when the surface is close to the observer but that gradient will tend to zero as the distance to the surface increases.

Mayhew and Longuet-Higgins (op cit.) showed that the pattern of vertical disparities provides information about what they called the two "viewing system parameters" – the absolute distance to the surface (d) and the angle of eccentric gaze (g). When a surface is viewed eccentrically, the image projected to one eye is larger in all dimensions than to the other eye (because the surface is closer) and Mayhew and Longuet-Higgins suggested that

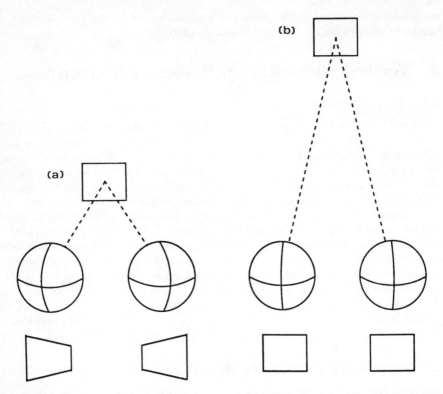

Figure 12: *A schematic representation of the binocular images created by a rectangular surface close to the observer (12a) which are trapezoidal in shape because the two sides of the rectangle are at different distances from each of the eyes. In contrast, a rectangular surface of the same angular extent at infinity will generate images which remain rectangular in shape (12b). Hence, there is a gradient of vertical size difference across the images from left to right which is minimal for a surface at infinity and maximal for a surface which is close to the observer.*

the human visual system might also be capable of detecting this interocular size difference in order to recover the angle of eccentric gaze (figure 12c). Because the magnitude of the size difference also varies with absolute distance (the size difference due to eccentric viewing of a surface at infinity will be negligible), it would be necessary to compute (d) before (g) could be recovered. Mayhew and Longuet-Higgins' theory is particularly attractive because it demonstrates that, in principal, there is sufficient visual information to determine absolute distance and eccentric gaze angle so that disparity scaling can be accomplished without the need to monitor non-visual, ocular motor information.

2.3 Experimental Evidence

Is there any evidence to suggest that the human visual system is capable of utilising these small differences in the vertical positions of images on the two retinae? To answer this question we first need to consider the four important predictions that are made by this theory. First, the appropriate manipulations of the pattern of vertical disparities to mimic those produced by a surface close to the observer (figure 12a) or at a large distance from the observer (figure 12b) should result in the surface being perceived either closer to the observer or farther away. Second, the appropriate manipulations of the pattern of vertical disparities to mimic those produced by a surface which is more or less eccentric should result in the surface being perceived as more or less eccentric. Third, an overall (isotropic) size difference between the stereoscopic images of a surface (mimicking eccentric viewing) should have no effect on the perceived slant of the surface with respect to the cyclopean line of sight but an anisotropic size difference such that the image is magnified vertically (but not horizontally) in one eye compared to the other (thereby creating vertical but not horizontal disparities) should generate the impression of surface slant with respect to the cyclopean line of sight. Fourth, appropriate manipulations of the pattern of vertical disparities to mimic those produced by a surface close to the observer (figure 12a) or at a large distance from the observer (figure 12b) should result in the percept of *more* depth *within* stereoscopic surfaces when they contain vertical disparities appropriate to the *greater* viewing distance. This prediction is made if the visual system used the estimate of (d) derived from vertical disparities to scale horizontal disparities in an analogous way to the scaling of horizontal disparities by the estimate of absolute distance derived from the vergence angle (Foley 1980).

To date, no one has reported that the perceived distance of a stereoscopic surfaces changes as a result of manipulations of the gradient of vertical disparities (prediction 1) *or* that the perceived eccentricity of a stereoscopic surface changes as a result of manipulations of the overall size difference (prediction 2).

2.4 The Induced Effect

The principal evidence for the vertical disparity hypothesis put forward by Mayhew and Longuet-Higgins comes from Ogle's original observation (1938) that when there is a difference in the vertical sizes of the two stereoscopic images (and the horizontal sizes are identical) the surface appears to slant out of the fronto-parallel plane from left to right. Ogle called this effect the "induced effect" because he believed that horizontal disparities were *induced* as a result of the visual system compensating (isotropically) for the vertical size difference.

Whilst the induced effect (the third prediction) is certainly consistent with the idea of the visual system extracting a viewing system parameter (g) and then using the parameter to rescale the values of the horizontal disparities, Rogers and Koenderink (1986) have proposed an alternative explanation. According to them, the visual system uses the amount of local deformation needed to map one eye's image on to the other to compute local surface slant directly. Their idea was based on Koenderink and van Doorn's (1976a) analysis which showed that local surface slant (up to a distance scaling factor) was specified by the amount of deformation between the two binocular images (see section 1.2). An isotropic size difference, as would be generated by eccentric viewing, does not alter the amount of deformation needed to map between the two images and therefore specifies local surface slant *independent* of the eccentricity of gaze. Therefore, whilst the induced effect, by itself, cannot be regarded as providing conclusive evidence for the deformation hypothesis, neither can it be regarded as providing conclusive evidence for Mayhew and Longuet-Higgins' vertical disparity hypothesis.

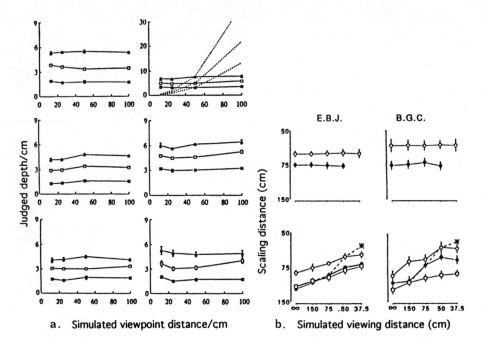

Figure 13: *The results of the two previous experiments on vertical disparity scaling reported by Sobell and Collett (1991) (13a) and Cumming et al. (1991) (13b). In both cases, it can be seen that manipulations of the gradient of vertical disparities appropriate to different viewing distances (x-axis) have no effect on the amount of perceived depth of a disparity discontinuity under various conditions (13a) or on the perceived shape of cylindrical surfaces (top two graphs of 13b). In contrast, it can be seen that manipulations of vergence angle do have an effect on the perceived shape of cylindrical surfaces (bottom two graphs of 13b).*

2.5   Disparity Scaling

Until recently, the fourth prediction of the vertical disparity hypothesis – that manipulations of vertical disparity should scale horizontal disparities – had not been adequately tested. In 1991, there were two reports, by Cumming, Johnston and Parker and Sobell and Collett, of experiments in which the perceived depth of stereoscopic surfaces was monitored as a function of the presence or absence of vertical disparities. Sobell and Collett asked subjects to judge the perceived depth difference in a surface containing a step change of disparity in the presence of vertical disparity gradients which were appropriate to viewing distances ranging from 12.5 to 100 cm. Vergence angle and accommodative cues were kept constant. Their results show that the amount of perceived depth was quite unaffected by the vertical disparity manipulations (figure 13a). Cumming *et al.* used a slightly different task. They argued that if disparities were scaled by an estimate of viewing distance derived from vertical disparities, then the perceived shape of a cylindrical surface should also be affected by manipulations of the gradient of vertical disparities. As was pointed out earlier, the horizontal disparities produced by two points on a surface will decrease with the square of the viewing distance or, putting it the other way around, for a constant horizontal disparity difference, the perceived depth should increase with increasing viewing distance. Thus a cylindrical surface with a given pattern of disparities over the surface should be seen as *more elongated* when the signalled viewing distance is *greater* and *more flattened* when the signalled viewing distance is *smaller*. In Cumming *et al.'s* experiments, the vergence angle was always kept constant and observers were asked to judge the perceived shape of the cylindrical surface as a function of the superimposed gradients of vertical disparity.

Cumming *et al.'s* results were also very clear cut. Manipulations of the gradient of vertical disparities to mimic viewing distances of between 37 cm and infinity had no effect on the perceived shape of the cylindrical surfaces (figure 13b). Manipulations of the vergence angle whilst keeping the vertical disparity gradient constant, on the other hand, did produce differences in the perceived shape of the surfaces, as would be expected from previous studies on vergence scaling (Foley and Richards 1972). The results of these two experiments strongly suggest that the human visual system does *not* use the information contained in the gradients of vertical disparity to scale horizontal disparities as would be predicted by Mayhew and Longuet-Higgins' theory.

2.6   Size Considerations

It is clear from the equations derived by Mayhew and Longuet-Higgins that the magnitude of the vertical disparity gradient generated by a surface at a given distance increases with increasing angle of vertical eccentricity (elevation) from the plane of regard. It is possible, therefore, that the failure of Cumming *et al.* and Sobell and Collett to find an effect of vertical disparity gradient on perceived depth was due to the relative small sizes of their displays (7-11 deg in *Cumming et al.'s* experiment and 25-30 deg in Sobell and Collett's). To test for this possibility, Rogers and Bradshaw (1992) examined the role of vertical disparity gradients with stereoscopic images which subtended 80 by 70 deg visual angle. One of the tasks they used was a forced choice comparison of the perceived depth in horizontal sinusoidal corrugations of 0.2 cycles/deg corrugation frequency and 10 *arc min* peak-to-trough depth. The corrugated surface subtended 25 by 20 deg of visual angle and was set in a larger flat, fronto-parallel surface which subtended 80 by 70 deg. Both the corrugated surface and the fronto-parallel surround were densely covered with random texture elements. Observers sat in a modified Wheatstone stereoscope with mirrors at 45 deg to the line of sight and viewed the rear-projected images on two translucent Mylar screens 57 cm from the observer's eyes. The vertical disparities of the stereoscopic images were accurately synthesised to mimic the viewing of a flat, textured fronto-parallel surface at either 28 cm or infinity. The "28 cm" and "infinity" stereopairs were each presented for five seconds in a randomised order. The observer's task was to judge whether the perceived depth in the corrugated surface which had a gradient of vertical disparities appropriate for 28 cm viewing distance had *more* or *less* depth than the same corrugated surface which had a gradient of vertical disparities appropriate for viewing at infinity. The stereoscopic images were correctly aligned so that the vergence angle was identical for both stereopairs and consistent with the physical distance of the projection screens of 57 cm. Thirty subjects took part in the experiment.

2.7   Results

The pattern of results obtained was quite clear and unequivocal over all the subjects tested. All subjects judged the amount of depth in the corrugations which contained a gradient of vertical disparities appropriate for 28 cm viewing to be *less* than the amount of depth in the corrugations which contained a gradient of vertical disparities appropriate for viewing at infinity. In separate trials, subjects were also asked to judge the perceived separation of the peaks (periodicity) as a function of the gradient of disparity present. If the human visual system uses the gradient of vertical disparities to provide an estimate of viewing distance, this estimate could also be used to scale the sizes of images on the retina. The visual angle subtended by an object of a given physical size varies inversely with the viewing distance and, in order to achieve size constancy, the visual system needs to scale retinal size with some estimate of absolute viewing distance. If the gradient of vertical disparity is used for size scaling as well as for disparity scaling, we would predict that the perceived separation of the peaks of the corrugations used in this experiment should also be affected by the vertical disparity manipulations. Although not as clear cut as the effect on perceived depth, the majority of subjects consistently judged the perceived separation of the corrugations to be greater when the gradient of vertical disparities was appropriate to a surface at infinity compared to when the vertical disparities were appropriate to 28 cm viewing.

2.8   A Modified Computational Theory

These results suggest that if the field of view is sufficiently large, the human visual system is able to exploit the different gradients of vertical disparity produced by surfaces at different distances. The computational theory of vertical disparities itself, however, should also be re-examined. Mayhew and Longuet-Higgins showed that the differences in the vertical position of images on the two retinae contained information about the two viewing system parameters of absolute distance to the fixation point (d) and the angle of eccentric gaze (g). That is, there is visual information about two parameters which, previously, were considered to have been only available from non-visual, ocular-motor sources. In order to derive their equations, Mayhew and Longuet-Higgins considered the projection of the 3-D scene on to two *planar* surfaces. If the two planar projection surfaces are converged (i.e. the binocular cameras are both directed towards a point closer than infinity), the projected images of a rectangular surface will indeed be trapezoidal in shape (figure 12a) and the amount of trapezoidal difference (the gradient of vertical disparities) will vary with the "vergence angle" between the two projection surfaces. Hence, if the spatial separation of the two vantage points is known, the absolute distance to the "fixation point" (d) can be computed. Note, however, that the magnitude of the gradient of vertical disparities does not vary with the distance of the surfaces within the scene – it is entirely a property of the "vergence angle" of the projection surfaces.

2.9   Spherical Projection

The retinae, however, are not planar projection surfaces and hence an analysis based on planar projection may not be entirely appropriate when considering the human visual system. A better approximation to the human retinae can be made by considering the consequences of spherical projection. With spherical projection, horizontal or vertical rotations of the projection surface (as would be produced by horizontal or vertical eye movements) cannot affect the gradients of *either* the vertical *or* the horizontal differences between corresponding points in the two images. Therefore, any gradients of vertical disparity cannot signal the vergence state of the two eyes. On the contrary, gradients of vertical disparity remain invariant with vergence angle. The absolute distance of a surface, on the other hand, does affect and indeed determines the gradient of vertical disparities under spherical projection. Similarly, whilst the difference in image size contains information about the angle of eccentric fixation of the two projection surfaces under planar projection, under spherical projection the difference of image size provides information about the eccentricity of the image with respect to the median plane and is unaffected by the horizontal and vertical positions of the two eyes.

It is not claimed that the modified computational theory based on spherical projection provides any *additional* information – necessarily there cannot be any additional information since the image points can be readily mapped from one projection surface to another. The difference lies in the fact that under spherical projection, the gradients of vertical disparity provide direct information about the *scene characteristics* of a (potentially) large number of different surfaces in the visual scene (the absolute distance to the surface and its eccentricity with respect to the median plane of the head), whilst under planar projection the gradients of vertical disparity provide direct information about a limited number of *viewing system parameters* (the vergence angle of the projection surfaces and the eccentricity of the camera axes). Thus, instead of yielding a single viewing system parameter of the vergence angle from which the absolute distance could be computed (and then used to scale horizontal disparities), the local gradient of vertical disparity produced by a surface under spherical projection provides direct information about the absolute distance to that surface which could be used, in principle, to scale horizontal disparities directly. Indeed, it is not even necessary to consider the process as sequential – the combination of vertical and horizontal disparities generated by a surface uniquely specifies the physical shape and depth structure of the surface.

2.10   The Appearance of Fronto-Parallel Surfaces

Empirical support for the idea that a combination of the pattern of horizontal and vertical disparities determines the perceived shape of surfaces comes from some preliminary observations on the appearance of fronto-parallel surfaces. Stereoscopic images were accurately synthesised to mimic the viewing of fronto-parallel, planar surfaces at (i) 28 cm and (ii) infinity. For a fronto-parallel surface at 28 cm, there is a changing gradient of horizontal disparities across the surface from left to right as a result of the fact that the surface deviates from the Vieth-Müller circle (the locus of zero disparity points in the plane of regard). For a fronto-parallel surface at infinity, on the other hand, there is no changing gradient of horizontal disparities as the surface falls close to the Vieth-Müller circle of infinite radius. However, *both* surfaces are seen as approximately fronto-parallel under natural viewing conditions, although Ogle (1950) and others have noted that there are small, systematic departures from the appearance of flatness of surfaces which are physically fronto-parallel. Why should both surfaces be seen as fronto-parallel when there are considerable differences in the pattern of horizontal disparities in the two cases? One possibility would be that information from the different vergence angle in the two conditions is used to correct the perceived shape of the surfaces. Whilst we cannot rule out this possibility completely, we have observed that correctly constructed fronto-parallel surfaces simulating viewing at *either* 28 cm *or* infinity, are both seen as approximately flat even when vergence angle is kept constant.

The alternative possibility is that vertical disparities may play a role and it is the combination of vertical and horizontal disparities that is responsible for the perception of a surface as flat and fronto-parallel. To test this idea, a stereoscopic surface was synthesised with vertical and horizontal disparities appropriate for viewing at 57 cm. Observers reported that this surface was flat and lying in a fronto-parallel plane. They were then presented with a second simulated surface with horizontal disparities appropriate for 57 cm viewing, as before, but with the gradients of vertical disparities that would be appropriate for a surface at *either* 28 cm *or* at infinity. All observers reported that the presence of vertical disparities appropriate to 28 cm viewing produced a very marked impression of a surface which was *concave* with respect to the observer and, conversely, that the presence of vertical disparities appropriate to infinite viewing produced the very marked impression of a surface which was *convex* with respect to the observer. These findings are consistent with the observations reported by Helmholtz (1909) that the presence of beads suspended on the strings of vertical threads used in judgements of the fronto-parallel horopter had a very significant effect on the appearance of frontoparallel surfaces. Seen together, these results provide clear evidence that manipulations of vertical disparities alone do have a significant affect on the appearance of stereoscopic surfaces. These effects may be result of mechanisms which (i) extract the gradients of vertical disparity between the two stereoscopic images, (ii) use them to provide an estimate of the absolute distance to the surface and (iii) use the estimate to scale horizontal disparities. Alternatively, and more parsimoniously, the human visual system may use the combination of vertical and horizontal disparities, perhaps in the form of the gradients of deformation over a surface, to compute surface slant and curvature directly.

2.11   Vertical Disparities and Binocular Stereopsis – Conclusions

In order to correctly recover the shape and depth magnitude of stereoscopic surfaces, it is necessary to have knowledge, explicitly or implicitly, about the absolute distance to the surface. The significance of Mayhew and Longuet-Higgins' computational theory is that it shows that there is *visual* information in the vertical differences between corresponding points in the two images which could be used to compute the absolute distance to a surface. Under spherical projection, which more closely approximates the projection of images on to human retinae, we have shown that the information about absolute distance is available directly from the gradients of vertical disparity. Moreover, we now have the first empirical evidence to show that manipulations of the vertical disparity gradient do effect the perceived depth, and periodicity of 3-D corrugated surfaces and the apparent curvature of fronto-parallel surfaces. An intriguing possibility, which has not been investigated, is that the human visual system may use the overall pattern of horizontal and vertical disparities to recover information about surface slant and curvature without the need to compute the viewing system parameters of absolute distance and eccentricity with respect to the head, explicitly.

References

Anstis, S.M., Howard, I.P., Rogers, B.J. (1978) A Craik-Cornsweet illusion for visual depth. Vision Res. 18, pp. 213-217

Barlow, B., Blakemore, C., Pettigrew, J.D. (1967) The neural mechanism of binocular depth discrimination. J. Physiol. 193, pp. 327-342

Bishop, P., Henry, G.H., Smith, C.J. (1971) Binocular interaction fields of single units in cat's striate cortex. J. Physiol. 216, pp. 39-68

Brookes, A., Stevens, K.A. (1989) The analogy between stereo depth and brightness. Perception 18, pp. 604-614

Burt P., Julesz, B. (1980) A disparity gradient limit for binocular fusion. Science 208, pp. 615-617

Cagenello, R.B., Rogers, B.J. (1988) Local orientation differences affect the perceived slant of stereoscopic surfaces. Investigative Ophthalmology and Visual Science 29, p. 399

Cagenello, R.B., Rogers, B.J. (1990) Orientation Disparity, Cyclotorsion and the perception of surface slant. Investigative Ophthalmology and Visual Science 31/4, p. 97

Cumming, B.G., Johnston, E.B., Parker, A.J. (1991) Vertical disparities and the perception of three dimensional shape. Nature 349, pp. 411-413

De Bruyn, B., Rogers, B.J., Howard, I.P., Bradshaw, M.F. (1992) Role of positional and orientational disparities in controlling cyclovergent eye movements. Investigative Ophthalmology and Visual Science 33/4, p. 1149

Foley, J.M. (1980) Binocular distance perception. Psychol Rev. 87, pp. 411-434

Foley, J.M., Richards, W.A. (1972) Effects of voluntary eye movements and convergence on the binocular appreciation of depth. Percept. & Psychophysics 11, pp. 423-427

Gibson, J.J. (1950) The Perception of the Visual World. Boston, Houghton Mifflin

Graham, M.E., Rogers, B.J. (1982a) Simultaneous and successive contrast effects in the perception of depth from motion-parallax and stereoscopic information. Perception 11, pp. 247-262

Graham, M.E., Rogers, B.J. (1982b) Interactions between monocular and binocular depth aftereffects. Investigative Ophthalmology and Visual Science 22, p. 272

Helmholtz, H. von (1909/1962) Physiological Optics. New York: Dover. English translation by J.P.C. Southall from the 3rd German edition of Handbuch der Physiologischen Optik. Hamburg, Vos, 1909

Howard, I.P. (1970) Vergence, Eye Signature and Stereopsis. Psychon. Monongr. Suppl. 3, pp. 201-219.

Howard, I.P., Zacher, J.E. (1991) Human cyclovergence as a function of stimulus frequency and amplitude. Exp. Brain Res. 85, pp. 445-450

Julesz, B. (1960) Binocular depth perception of computer generated patterns. Bell System Technology Journal 39, pp. 1126-1162

Kertesz, A.E., Sullivan, M.J. (1978) The effect of stimulus size on human cyclofusional response. Vision Res. 18, pp. 567-71

Koenderink, J.J. (1985) Space, form and optical deformations. In: D.J. Ingle, M. Jeannerod, and D.N. Lee (eds.) Brain mechanisms and spatial vision. Dordrecht, Martinus Nijhoff

Koenderink, J.J. (1986) Optic flow. Vision Res. 26, pp. 161-180

Koenderink, J.J., van Doorn, A.J. (1975) Invariant properties of the motion parallax field due to the movement of rigid bodies relative to the observer. Optica Acta 22, pp. 773-791

Koenderink, J.J., van Doorn, A.J. (1976a) Local structure of movement parallax of the plane. Journal of Optical Society of America 66, pp. 717-723

Koenderink, J.J., van Doorn, A.J. (1976b) Geometry of binocular vision and a model for stereopsis Biological Cybernetics 21, pp. 29-35

Longuet-Higgins, H.C., Prazdny, K. (1980) The interpretation of moving retinal images. Proceedings of the Royal Society London, Series B/208, pp. 385-387

Marr, D., Poggio, T. (1976) Cooperative computation of stereo disparity. Science 194, pp. 283-287

Marr, D., Poggio, T. (1979) A computational theory of human stereo vision. Proceedings of the Royal Society London, Series B/204, pp. 301-328

Marr, D. (1982) Vision. Freeman, San Francisco, CA

Mayhew, J.E.W., Longuet-Higgins, H.C. (1982) A computational model of binocular depth perception. Nature 297, pp. 376-379

Ogle, K.N. (1938) Induced size effect I. A new phenomenon in binocular space perception associated with the relative sizes of images in the two eyes. AMA Arch. Ophthal. 20, pp. 604-623

Ogle, K.N. (1950) Researches in Binocular Vision. New York, Hafner

Ogle, K.N. (1955) Stereopsis and vertical disparity. AMA Arch. Ophthal. 53, pp. 495-504

Pollard, S.B., Mayhew, J.E.W., Frisby J.P. (1985) PMF: A stereo correspondence algorithm using a disparity gradient limit. Perception 14, pp. 449-470

Rogers, B.J., Graham, M.E. (1979) Motion parallax as an independent cue for depth perception. Perception 8, pp. 125-134

Rogers, B.J., Graham, M.E. (1982) Similarities between motion parallax and stereopsis in human depth perception. Vision Res. 22, pp. 216-270

Rogers, B.J., Graham, M.E. (1983) Anisotropies in the perception of three-dimensional surfaces. Science 221, pp. 1409-1411

Rogers, B.J., Graham, M.E. (1984) Aftereffects from motion parallax and stereoscopic depth: similarities and interactions. In: L. Spillmann, and B.R. Wooten (eds.) Sensory experience, adaptation and perception. New Jersey, Lawrence Erlbaum

Rogers, B.J., Graham, M.E. (1985) Motion parallax and the perception of three dimensional surfaces. In: D.J. Ingle, M. Jeannerod, and D.N. Lee (eds.) Brain mechanisms and spatial vision. Dordrecht, Martinus Nijhoff

Rogers, B.J., Howard, I.P. (1991) Differences in the mechanisms used to extract 3-D slant from disparity and motion parallax cues. Investigative Ophthalmology and Visual Science 32/4, pp. 152

Rogers, B.J., Bradshaw, M.F. (1992) Differential perspective effects in binocular stereopsis and motion parallax. Investigative Ophthalmology and Visual Science 33/4, pp. 1333

Rogers, B.J., Koenderink, J.J. (1986) Monocular aniseikonia: A motion parallax analogue of the disparity induced effect. Nature 332, pp. 62-63

Sobel, E., Collett, T. (1991) Does vertical disparity scale with the perception of stereoscopic depth. Proc. R. Soc. London B/244, pp. 87-90

Tyler, C.W. (1973) Stereoscopic vision: cortical limitations and a disparity scaling effect. Science 181, pp. 276-278

Tyler, C.W. (1974) Depth perception in disparity gratings. Nature 251, pp. 140-142

Experiments on Stereo and Texture Cue Combination in Human Vision Using Quasi-Neutral Viewing

John P. Frisby and David Buckley

AI Vision Research Unit, University of Sheffield

It has long been known to psychophysicists, from numerous demonstrations and experiments, that the visual system is not so much a single sense as a collection of many subsystems or modules. Each module is capable of extracting useful information about the attributes of objects and scenes from just one kind of data (or cue) carried by the optic array. Recent developments in neurophysiology have been consistent with this general picture. Many visual cells are sharply tuned to particular aspects of visual stimulation and they are organised into numerous different visual regions, each one appearing to be specialised for a particular form of visual information processing (e.g. for motion, for colour, etc.).

This picture of biological visual systems designed on a quite strongly modularised basis has been nicely complemented by a great deal of work in computer vision over the past two decades elucidating the computational structure of the tasks facing individual modules. This research has achieved some real successes. For example, numerous methods are now known for acquiring useful depth information from a variety of cues considered on their own, leading to a multitude of so-called *shape-from-...* algorithms (e.g. from *...stereo, texture, contour, shading, motion*). But sadly, the robustness of these algorithms in dealing with the vagaries of natural images presently leaves much to be desired. In this respect they fall well below the performance level of the human visual system, a fact which continues to hamper the use of machine vision systems for industrial purposes.

One way to improve the robustness of machine vision systems is to work towards an improved understanding of how to design modules specialised for a single cue but better able to cope with the form in which that cue appears in natural images. That obviously sensible approach is being adopted in many computer vision laboratories.

A separate and complementary approach is to seek methods for dealing with cues not in isolation but in combination. The underlying idea here is that the superbly robust performance of mammalian visual systems may reside in the ways they have evolved for exploiting the patterns in which different cues occur together in natural images of natural scenes. Perhaps biological visual systems treat cue combination not so much as a problem to be solved but rather as an opportunity to be seized.

If there is anything in this viewpoint, the task for visual scientists is to discover and exploit benefits that arise from processing cues not in isolation but jointly. The tools for tackling this problem straddle the fields of psychophysics, neurophysiology and computer vision. INSIGHT work has sought guidance from the structure and design principles exhibited by biological visual systems towards creating better computer vision systems, while at the same time attempts in INSIGHT to create the latter should lead to an improved understanding of why biological systems are designed the way they are. For reviews of the recent depth cue integration literature, see Bulthoff and Mallot (1987, 1988), Bulthoff (1991), and Aloimonos and Shulman (1989).

In the INSIGHT work described here, the focus was on developing psychophysical techniques to investigate the way stereo and texture cues interact in human vision of real 3D surfaces. Our concern to study real surfaces is in keeping with the general INSIGHT theme of vision for a natural environment. This work will be presented in three sections.

1 Judging the Amplitudes of 3D Ridges

Our first sequence of INSIGHT experiments measured perceptions of the amplitude of 3D ridges (figures 1-3) using a cue conflict paradigm in which the stereo cue to ridge amplitude was pitted against a combination of texture and outline depth cues (the latter two always signalling the same ridge amplitude). Cue conflict has often been used before to investigate disparity processing mechanisms (for a review, see Stevens and Brookes 1987, 1988) but there were several distinguishing features in our use of this paradigm.

First, we set out to investigate what we hoped would prove to be small cue conflicts. The largest conflict studied was a difference of 6 cm in ridge amplitude and generally it was much less than this. This maximum meant that the difference between slants of the steepest tangent planes to the ridges signalled by each cue was no greater than about 20°. We have previously discovered evidence that stereo and texture cues are pooled for this range of horizontal planar surface slants whereas larger conflicts can give rise to a species of rivalry in which one or other cue dominates at any given moment (Buckley 1988; Buckley *et al.* 1988; Buckley *et al.* 1991). This outcome indicates the existence of mechanisms that pool data about 3D shape from different cues only if those data are reasonably similar, and we aimed our experiments at studying possible pooling mechanisms of this type by keeping cue conflicts small. (The limits on 'similarity' implemented by such mechanisms would presumably be determined by the noise expected on each channel. An error model for the recovery of surface information from each cue would be required to arrive at a principled choice of 'small' cue conflicts but such data are not available. Nor did we think it a sensible starting point to try to obtain them before establishing the general character of the cue integration observed for the ridge stimuli.)

A second distinguishing feature of our experiments was that they studied cue integration for ridges with both vertically and horizontally oriented axes. We have previously discovered a substantial surface orientation anisotropy in cue integration for planar surfaces. Using stereograms, the perceived slant of planes rotated around the horizontal axis was found to be roughly the average of the stereo and texture cues for small slant cue conflicts (range 0-20°), whereas for planes rotated around the vertical axis the texture cue tended to be dominant (Buckley 1988; Buckley *et al.* 1988). This finding adds further weight to the evidence for a vertical/horizontal anisotropy for certain aspects of stereo processing (Anstis *et al.* 1978; Rogers and Graham 1983; Rogers and Cagenello 1989; Gillam *et al.* 1984; Gillam *et al.* 1988). Third, Stevens and Brookes (1987, 1988) suggested, from cue conflict studies, that stereo might in general be a stronger cue where the "surface exhibits curvature or edge discontinuities, i.e. where the second spatial derivatives of disparity are non-zero". Gillam *et al.* (1988) proposed a similar idea to explain their observations of long perceptual latencies for vertical surfaces by noting that "Relative disparity, unlike absolute disparity, does not change across a [planar] surface slanted around a vertical axis (p. 173)." Hence, they reasoned, long latencies for vertical surfaces might be explicable if the visual system generally relied on *changes* in relative disparity, needing highly non-local comparisons to extract surface slant if relative disparity information remained constant (as it does for vertical but not horizontal planar slants). This is an interesting notion and well worth examination even if the latency data that prompted it are controversial (Frisby *et al.* 1992b). We set out to test it by exploiting the finding that texture strongly dominates stereo for vertical planar stimuli (Buckley 1988). This result permitted us to ask the question: could stereo be made to play as strong a role as texture for vertical surfaces if disparity cues were provided with non-zero spatial derivatives? This was achieved by comparing cue integration for vertical and horizontal 3D ridge surfaces with parabolic and triangular depth profiles.

Fourth, we were interested in checking whether observations made using stereograms were replicable using *real* 3D ridges. Despite their many advantages, it is not always remembered that stereograms are intrinsically depth cue conflict stimuli in which accommodation is pitted against disparity for at least some regions of the field of view. The likelihood of this being an important factor is increased by the synkinesis that exists in the control of vergence and accommodation, as will be discussed at greater length later. Hence we used the cue conflict paradigm to investigate whether the operation of disparity mechanisms is different for natural viewing and stereogram viewing of putatively the same 3D surface shape.

The main features of the methodology were as follows (full details in Buckley and Frisby 1992).

Stereogram Display Device

Observers were seated in a darkened room and viewed stereograms displayed as red/green anaglyphs on a high quality RGB monitor. The screen was viewed from about 57 cm through red and green filters mounted in a headrest. The most distant edges of the 8 cm×8 cm ridges were arranged to appear in the plane of the screen and so only for those locations were the disparity cues in synchrony with the accommodation cues.

Surface Texture

The surface texture elements used throughout were circles of 2 cm diameter on the portrayed scene surface (not on the screen). These circles projected into the two stereo images as ellipses whose shape was determined by the size of the stereo and texture cues present in any given condition (figure 1). Circles were chosen because they present line segments of all orientations.

Arrangements of circles on the surface were of two types: 'regular' or 'jittered' (figures 1 and 3 respectively). The latter were used to check whether the intrinsic stereo ambiguity of regularly-repeating patterns, which means they can be fused in many different ways (the *wallpaper illusion*), played any part in our experiments. We doubted this was a serious hazard for our stimuli because there were always two sources of disambiguating information: the outline edges of the ridges, and a short (1 cm) straight line inserted in each stimulus at the apex of the ridge and oriented along the ridge axis specifically to

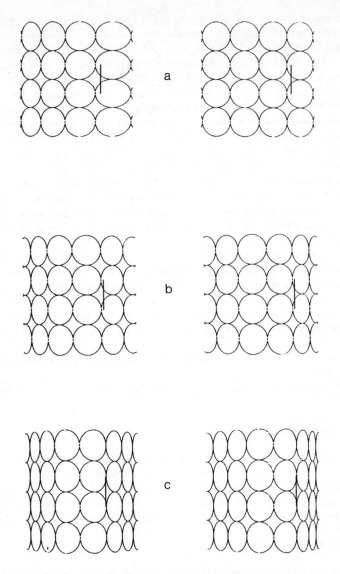

Figure 1: *Ridge stereograms with parabolic depth profiles similar to those used in the experiment described in the text, arranged for cross-eyed fusion. The requirements of this form of viewing mean that the stimuli are not exact replicas of those used in the experiment but they are sufficiently similar to give a realistic impression of what subjects saw. The texture and contour cues are appropriate for a viewing distance of 57 cm. The short straight lines are to help prevent incorrect fusions (see text, but note that in the experiment they were placed at the peaks of the ridges). (a-c) Vertical ridges. (d-f) Horizontal ridges. In all six stereo-pairs the stereo cue remains constant (i.e. portrays the same ridge amplitude). In (b) and (e) the stereo and texture/outline cues are consistent. In a and d the texture/outline cues signal a lower ridge amplitude than stereo, which is itself set to the same amplitude as in (b) and (e).*

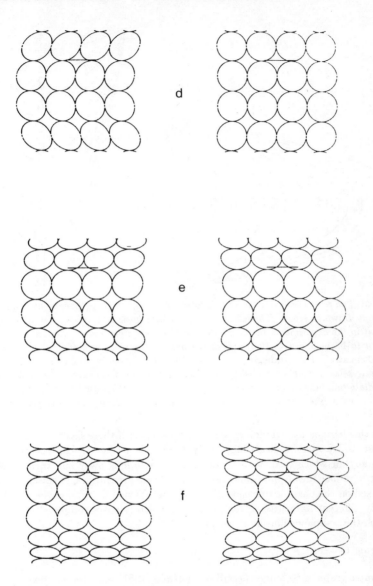

Figure 1: *(Cont.) Conversely, in (c) and (f) the texture/outline cues are signalling a higher ridge amplitude than stereo. If the reader observes the same type of vertical/horizontal anisotropic cue interaction as the subjects in the experiment, then the texture/outline cue will be seen to have a marked effect on the perceived amplitude for the vertical but not the horizontal ridges. Some highly practiced observers of difficult stereograms, however, including the first author, find that although they may experience strong domination by texture/outline cues on initial exposure to the vertical stereograms, with careful and sustained viewing they see as much depth build up over time as occurs from the outset for the equivalent horizontal stimuli.*

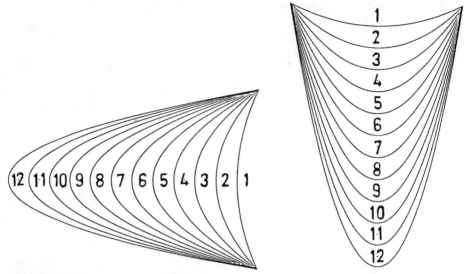

Figure 2: *The response scales used by the observers in all the experiments involving parabolic ridges (not drawn to scale). The left scale was used for horizontal ridges, the right for vertical. Note that in the experiments the bases of the scales were 8 cm across and each end of the base was fixed to be in line with the edges of the base of the surface in the stimulus, which were about 10 cm away. The parabolas had amplitudes of 1 to 12 cm in steps of 1 cm. The observer's task was to select a number guided by the scale that reflected perceived ridge amplitude, using intermediate numbers as appropriate. For Experiment 3, which used triangular ridges, similar scales were used except that the parabolas were replaced by triangular ridge profiles.*

prevent this happening. Observers were asked to report if these short lines were ever seen as double, which would indicate incorrectly fused circle elements. This occurred only very occasionally, in which case the observer was asked to look away and then make a fresh attempt to re-fuse the display. But in addition to these precautions, in certain experiments we checked on the possible intrusion of a wallpaper illusion by testing whether the results were any different for a jittered texture layout which removes completely the ambiguity on which the wallpaper illusion depends.

Stereogram Generation

A computer graphics technique described by Ninio (1981) was used to create stereograms prior to the experiments. His method takes points lying on the desired binocularly-viewed scene surface and then uses ray casting through the optical centres of the eyes to compute perspective projections of the left and right stereo images suitable for the geometry of the stereoscopic apparatus. If texture elements are evenly distributed over the desired scene surface then this technique ensures that image projections contain geometrically-correct texture gradient cues to the required surface shape. The texture gradient and outline cues in the cue conflict conditions were controlled in the left eye's view only, with the imposition of the required stereo disparity cue then determining the positions of matching elements in the right eye's image.

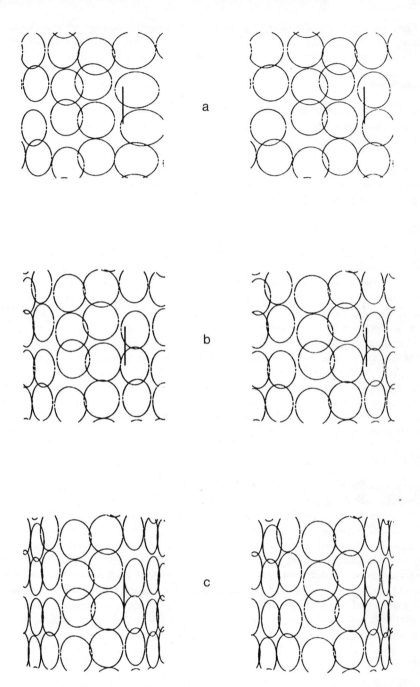

Figure 3: *As for figures 1a-c but with jittered texture.*

The accuracy of the computer program for generating the stereograms was checked by comparing its output with video pictures of a sample of real ridges possessing the required surface texture and outline cues. No discernable discrepancies were found.

Outline Cues

Youngs (1976) found that outline shape had a strong effect on the perceived slant of a simple untextured stereogram of a quadrilateral: stereograms containing converging outlines were seen as more slanted than ones with parallel contours (see also Clarke *et al.* 1956). This is a strong effect which also occurs for textured stereograms (Buckley 1988). In the present experiments, the outline cue was arranged to convey the same amplitude information as the texture cue, and hence it too was always manipulated only in the left image. Thus the cue conflict was always stereo vs. texture/outline.

Ridge Amplitude Judgements

The observer was asked to judge the amplitude of the surface by matching it to one of a series of numbered depth profiles displayed in response scales (figure 2; cf. Todd and Akerstrom 1987). These were mounted on a large matt black frontoparallel screen which surrounded the monitor. Observers were encouraged to use intermediate numbers on the scale if they wished to record intermediate judgements of perceived amplitude.

It is worth noting at this point that vergence eye movements were needed to fuse all the stimuli, given standard definitions of Panum's fusional limits (Boff and Lincoln 1988), as the disparity ranges of the various amplitudes were as follows (in ° visual angle): 3 cm $= 0.36°$, 5 cm $= 0.63°$, 6 cm $= 0.77°$, 7 cm $= 0.91°$, 8 cm $= 1.06°$, and 9 cm $= 1.22°$. Observers were free to scan the stimuli as they wished. It is reasonable, therefore, to assume that vergence movements were constantly being made, with data arising from local fusions established within Panum's limit contributing to the overall fused percept of the stimulus.

Real Ridge Stimuli

All experiments began by training the observer in the use of the amplitude response scales for a sample of accurate full-scale models of the kind of parabolic ridges portrayed in the experimental stereograms, with amplitudes 2, 4, 6, 8, and 10 cm. These 'real ridge stimuli' were made by bending cardboard bearing a print-out of the texture used for the experimental stimuli over a wooden former that was itself invisible to the observer. The only obvious differences between the resulting texture/outline cue and that for the anaglyphs were: (a) for these real stimuli the circles were dark on a white background, whereas contrast was reversed in the anaglyphs; and (b) the outline cue was created by the real edge of the white paper being seen against the black surround, whereas for the anaglyphs it was a subjective contour formed by the interrupted circle outlines (figure 1). Care was taken with the lighting of these real ridges to avoid shading depth cues.

Observers

The 26 naive observers were unpaid volunteers aged between 22 to 26. All had normal or corrected to normal vision. Each observer served for only one experiment, and each was screened for good stereopsis using the Titmus Randot Test (criterion for inclusion was stereoacuity of 30 sec arc or better).

Experimental Design

All experiments used a fully repeated measures design, with counter-balanced pseudo-randomly ordered presentations spread over a sequence of two or three experimental sessions, each lasting up to about 40 minutes. Two judgements of amplitude were collected per stimulus condition and the means of these were analysed using anovas. Monocular (left eye) presentations of the left halves of the eight stimuli whose cues were all in accord were also included. The experiment began with a short training session using feedback for both vertical and horizontal real ridge stimuli with amplitudes of 2, 4, 6, 8, and 10 cm to a criterion of two correct scale choices with no intervening errors for each of the training surfaces.

Results

A series of five experiments were carried out using the above methodology. Full results are provided in Buckley and Frisby (1992). The findings from two experiments are used here to illustrate our main conclusions.

Figure 4 shows the results from an experiment which used stereograms of ridges with parabolic 3D profiles, vertical and horizontal ridge orientations, regular textures, and ridge amplitude cues of 3, 5, 7 and 9 cm. All permutations of the four levels of stereo and texture/outline cues were included, producing 16 stimuli for each ridge orientation.

Figures 4a and 4b display the group means (N=6) for the vertical and horizontal ridges respectively. Amplitude judgement is plotted on the ordinate, stereo cue amplitude is on the abscissae and the parameter is the level of the texture/outline cue. Alongside each line in brackets is the slope of its best-fitting straight line. These give an indication of the power of the stereo cue at each level of the texture/outline cue (see later). Inter-subject differences can be judged from the $+/- 1$ s.e. error bar, which uses the average of the s.e.s for the 16 data points.

The dashed lines link together points arising from conditions in which the three cues were consistent. Points on the solid lines have the same level of the texture/outline cue and comprise one point from a consistent cue pairing and three from inconsistent pairings. If the observers had judged the consistent conditions veridically (i.e. as taught to do so in training on real ridges with consistent cues), then the dashed lines should have had a slope of 1 and they should have passed through the 9/9, 7/7, 5/5 and 3/3 points. In fact, the consistent cue pairings produced judgements quite close to veridical but with some undershoot for shallow vertical ridges and some overshoot for steep horizontals ridges, producing slopes for both consistent lines of about 1.2.

Separate anovas for the two ridge orientations showed that the stereo and texture/outline cues produced significant main effects in both cases. The vertical ridges, however, unlike the horizontals, also showed a significant interaction between the two cues. The smallest significant F for these tests was $F_{3,5} = 11.38$ (p < 0.01).

These anovas reflect the very different pattern of cue integration evident in 4a,b for the two ridge orientations. Thus, stereo was by far the stronger cue for the horizontal ridges (slopes >= 0.73), although the separations between the lines for the various values of the texture/outline cue show this cue had some effect on perceived ridge amplitudes. For vertical ridges, the picture is one of stereo being completely overwhelmed by the most shallow (3 cm) texture/outline cue, and then of a progressively greater influence of stereo (steeper slopes) as the texture/outline cue itself signalled higher ridge amplitudes.

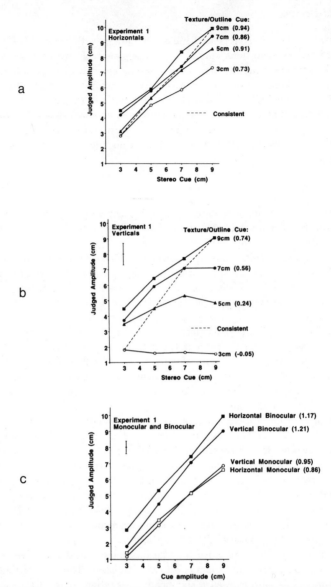

Figure 4: *Group means from Experiment 1 (N=6). (a) vertical and (b) horizontal parabolic ridges. The numbers in parentheses are the slopes of each line. The slopes of the dashed lines (see text) that link the consistent stereo/texture points were 1.21 in a and 1.17 in (b). The error bar at the top left corner of each graph is the mean of the 16 standard errors calculated for each of the means shown from the individual subject means. This bar therefore reflects differences between subjects and not the error variation used in the various anovas cited in the text, as they were repeated measures designs. (c) shows the means for the vertical and horizontal monocular judgements, together with, for easy comparison, the means of the binocularly viewed stimuli from (a) and (b) which had consistent texture/outline and stereo cues. In (c) the error bar is for the monocular means only.*

The question arises as to what the data would have looked like if they had been produced by an equal weighting of the two cues. One approach to answering that question is to generate predictions for the em inconsistent cue conditions on the assumption that the judged amplitudes for the consistent pairings were determined equally by stereo and by texture/outline. As the dashed lines linking the consistent pairings had slopes of about 1.2, this approach generates equal-weighting predictions of slopes of half that size for the solid lines (i.e. of about 0.6 for the present experiment). While the assumption on which this approach is founded can of course be questioned, it nevertheless provides a helpful baseline for judging the meanings of the slopes actually observed. It further emphasises the strong domination by stereo for the horizontal ridges (slopes generally much greater than 0.6).

For the vertical ridges the pattern was less straightforward. They showed complete domination by texture/outline for low levels of that cue. Stereo played an increasing role as the texture/outline cue itself was increased, until for the 9 cm texture/outline conditions stereo had a greater effect (slope = 0.74) than predicted on the assumption of equal weighting (0.6). The slope of 0.74 reflects the fact that, when the stereo cue was signalling 3 cm, the 9 cm texture/outline cue failed to 'pull up' perceived ridge amplitude as much as predicted given the equal-weighting hypothesis.

We find it remarkable that the vertical ridge conditions with a 3 cm texture/outline cue *showed no appreciable effect of the stereo cue regardless of its size despite it being seen as non-planar*. Remarks volunteered by some observers when viewing the 3 cm texture/outline *vs* 9 cm stereo vertical ridge condition suggested that it caused "eye strain". Perhaps those observers found it difficult to produce the vergence shifts required for fusion of the disparate texture elements when the texture/outline cue was working strongly in the opposing direction, i.e. encouraging vergence positions close to the monitor screen, which is where the accommodation/vergence synkinesis was also directing vergence. Note that no reports of difficulties were reported for any vertical ridges other than the 3 cm texture/outline vs 9 cm stereo conflict, nor for any horizontal stimuli. Despite these reports of difficulties from some observers, as far as we could judge all managed to fuse the vertical 3 cm texture/outline *vs* 9 cm stereo stereograms successfully before making their judgements. Observers were reminded throughout that they should achieve correct fusion before responding (see precautions described earlier).

The experiments used relatively inexperienced observers of stereograms. Inspection of their individual data shows they all produced the general pattern of results shown in figure 4.

We turn now to the monocular conditions. They were included to check that the texture/outline cue was effective when it was presented without stereo. Figure 4c displays the group means for the monocular conditions, along with the comparable (i.e. consistent cue) binocular conditions. The main point to observe in 4c is that the texture/outline cue was sufficient on its own to generate 3D ridge perceptions even though the overall means for the monocular conditions were well below those for the equivalent binocular ones ($F_{1,5}$ = 62.48, p<.001). The other point worthy of mention is that there was no significant 3-way interaction between the factors of vertical/horizontal ridges, monocular/binocular viewing, and cue amplitude. This suggests that the ridge orientation anisotropy observed for cue integration in the cue-conflict stereograms was not caused simply by differences in the power of the texture/outline cue as a function of surface orientation, although it is unsafe to assume that the performance of a 'monocular depth cue' under monocular viewing can be extrapolated straightforwardly to its role under binocular viewing.

The weakness of the stereo cue for vertical surfaces when opposed by a conflicting texture/outline cue is in keeping with previous demonstrations of stereo being relatively poor for vertical surfaces. However, the important new feature is the fact that this anisotropy was also observed here *even though surfaces were used with non-zero second spatial derivatives.* This is strong evidence against the theories of the stereo anisotropy suggested by Gillam *et al.* (1988) and Stevens and Brooks (1987, 1988).

Other experiments in the series of five explored *inter alia*: (a) a reduced cue amplitude range of 5-8 cm (a further precaution against using extreme cue conflicts); (b) jittered textures (a precaution against the wallpaper illusion); and (c) triangular ridge profiles (which inject a sharp 3D depth discontinuity). Despite these variations, all studies demonstrated the same general pattern of results described here.

Two of the five experiments compared real ridges with stereograms of ridges. Figure 5 shows the group means (N=6) from one of these studies which was similar to that just reported except that it used only vertical ridges and only texture cues for 3 cm and 9 cm ridge amplitudes (the fuller sampling of the stereo cue avoided subjects adopting a restricted set of responses).

This experiment also varied texture density as a control against the fact that the texture cue for a 9 cm amplitude ridge introduces into each image roughly twice as many circles as does the texture cue for a 3 cm ridge, simply because of its large surface area (recollect that in all cases the circles *on the surface* were 2 cm in diameter). Density was manipulated by choosing appropriate sizes for element diameters to achieve the required densities of 20 and 40 elements per stimulus. It turned out that there was an effect of density but it was very small, with the denser stimuli being judged 0.36 cm higher in amplitude overall ($F_{1,5}$ = 12.48, p<0.05). Moreover, there were no significant interactions between density and other factors. We conclude that a density effect can be neglected in discussing the main features of this series of experiments.

The pattern of results shown in figure 5 for the vertical stereograms replicates that from the previous experiment (recollect that horizontal ridges were not included in the present experiment). Thus for the stereograms (5*a*), the cue interaction term was significant ($F_{3,15}$ = 7.09, p<.05): the 3 cm texture/outline cue strongly dominated all levels of the stereo cue, whereas cue integration was observed for conditions including the 9 cm texture/outline cue. For the real ridges (figure 5*b*), on the other hand, stereo dominated the texture/outline cue throughout ($F_{3,15}$ = 80.21, p<.001) and the cue interaction term was not significant. In a separate experiment, whose results we do not report here, data for horizontal real ridges closely resembled those depicted in figure 5*b* for vertical real ridges, i.e. the ridge orientation anisotropy shown in figure 4 for the stereograms did not occur for the real ridges.

In view of the spontaneous reports from some observers in the first experiment regarding "eye strain" from the 3 cm texture/outline vs 9 cm stereo condition, all observers in the present one were explicitly asked whether they experienced difficulties with this stimulus. Three (out of 6) said they did so, using such remarks as they found this stimulus "very difficult", "very unstable", "difficult to fuse", "made me feel that I'm looking in the wrong place". No difficulties were reported for the other stereograms, nor for the real ridges. The implications of this outcome are discussed below.

The main conclusion we draw from this experiment is that *stereo is not a weak cue when pitted against texture/outline in real vertical ridges.*

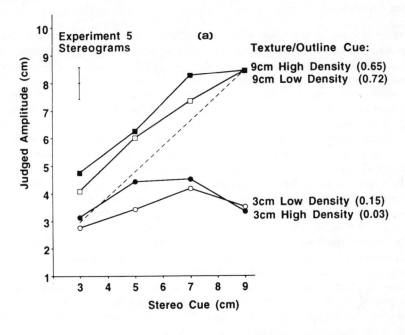

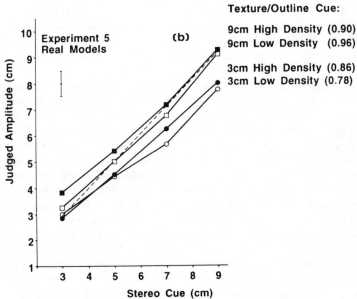

Figure 5: *Results (N=6) of the second experiment described in the text (fifth in the series sampled here) which used vertical ridge stereograms (a) and vertical real ridges (b), in the same format as in Figure 4. The slopes of the dashed lines (consistent stereo/texture pairings) are 0.93 in a and 1.04 in (b).*

General Discussion

This series of experiments, which used a total of 26 naive observers, produced clear and consistent evidence of a vertical/horizontal cue integration anisotropy in stereograms of 3D ridges, for both parabolic and triangular depth profiles. When using stereograms of horizontal ridges, stereo strongly dominated the texture/outline cue, whereas the reverse was true of stereograms of shallow vertical ridges (up to about 5-6 cm). This anisotropy was not seen for real ridges, for which stereo was the strongly dominant cue throughout.

A key objective of the present work was to test whether, in human vision, stereo for vertical surfaces would be strengthened when disparity cues were arranged to have non-zero second spatial derivatives. The answer to this question is negative, for our cue conflict paradigm. The evidence is quite conclusive that, even though disparity cues in stereograms may be strengthened in other settings when they portray non-planar surfaces, such an effect is not sufficiently strong for them to overcome the dominance of texture/outline for stereograms of vertical ridges with amplitudes up to about 5-6 cm.

The explanation we offer for our findings runs as follows.

The human visual system experiences certain problems in extracting and/or interpreting disparity cues in stereograms for vertical surfaces, be they planar or non-planar. These problems (whose nature we will not discuss in detail here) tend to make both vertical planes and vertical ridges appear shallower than they 'should' do. If a texture/outline cue is present which supports the perception of a shallow vertical ridge then this is in keeping with the tendency to 'under-utilise' the disparity information, producing in our paradigm the result of domination by the texture/outline cue.

Equally, if a higher-amplitude ridge is indicated by the texture/outline cue than by stereo, the tendency of stereo to be interpreted as relatively shallow 'pulls down' perceived ridge amplitude. Paradoxically, this produces in our paradigm the result of stereo cue domination insofar as it steepens the slope of the 9 cm texture/outline cue line in figures 4 and 5.

We further suggest that the reason why the *real* ridges did not display anisotropy in cue integration is that they presented oculomotor information (vergence and accommodation) that differentially assisted stereo due to the synkinesis that exists between disjunctive eye movements, accommodation and disparity. In the real ridges, the disparity cues were, by virtue of the means used to create them (i.e. cardboard 3D models), everywhere in synchrony with the oculomotor cues. In the stereograms, disparity was everywhere in conflict with accommodation, except at the most distant edges of the stereogram ridges which lay in the plane of the monitor screen. The size of this conflict in the stereograms increased with the size of ridge amplitude signalled by the stereo cue.

In support of this explanation of the real ridge data, we note the difficulties that observers experienced with stereograms presenting a 3 cm vs 9 cm conflict between texture/outline and stereo. Their introspections suggest that they had difficulties in generating the vergence movements necessary for fusion. We think they overcame these difficulties prior to making their amplitude judgements (because of the checks in our experimental procedure). Nevertheless, these introspections draw attention to the importance of taking into account the synkinesis between disparity and oculomotor information, and the associated fact that stereograms are intrinsically cue conflict stimuli.

This may be why the vertical/horizontal stereo anisotropy described in the psychophysical literature has no obvious counterpart in the everyday perception of normal scenes. Walls and doors do not appear obviously non-veridical when seen from angled viewpoints. The psychophysical laboratory findings may be a phenomenon only of the impoverished

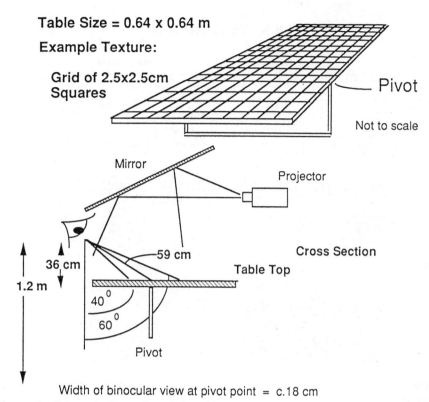

Figure 6: *Schematic illustration of the table apparatus, for a viewing distance ((d), eye to pivot) of 1.95m. See text for explanation.*

visual environment that generated them, in which, typically, disparity cues are out of synchrony with oculomotor cues because of the use of stereograms.

The question as to why disparities arising from vertical and horizontal surfaces should be treated differently by the human visual system is left open by these results. The present experiments are, however, evidence against explanations which predict that the anisotropy should be lost if disparity cues are arranged to carry non-zero second order spatial derivatives.

2   Judging the Slants of Ground Plane Surfaces

The experiments just described demonstrated that quite different results can be obtained when stereo mechanisms are explored using stereograms and real surfaces. Given IN-SIGHT's concern with vision for a natural environment, we decided to concentrate all our investigations on stereo/texture cue integration using real surfaces, specifically ground plane surfaces. The ecological significance of the ground plane has often been noted (Helmholtz, Gibson) and yet surprisingly little work has been done using ground planes as experimental stimuli. An interesting exception is Hanny and von der Heydt (1982) who explored Helmholtz's idea that the tilt of the vertical horopter was such as to place it in the ground plane. They reported neurophysiological data suggesting plasticity in tilt

SIX TEXTURES

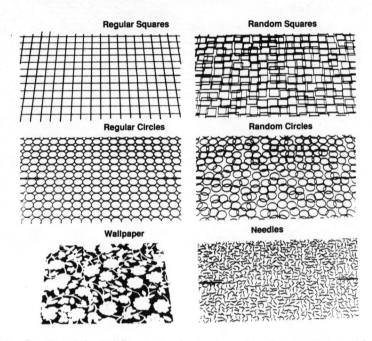

Figure 7: *Samples of the 6 different stimulus types used in the experiments with the table apparatus.*

of the vertical horopter: it was tilted backwards for kittens reared in the dark except for periods of exposure to light in a box with only its floor illuminated, but tilted forward for kittens whose exposure box had only the ceiling illuminated. In each case the tilt could be interpreted as an adaptation to facilitate disparity processing of a prominent environmental feature.

The apparatus we developed to explore disparity mechanisms tuned to the ground plane is illustrated in figure 6. An observer standing at the edge of a table looks down on it through a tilted headrest that occludes the table's edges and provides a field of view of about 20×20° v.a. The table-top is supported on a pivot which allows it to be slanted in the 'uphill/downhill' direction from the observer's vantage point to angles +/-10 degs from 'horizontal' (defined as parallel to the ground plane on which the observers stood; this plane had a slant angle of about 53° from fronto-parallel along the optic axis through the central spot). Viewing distance (d) to a small illuminated spot on the table-top in the centre of the field of view was set to either 1.95m or 0.59m. This was achieved by using two different tables, one for each distance, while maintaining the same headrest and hence viewing location. The observer's task is to judge the slant of the table's surface. About 70% of our volunteer naive observers have proved able to do this subjective estimation task quite reliably following about 10 minutes of training with feedback.

The advantage of this simple apparatus is that because slant cues are created by natural viewing of a real surface, disparity, accommodation and vergence are in synchrony wherever the observer looks. This is not the case in stereograms. Moreover, it is feasible in this apparatus to pit a conflicting texture slant cue against the stereo table-slant cue by projecting suitable artwork on to the table-top from a projector overhead, making due allowance in generating the projected patterns (using computer graphics) for the affect of the physical slant of the table on the images of the texture. We have used a variety of different textures in this apparatus (figure 7).

Texture Components

In the experiments to be described, we investigated the roles played by different components of texture depth cues, a question studied by many others (e.g. Cutting and Millard 1984; Stevens 1981; Blake and Marinos 1991). Specifically, we have explored the relative effectiveness of perspective (p) and compression (c) components of texture in combating conflicting stereo cues. Figure 8 and its legend define what we mean by p and c, and also introduces the labels Cpc, Ipc, Ip and Ic to refer to various mixtures of stereo and texture components.

We have used the table apparatus in a series of 7 experiments (using over 40 subjects), with the main results as follows.

Stereo & Texture Cue Combination

Figure 9 provides group means (N=4) obtained by pooling data for the 6 textures shown in figure 7 as there were no significant differences created by texture type. For clarity, figures $9a$ and $9b$ show the data for just the Cpc and Ipc conditions. In $9a$ ($d = 1.95$m), the roughly horizontal slope of the data from the Ipc conditions (stereo inconsistent with both p and c texture components) indicates roughly equal weighting of stereo and texture: in these two Ipc conditions, one cue was set to 0 degrees, the other to -10 degrees, and the mean judged slant in both cases was about -5 degrees. On the other hand, in $9b$ ($d = 0.59$m), greater weighting was given to stereo ($F_{1,6} = 20.74$, p $<.01$) as can be seen from the non-zero slope of the Ipc data line.

This result is consistent with the idea that the human visual system combines information from different depth cues according to the errors to be expected from each one, in that stereo errors increase with increased viewing distance so that stereo should receive more weight for the nearer of the two values of d used here. Alternatively, or perhaps in addition, the fact that texture had less influence for the nearer distance may reflect a contribution from accommodation cues working in support of those for stereo, in so far as the accommodation cue might also be more effective for the nearer viewing distance.

Combination of Texture Components

Figures $9c$ and $9d$ show the same results as $9a$ and $9b$ but added to them are data from the inconsistent cue conditions in which just one or other of the p and c texture components were set to be inconsistent with stereo (Ip and Ic respectively). For both values of d, the effectiveness of any given texture density cue could be predicted from a linear combination of the effects of its p and c components ($Ipc = Ip + Ic$). However, the details differed for the two d values, with greater weight given to c than to p at the nearer d.

This finding of a linear combination of p and c might be accounted for by spatial frequency tuned slant-from-texture mechanisms (e.g. Turner *et al.* 1991) that pool power in all orientations, such that they behave isotropically. We plan to develop a computational model exploring this idea.

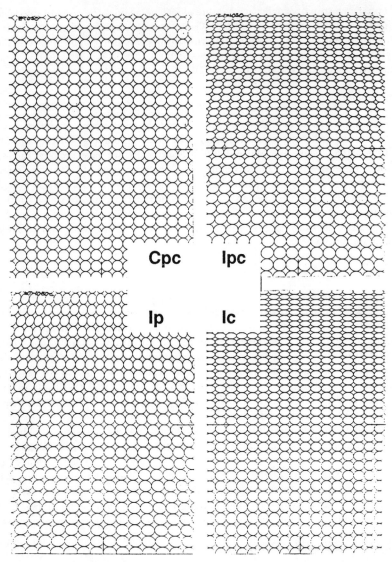

Figure 8: *Illustrations of stimuli showing the definitions of perspective ('p') and compression ('c') components of texture used here. (a) A regular grid. (b) The width of the texture elements in (a) changed to signal a receding slant: this is a change in the 'p' component. (c) The height of the texture elements in (a) changed to signal a receding slant – this is a change in the 'c' component. (d) Both 'p' and 'c' components changed. We use the label 'pc' to refer to conditions in which both perspective and compression components of texture are signalling the same slant; whereas 'p' or 'c' on their own mean that just the 'p' or the 'c' texture component was set to a given value, with the other component being consistent with stereo. Hence, if these stimuli were projected on to the table apparatus when it was set to be horizontal, 'Cpc' would mean both texture components were consistent with stereo, 'Ipc' that both were inconsistent, and 'Ip' and 'Ic' that just one or other were inconsistent with stereo.*

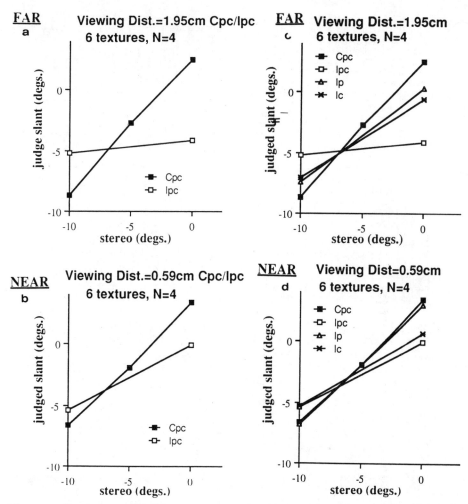

Figure 9: *Group means (N=4) from experiments described in the text using the table apparatus pooled over 6 textures. For clarity, s.e.s are not shown but they were about 1 degree. The solid squares show conditions in which the texture and stereo cues were Consistent: these are coded Cpc (see figure 8 for terminology). The open squares show Inconsistent (Ipc) conditions in which a 10° conflict existed between the stereo and texture components. The triangles indicate Inconsistent 'p' conditions, the crosses Inconsistent 'c' conditions. All the data are shown in (c) and (d) but for clarity just the (same) 'Cpc' and 'Ipc' data are shown in (a) and (b).*

These results conflict with those of Cutting and Millard (1984). They required observers to make surface judgements from texture displays on a computer monitor, and they found that the p component was important for planarity judgements and the c component for curvature. Our results do not reveal any special significance of p for our planar slant task when $d = 1.95$m, and they suggest dominance by c when $d = 0.59$m. These are sharply differing findings from Cutting and Millard. They may reflect the fact that we used real surfaces rather than pictorial displays. If so, it underlines the need to be cautious about extrapolating conclusions from artificial to natural settings. We are engaged

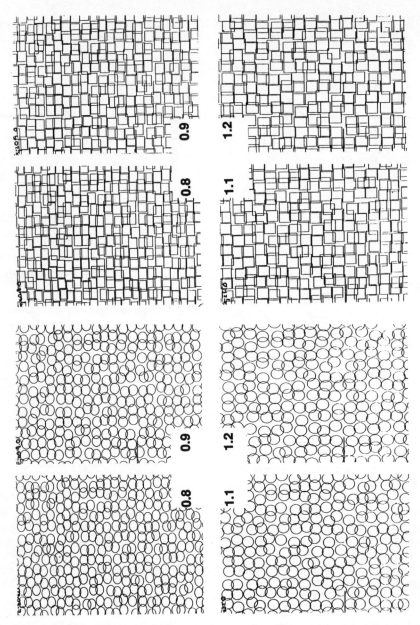

Figure 10: *Textures comprised of ellipses and rectangles. The numbers 0.8, 0.9, 1.1 and 1.2 are defined on the abscissa of the graph shown in figure 11a.*

in further experiments to check these findings for planar slant judgements and we are developing our table apparatus to permit investigations of 3D curvature perception. This work will continue to be guided by the computational analyses of our INSIGHT partners, Gårding and Eklundh (see Chapter 14).

Effect of Texture Element Shape

Using the table apparatus, we have compared textures of ellipses with textures of circles, and rectangles with squares (figure 10). The theoretical background to this work was an attempt to test whether the human visual system implements slant-from-texture methods, such as the WISP model of Gårding (1991; see Chapter 14), which find the plane in which the texture elements are weakly isotropic. Ellipses and rectangles were defined as contractions or expansions (0.8 to 1.2) from circles/squares (1.0) along an axis extending away from the observer. The abscissa of figure 11 a illustrates our meaning of the 0.8 to 1.2 scale, and it shows that there were four elliptical and four rectangular texture types, with the circles/squares at the centre of each series (each series, therefore, having five members). Note that the 0.8 to 1.2 values refer to scene elements, not to the shapes of their projections into images (which is why, in figure 10, the textures are displayed from a viewpoint perpendicular to the surface).

Table 1: *Table of group means (N=5) of slant judgements for all conditions in an experiment using elliptical and rectangular texture elements of the type shown in figure 10.*

Stereo	Ellipse					Rectangle				
Slant	Contraction			Expansion		Contraction			Expansion	
(Deg)	0.8	0.9	1.0	1.1	1.2	0.8	0.9	1.0	1.1	1.2
0.0	-0.3	1.8	1.2	2.4	3.3	1.9	1.5	0.9	0.7	1.7
-2.5	-2.1	-1.3	-2.3	-0.5	0.6	-0.5	-1.1	-1.6	1.0	1.0
-5.0	-4.5	-4.0	-3.8	-2.4	-1.8	-3.5	-3.5	-3.5	-3.1	-1.7
-7.5	-5.4	-4.7	-5.5	-5.0	-3.8	-7.8	-7.2	-6.1	-5.7	-6.2
-10.0	-9.0	-8.3	-8.2	-7.3	-6.3	-10.7	-9.5	-9.6	-9.4	-8.0
Means	**-5.0**	**-4.3**	**-3.3**	**-3.7**	**-2.6**	**-1.6**	**-4.1**	**-4.0**	**-4.0**	**-3.3**	**-2.7**

For each texture, five stereo slants were used by varying table angles in 2.5° steps from 0° to -10° (see table *1*). These conditions were randomly ordered over two experimental sessions, with two judgements made per condition by each of the five observers. The group mean data pooled over the various levels of the stereo slant cue (bottom row of table *1*) are shown in figure 11 a.

A visual system has (at least) two ways of interpreting an image comprised of a field of ellipses or rectangles. It could assume the image texture elements arise from certain elliptical/rectangular shapes in the scene and find a slant-from-texture consistent with the particular elliptical or rectangular scene texture element chosen. We will call this the *E/R assumption*. (We do not discuss further here how a particular value could be chosen for the assumed elliptical/rectangular scene shapes.)

Alternatively, it could assume that the elliptical/rectangular image elements arise as the projections of circles/squares and find a slant-from-texture accordingly. We will term this the *C/S assumption*. (Or, of course, the visual system in question might pool data arising from different texture modules each operating according to one or other of these different assumptions.)

We have enquired what the human visual system does when faced with fields of ellipses and rectangles of the kind shown in figure 10 using the table apparatus with $d =$ 1.95m. The design of the experiment was such that the texture elements always provided slant information consistent with the stereo cue (physical slant of table) *given the E/R*

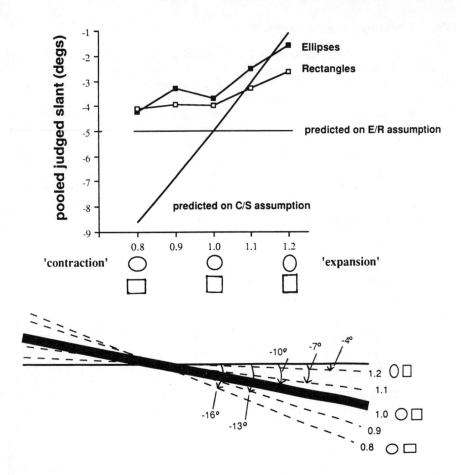

Figure 11: *(a) Plot of means shown in the bottom row of the table obtained by pooling over different levels of stereo slant cue. For clarity, s.e.s are not shown but they were about 1 degree. (b) Illustration of conflicts between the texture and stereo cues, given the C/S assumption (see text). The solid line illustrates the case when the stereo cue was -10° (created by the setting of the table angle). The dotted lines show the slants signalled by the various texture shapes (0.8–1.2) given the C/S assumption that these shapes are interpreted as circles and squares in the scene. Assuming equal weighting of the stereo and texture cues, the predicted perceived slant for any given pairing would therefore fall midway between the value for stereo (-10°) and the value shown for the particular texture element shape. Figure 11.a shows predictions (pooled over all five levels of the stereo cue) for both the C/S and E/R assumptions.*

assumption. Hence, on the C/S assumption (i.e. circles/squares in the scene), the elliptical/rectangular texture elements were in conflict with stereo to varying degrees (an example of this situation for a table slant of -10° is shown in figure 11b).

The results, averaged over all stereo slants, are shown in table 1 along with predictions for each assumption. The contraction/expansion factor was significant but the data are separated into two regions. For the expanded shapes (1.1 and 1.2) the data seem to support the C/S assumption, whereas for the contracted shapes (0.8 and 0.9) they are more consistent with the E/R assumption.

However, the results for the 0.8 and 0.9 conditions might reflect the fact that the method used to create the stimuli led to the overall density of elements being confounded with element shape. That is, the method for stimulus generation began by creating a field of boxes of the shape required for a given expansion/contraction, holding box width constant. For the elliptical stimuli, an ellipse was then drawn to touch the sides of this box. Finally, the centres of the ellipses were jittered to create the texture fields displayed in figure 10, and the box outlines suppressed. A similar procedure was used for the rectangles, except that they were formed directly by the box outlines. This procedure produces a denser layout the more contracted the elements.

Inspection of the textures shown in figure 10 reveals that this density effect was modest and we hoped that it would be unimportant. We allowed its presence because we were more concerned to make the various textures have roughly equal numbers of contour intersections per element, and our procedure controlled for that factor. But it is conceivable that the effect of this greater density may have been to strengthen the contribution of a shape-from-texture module operating on the compression component of the overall texture density gradient without regard to the expansion/contraction of texture element shapes. If so, this module may have led to greater texture support for the stereo cue for the contracted element conditions. The potential role of this density factor may repay further study.

The differences between the rectangles and ellipses for the 1.1 and 1.2 conditions (table *1* and figure 11 *b*) were not significantly different although the trend was for a weaker effect for rectangles. This might reflect the fact that whereas it will generally be 'ecologically valid' to assume that ellipses in the image arise from circular scene elements, it would not be safe to assume that image rectangles generally arise from squares. This was the underlying reason for including the rectangles, and we are surprised that they produced even moderately similar data to the ellipses.

To sum up on (iii), the data suggest that the human visual system, over a fair range of the stimuli used, interpreted the texture elements as the projections of circles or squares. This evidence is in keeping with models of slant-from-texture, such as the Gårding's WISP model (Chapter 14), which find the plane in which the texture elements are (weakly) isotropic. On the other hand, this interpretation is not sufficient to account for all the characteristics of the data.

The table apparatus seemed in these experiments to be proving of value as a general technique for measuring the strengths of putative texture effects by seeing how well they compete against stereo in a quasi-natural viewing arrangement. We thus decided to use it for other INSIGHT work addressing the stereo calibration problem.

3   Calibration of Disparities

Differences between left and right stereo images can be described in a vector field obtained by superimposing the left image on top of the right image and connecting up matching points. This *binocular disparity field* is determined partly by the three-dimensional structure of the viewed scene and partly by the positions of the eyes (i.e. vergence, gaze, and elevation angles). Knowledge of the latter is therefore required to interpret disparities for the purpose of recovering scene structure. If oculomotor signals are used by the human visual system as a source of the required eye position information then manipulating vergence angle should have an effect on three-dimensional percepts from disparities.

Experiments testing this prediction have a long history (Helmholtz 1924), with conflicting results. We have extended such investigations by measuring the effect of wearing prisms on the slant estimation task presented by the table apparatus. As the observer scans the table-top, the disparity vector field changes due to its dependence on the positions of the eyes (Mayhew 1992). The observer's percept, however, remains constant over varying fixations. This is an example of the classic phenomenon of stereoscopic depth constancy (Ono and Comerford 1977). Somehow the visual system can interpret disparity vector fields taking into account components due to eye position. How it does so is an unsolved problem (see Rogers, Chapter 10 of this book). In computer vision, the equivalent problem is referred to as stereo camera calibration (e.g. Tsai 1986).

Where might the human visual system obtain the information about eye position required to interpret the disparity field correctly? One obvious possibility is from oculomotor sources, such as data from stretch receptors in the eye muscles or from internal messages sent to control the eye muscles (so-called 'corollary discharges' or 'efferent copies').

There is an extensive literature showing that large oculomotor changes can have effects on judgements of size (e.g. Heineman *et al.* 1957; Regan, Erkelens and Collewijn 1986; Cummings *et al.* 1991). For the purposes of disparity calibration, however, exquisite sensitivity to eye movement position is required as disparity components due to depth are often small in comparison with those due to eye positions. Helmholtz (1924) found that small changes in vergence angle created with prisms produced little if any effect on depth percepts if stimuli were used that provided information on both the vertical and horizontal components of disparity vectors (i.e. the relative depths of vertical threads with beads attached were not much changed by wearing prisms, but if the beads were removed then an effect from the prisms became manifest).

We wished to extend such enquiries to ask whether relatively small alterations to vergence angle have the predicted effects when the observer is faced with a reasonably large field of view (20×20° visual angle) of a richly-textured object for which the texture information could be manipulated to be in synchrony or in conflict with stereo.

Small vergence changes were created using pairs of clinical Fresnel 2-dioptre base-in and base-out prisms mounted in spectacles. The effect of these prisms was such that if the observer fixated a spot on the table in the centre of the field of view whose real distance was 59 cm (vergence angle 6.2°) then, when wearing the 2D-BI and 2D-BO prism pairs, this distance as signalled by vergence angle would have become about 94 cm and about 43 cm respectively (vergence angles of 3.9° and 8.5°). A control (no prism) viewing condition was provided by spectacles containing plain acetate.

The first experiment (Frisby *et al.* 1991) began by training the observer to make table angle judgements using a random needle texture (figure 7) and the plain acetate spectacles. The textures of the experimental stimuli were grids formed by thin white lines on a dark ground. There were seven table angles (+10°, +7.5°, +5°, 0°, -5°, -7.5°, -10°) for which stereo and texture were arranged to be consistent. In all of them the grid lines formed 2.5×2.5 cm squares on the table surface. The general experimental procedure followed that used previously for the table rig experiments. Following training, each observer was run over two sessions, each lasting about 30 mins and each split into two halves during which all stimuli were shown twice in a different pseudo-random order for each subject. Half the observers wore prisms (either 2D-BI or 2D-BO) in the first half-session and then spectacles containing plain acetate as a control in the second half-session; in the second session this order was reversed, and the prims type changed from that used in the first session. The remaining observers had the prism/no prism conditions counter-balanced appropriately.

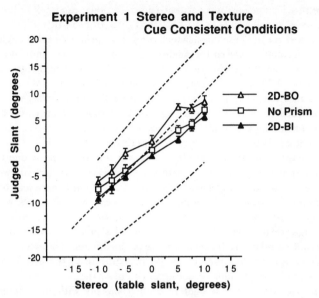

Figure 12: *Group means (N=8) for the consistent cue conditions of the first prism experiment (see text). Pooled data are shown as there were no significant differences between the 4 experienced and the 4 naive observers. The bars show standard errors. The dotted lines show predicted results if disparities were calibrated by vergence angle alone: these values were obtained by using the value for d given by vergence angle in Mayhew's disparity equations (see text: upper dashed line – 2D-BO prisms; middle – No prisms; lower – 2D-BI prisms). The obtained prism effects are on average about 13% of those predicted.*

The observers were four experienced psychophysicists who knew the experimental hypothesis, and four naive volunteers who were not informed they were wearing prisms. No observers were told whether the spectacles worn at any given moment contained prisms or plain acetate. To minimise the chances of significant prism adaptation, observers were told to close their eyes at all times other than when they were performing the slant estimation task, and all stimulus presentations were restricted to 5 secs by automatically controlled shutters in the headrest. Lamps in the headrest maintained the state of the observers' light adaptation during inter-stimulus intervals when they had their eyes closed.

The results are shown in figure 12, along with predictions of the oculomotor hypothesis. The latter were calculated by substituting values for distance to fixation (d) into Mayhew's (1982) formula for the horizontal components of disparities, making due allowance for the effect on d of the prisms (see also Mayhew and Longuet-Higgins 1982; Porrill *et al.* 1987).

Although the prisms had a significant effect on slant judgements ($F_{1,7} = 10.55$, p<.05), it can be seen that this effect was slight and much smaller than values predicted by Mayhew's formula (being only about 13% of the latter). Measurements of photographs of the table-top taken through the spectacles suggested that the observed prism effect might even have been caused solely by a small amount of prism distortion (Ogle 1960) injected by the Fresnel prisms (distortion was kept low by using only relatively weak prisms, by limiting the field of view to 20×20°, and by using the near viewing distance of $d = 59$ cm. We conclude that no convincing evidence emerged to indicate that vergence angle played much role in the calibration of disparities for this task.

There were in addition four inconsistent cue conditions, randomly-intermingled with the consistent ones, as follows: +10°T/0°S, -10°T/0°S, 0°T/S+10°, 0°T/S-10° (T= Texture cue, S = Stereo cue). These inconsistent conditions produced table angle judgements which were a compromise between the angles signalled by each cue individually, but with greater weight given to stereo (figure 13; cf. figure 9). This cue integration effect was significant ($F_{1,7} = 14.42$, p<.01). For present purposes, however, the most interesting feature of these results is that wearing prisms had no significant effect on the pattern of cue integration which is further evidence against oculomotor information having played much part in determining the results of the present experiment.

One issue that needs to be considered in experiments of this kind is the phenomenon of oculomotor adaptation: might an effect that was initially present due to the prisms disappear as observers somehow recalibrate their vergence system while wearing them? We think this factor is unlikely to have been important in our experiment because of two precautions: stimulus exposures were kept brief (5 secs), and observers were instructed to keep their eyes closed in between stimulus presentations. Also, the results showed no discernible change in the effect of the prisms over the time they were worn (e.g. there were no significant correlations between the size of the difference between first and second judgements for each condition and the length of exposure to the prisms between those judgements created by the random ordering of conditions).

The present data seem at first sight to conflict sharply with those recently reported by Cummings *et al.* (1991) who found evidence giving some support to the oculomotor hypothesis (see also Chapter 14). They found that manipulating vergence angle using mirrors while viewing random-dot stereograms of horizontal cylinders produced changes in the perception of cylinder curvature. However, this effect was only about 25% of that expected if the observers had been able to make veridical judgements of cylinder curvature from disparities scaled by vergence angle, which led them to conclude that factors other than vergence angle must have played a part. Moreover, their effect was only about 10% of that expected if vergence angle was used for calibration over the range of vergence angles used in our experiment (the full vergence range created with mirrors by Cummings *et al.* extended from 37.5 cm to infinity, i.e. to parallel optic axes). This figure is very similar to that found by us: our own of about 13% , particularly when note is taken of the fact that for their calculations (unlike ours) they assumed that effective disparity values would be scaled down by micropsia induced by vergence angle.

Another question that arises is whether the strong monocular depth cues provided by the grid textures diminished the effect of the prisms. We have tested whether this was so in a second experiment comparing results obtained with the grids with a texture of small isolated random dots (5 dots/deg^2 vis angle). Dot textures of this type are good for stereo but they are a weak texture cue for surface slant (Stevens 1981), as we confirmed by finding that monocular inspection of them in the apparatus made it almost impossible to judge table angle.

This second experiment was similar to the one just described except that inconsistent cue conditions were excluded. Also, the opportunity was taken to test the generality of the previous findings by using the method of adjustment: the observer's task was now to adjust the table-top "to appear horizontal" by manipulating a knob (details in legend to figure 14). Otherwise the procedure was as before, with the same 4 experienced psychophysicists and 4 new naive volunteers were trained to set the table angle to horizontal. Finally, to exclude any opportunity for build up of prism adaptation, spectacles were changed randomly over successive trials.

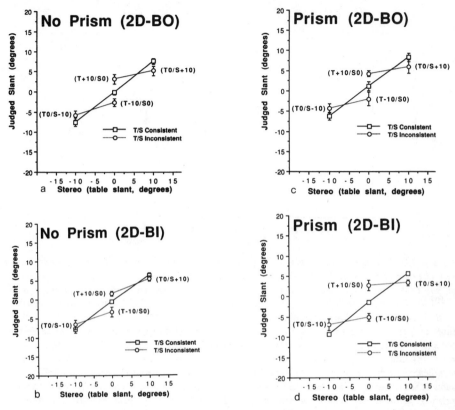

Figure 13: *Group means (N=8) for the inconsistent cue conditions of the first prism experiment compared with consistent ones. The codes shown for the inconsistent conditions are explained in the text.*

Group mean results are shown in figure 14. There was a significant effect due to the prisms ($F_{2,14} = 24.10$, p<.0001) but again one of small size. There was a significant difference between the overall means for the grid and the dots textures ($F_{1,7} = 15.24$, p<.01). There was however no significant interaction between the prism and the texture factors, reflecting the same (weak) effect of prisms on both.

Subsequent experiments (Frisby *et al.* 1992a) have indicated that similar results are obtained when ophthalmic lenses are used to create an accommodation demand in keeping with the vergence angle induced by the prisms.

If oculomotor information on vergence was of little importance for our table angle judgement task, how then was the disparity field calibrated? For the grid textures, perhaps surface slant estimates from texture gradients provided sufficient information. We are presently pursuing that idea by developing computational models of how slant-from-texture might be used to calibrate stereo (Porrill *et al.* 1991).

The dots, on the other hand, provided such weak slant perceptions that it is unlikely that for them texture methods played an appreciable role. The only method of calibration we know of that might have sufficed for the dots, given the small effect of the prisms, relies on the vertical components of the disparity vectors. Vertical disparities are unaffected (to

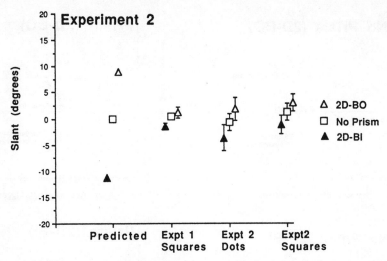

Figure 14: *Settings of the table to horizontal for grid and random dots textures from the second prism experiment described in the text. The texture cues were provided by artwork laid on the table (not projected on to it as previously; there were no texture/stereo cue conflict conditions in this experiment). Settings to horizontal could be made only when the shutters in the headrest were down, thereby avoiding motion cues. Observers proved able to set the table to 'perceived horizontal' over a sequence of up to 7 stationary exposures of the table-top, for any given condition. Three settings were made for each condition, with all settings being randomly-intermingled. Data are again shown averaged over the 8 subjects as there were no significant differences between the experienced psychophysicists and the naive volunteers. Bars show standard errors. Observed and predicted data points for equivalent conditions in the first prism experiment are shown for comparison. The data from the grid conditions in the two experiments are not significantly different ($F_{1,14} = 0.56$).*

first order) by 3D scene structure, here table angle, but they are affected by alterations in eye positions. This fact allows vertical disparities to be used for estimating the viewing geometry parameters of distance to fixation, gaze angle and elevation angle (Mayhew 1982; Mayhew and Longuet-Higgins 1982; Peek *et al.* 1984; Porrill *et al.* 1987). The susceptibility of the human visual system to an illusion called the induced effect, which occurs when one eye's view is magnified vertically, is consistent with use of vertical disparities for stereo calibration (see Rogers, Chapter 10 of this book, for further discussion).

Cummings et al, however, concluded that the human visual system does not use vertical disparities for this purpose because their cylinder curvature judgement task was unaffected when vertical disparities were changed. Sobel and Collett (1991) also failed to find evidence, using a depth interval judgment task, that the human visual system employs vertical disparities for calibration. Rogers and Bradshaw, on the other hand, recently observed that changing vertical disparities can have an effect when the field of view was very large (about 80×80°; see Chapter 10). And Helmholtz provided various demonstrations of the importance of the vertical dimension in stereopsis, such as the vertical threads and beads experiment described above. Clearly, the present picture is confused.

One reason for conflicting results is perhaps the fact that stereo interacts quite strongly with other cues, as shown here for texture and outline cues. Hence manipulations of vertical disparities might conceivably be rendered ineffective by contradictory visual information. In any event, in the quasi-natural viewing circumstances used here, vertical disparities were certainly available, and for the dots perhaps they were the only source of the required calibration data.

4 Overview

The most important conclusions from this series of experiments are as follows:

(a) Care is required in drawing conclusions from artificial stimuli, such as stereograms, about how the visual system deals with real surfaces, as substantially different patterns of results can be obtained in the two cases. It needs to be remembered that stereograms are intrinsically cue-conflict stimuli.

(b) The vertical/horizontal stereo anisotropy observed using stereograms is not eliminated by using parabolic or triangular surfaces, which is evidence against theories of the anisotropy which suppose that stereo is strengthened as a cue when second spatial derivatives of disparity are non-zero.

(c) Stereo and texture cues jointly determine the perceived slants of real ground plane surfaces, but with greater weight given to stereo for nearer viewing distances.

(d) The strength of a given texture cue in competition against stereo for the purposes of ground plane slant perception can be predicted from the sum of the strengths of its perspective and compression components measured separately.

(e) Altering vergence angle using prisms has little effect either on ground plane slant perceptions or on the pattern of stereo/texture cue combination, which suggests that oculomotor information is unlikely to be the main source of stereo calibration data for this task.

Acknowledgements

This work was funded by an Esprit BRA grant within the INSIGHT consortium. Our thanks to Catherine Toase and Eleanor Spivey for help in conducting some of the experiments. We thank John Mayhew, John Porrill and Stephen Pollard for many useful discussions. Philip McLauchlan gave advice on how to produce the textures for the real ridges. Stephen Pollard and John Porrill were of great assistance in writing and debugging the computer graphics program used for the stereograms. Michael Port helped build the apparatus and Len Hetherington made the slides.

References

Aloimonos, J., Shulman, D. (1989) Integration of visual modules: An extension of the Marrian paradigm. San Diego, CA: Academic Press

Anstis, S.M., Howard, I.P., Rogers, B. (1978) A Craik-O'Brien- Cornsweet illusion for visual depth. Vision Research 18, 213-217

Blake, A., Marinos, C. (1991) Shape from texture: estimation, isotropy and moments. Artificial Intelligence 45, 323-380

Boff, K.R., Lincoln, J.E. (1988) Engineering Data Compendium: Human Perception and Performance. Harry G. Armstrong Aerospace Medical Research Laboratory, Wright-Patterson Air Force Base, Ohio, Volume 1, Section 1.8

Buckley, D. (1988) Processes of 3D Surface Reconstruction in Human Vision. PhD, University of Sheffield, U.K.

Buckley, D., Frisby, J.P. (1992). Interaction of stereo, texture and outline cues in the shape perception of 3D ridges. Submitted

Buckley, D., Frisby, J.P., Mayhew, J.E.W. (1988) Interaction of texture and stereo cues in the perception of surface slant: evidence for surface orientation anisotropy in cue integration. Perception 17, 384

Buckley, D., Frisby, J.P., Spivey, E. (1991) Stereo and texture cue integration in ground planes: An investigation using the table stereometer. Perception 20, 91

Buckley, D., Frisby, J.P., Mayhew, J.E.W. (1989) Integration of stereo and texture cues in the formation of discontinuities during three-dimensional surface interpolation. Perception 18, 1-26

Bulthoff, H.H. (1991) Shape from X: psychophysics and computation. In: Landy, M.S. and Movshon, J.A. (eds.) Computational Models of Visual Processing, Cambridge, MA: MIT Press, pp. 305-330

Bulthoff, H.H., Mallot, H.A. (1987) Interaction of different modules in depth perception. In: Proceedings of First International Conference on Computer Vision, pp. 295-305

Bulthoff, H.H., Mallot, H.A. (1988) Integration of depth modules: Stereo and shading. Journal of the Optical Society of America A/5, 1749-1758

Clarke, W.C., Smith, A.H., Rabe, A. (1956) Retinal gradients of outline distortion and binocular disparity as stimuli for slant. Canadian Journal of Experimental Psychology 10, 1-8

Cutting, J.E., Millard, R.T. (1984) Three gradients and the perception of flat and curved surfaces. Journal of Experimental Psychology 113, 198-224

Cummings, B.G., Johnston, E.B., Parker, A.J. (1991) Vertical disparities and perception of three-dimensional shape. Nature 349, 411-413

Frisby, J.P., Buckley, D., Spivey, E., Hill, L., Mayhew, J.E.W. (1991) Is slant perception from stereo calibrated from vergence cues? Perception 20, 91

Frisby, J.P., Buckley, D., Hill, L. (1992) Calibration of disparities in the human visual system. Submitted

Frisby, J.P. Bradshaw, M., Buckley, D., Crawford, M. (1992) Short perceptual latencies for discriminating the slant of simple planar stimuli rotated around a vertical axis. 15th European Conference on Visual Perception Abstract in Perception. (In press)

Gårding, J. (1991) Shape from surface markings. CVAP 85, Royal Institute of Technology, University of Stockholm. PhD Thesis

Gillam , B., Flagg, T., Finlay, D. (1984) Evidence for disparity change as the primary stimulus for stereoscopic processing. Perception and Psychophysics 36, 559-564

Gillam, B., Chambers, D., Russo, T. (1988) Postfusional latency in stereoscopic slant perception and the primitives of stereopsis. Journal of Experimental Psychology: Human Perception and Performance 14, 163-175

Hanny, P., von der Heydt, R. (1982) The effect of horizontal-plane environment on the development of binocular receptive fields of cells in cat visual cortex. Journal of Physiology 329, 75-92

Heineman, E.G. Tulving, E., Nachmias, J. (1957) The effect of oculomotor adjustments on apparent size. American Journal of Psychology 72, 32-45

Helmholtz, H. von (1924) Handbook of Physiological Optics. 3rd edition translated by J.P.C. Southall for the Optical Society of America

Marr, D. (1981) Vision W.H. Freeman & Co.: San Francisco

Mayhew, J.E.W. (1982) The interpretation of stereo-disparity information: The computation of stereo orientation and depth. Perception 11, 387-403

Mayhew, J.E.W. (1992) The adaptive control of a four-degrees-of-freedom stereo camera head. Proceedings of the Royal Society of London, series B, (In press)

Mayhew, J.E.W., Longuet-Higgins, H.C. (1982). A computational model of binocular depth perception. Nature 297, 376-379

Ninio, J. (1981) Random-curve stereograms: a flexible tool for the study of binocular vision. Perception 10, 403-410

Ogle, K.N. (1960) Optics (Second edition. Charles C. Thomas, Springfield, Illinois)

Ono, H., Comerford, J. (1977) Stereo Depth Constancy. In W. Epstein (ed.) Stability and Constancy in Visual Perception Mechanisms and Processes, (pp. 91-128). New York: Wiley

Peek, S.A., Mayhew, J.E.W., Frisby, J.P. (1984) Obtaining viewing distance and angle of gaze from vertical disparity using a Hough-type accumulator. Image & Vision Computing 2, 180-190

Porrill, J., Mayhew, J.E.W., Frisby, J. P. (1987) Cyclotorsion, conformal invariance and induced effects in stereoscopic vision. In Frontiers of Visual Science: Proceedings of the 1985 Symposium 90-108 (National Academy Press, Washington DC)

Porrill, J., Gårding, J., Eklundh, J.O., Frisby, J.P., Buckley, D., Pollard, S.B., Mayhew, J., Spivey, E. (1991) Using shape-from-texture to calibrate stereo. Perception 20, 90

Regan, D., Erkelens, C.J., Collewijn, H. (1986) Necessary conditions for the perception of motion in depth. Investigative Ophthalmology & Visusal Science 27, 584-596

Rogers, B.J., Graham, M.E. (1983). Anisotropies in the perception of three-dimensional surfaces. Science 221, 1409-1411

Rogers, B.J., Cagenello, R. (1989) Disparity curvature and the perception of three-dimensional surfaces. Nature 339, 135-137

Sobel, E.C., Collett, T.S. (1991) Does vertical disparity scale the perception of stereoscopic depth? Proceedings of the Royal Society of London, series B/244, 87-90

Stevens, K.A. (1981) The information content of texture gradients. Biological Cybernetics 42, 95-105

Stevens, K.A., Brookes, A. (1987) Depth reconstruction in stereopsis. In Proceedings of the First International Conference in Computer Vision, pp. 682-686

Stevens, K.A., Brookes, A. (1988) Integrating stereopsis with monocular interpretations of planar surfaces. Vision Research 28, 371-386

Todd, J.T., Akerstrom, R.A. (1987) Perception of three-dimensional form from patterns of optical texture. Journal of Experimental Psychology: Human Perception and Performance 13, 242-255

Tsai, R.Y. (1986) An efficient and accurate camera calibration technique for 3D machine vision. Proceedings IEEE Computer Vision and Pattern Recognition 86, 364-374

Turner, M.R, Gerstein, G.L., Bajcsy, R. (1991) Underestimation of visual texture slant by human observers: a model. Biological Cybernetics 65, 215-226

Youngs, W.M. (1976) The influence of perspective and disparity cues on the perception of slant. Vision Research 16, 79-82

The Analysis of Natural Texture Patterns

Roger J. Watt

Department of Psychology, University of Stirling

1   Introduction: Natural Textures

Many naturally occuring surfaces have very obvious and distinctive texture patterns. Sometimes this texture is actually engraved into the surface and the visual quality is thus similar to the tactile quality of the surface. Occasionally it is marked onto the surface and is thus only a visual quality. This chapter is concerned with an examination of the principal issues involved in a study of the human visual perception of naturally-occurring texture patterns. A definition of texture at this stage would prove elusive for reasons that will become apparent later on. However, natural textures have certain characteristics that allow us to identify them as such. Just what are these characteristics? A possible starting point for an analysis of natural textures is to attempt to discover what these properties are that characterize them.

1.1   Synthetic Patterns

An instructive way to examine what characterizes natural textures is to attempt to make a process that can synthesize patterns that could pass for natural textures. In order to avoid having to decide whether synthesized patterns can be regarded as natural texture or not, a series of binary choices will be made, so that for each putative characteristic, the decision is simply whether the resultant images are more or less like natural textures. Although far from ideal and objective, this makes the process much easier.

An approach is to build a pattern from simple texture elements. To start with we can use two types of elements, short, straight lines, or small, complete circular discs. Quite clearly, given the choice between an image with only one such element and another with many elements, the more complex image will be chosen as the more like a natural texture. One of the most striking aspects of natural textures is the degree of randomness that they exhibit. Given the choice between regular arrays of texture elements and randomly placed texture elements, the obvious preference is for random placing. The repetition of texture elements has to be done with random placing.

The next step is to allow the form of the elements to vary randomly in some or all of the dimensions of their structure. Short lines might be allowed to vary in their length and orientation according to a probability density function. Circles might be allowed to

Figure 1: *Various types of synthetic texture image are shown in this figure. They are suitable as pseudo-natural textures to varying degrees. In general the more complex and random the pattern, the more like a natural texture it tends to be.*

vary in their size and their colour. In each case, the new pattern is more acceptable than the old as a natural texture.

The placing of elements in the patterns need not be completely random: some degree of order can be imposed by constraining where the elements might be placed. For example it is possible to require that the elements do not overlap each other, or to require that elements adhere to each other.

Some texture elements, such as short line segments, can be made more natural in their appearance by joining them together. The circles can be strung together in a chain with a similar enhancing effect.

Most natural images are notable in having a lack of areas of uniform intensity. This is equally true for natural textures, and the textures considered so far are all deficient in that they are either binary (two grey values) or are made up of elements that are uniform in intensity. This is easily amended by using elements that are made from varying intensity profiles. When this is done, there is however, the difficulty that where elements overlap, the values from the various elements have to be considered. A simple procedure is to add together all the elements at each point in the texture image. This can be used to produce Brownian fractal patterns, which look very much like lichen growth on a rock. An alternative is to allow some elements to occlude others, which can produce effects like pebbles on a beach.

The main characteristics that have been considered here are as follows. A pseudo-natural texture can be made by randomly placing texture elements across an area of an image. The elements should be allowed to exhibit some variation in their size, orientation and form. The elements can be connected together to form longer structures. Finally, the elements should not be restricted to being uniform in their grey-level.

Most natural surfaces are rough and complex. Another approach to generating textures is to note that natural textures are generally characterized by being fractal, at least over a restricted range of scales. This statement, although true, has many implications, and the converse, that all fractal patterns are natural textures, is certainly not true. The statement is often taken only to be an observation that the power spectrum of natural images, and especially the power spectrum of natural textures shows a simple relationship between energy and spatial frequency, so that power varies inversely with spatial frequency raised to a power of between one and two. For a pattern to be fractal, the relationships in the phase spectrum also need to show scale invariance and cannot be disregarded. The difficulty with the proposition is that there are many images which we would not want to regard as being textured in any sense of the word, but which also have this specified power spectrum. Moreover, if we attempt to synthesize images with the requisite power spectrum and random phase spectrum, then what results most often is not a natural-looking image. This approach seems to be doomed.

1.2   Natural Textures

Most natural surfaces are rough and patterned and different types are characterized by particular types of pattern. The patterns on natural surfaces can be of many different sorts, but there are two general sources: they can be markings, reflectance changes; or they can be engravings which cause the surface to have illuminance changes. In both cases, the result is spatial variation in the luminance of the surface and therefore variation in the intensity of light formed in an image of that surface. The markings and engravings that are found on natural surfaces are usually made by processes that have a stochastic factor, so that the surfaces of basically similar objects, such as two rocks, are not identical,

Figure 2: *A sample of different naturally occuring surfaces are shown in this figure. For each of these images, it is possible to devise tasks that would require predominantly statistical or predominantly geometric information about the pattern. In this respect, the images can be said to be natural textures.*

but are only similar in the sense of having been made of the same material and exposed to the same types of force. The processes that make natural surfaces often work at a range of scales. For example, the processes of erosion have a number of different scales – glaciation creates rock structures at the scale of kilometres, frost creates rock structures at the scale of metres, and wind weathering creates structures on rock at much finer scales still. These factors ensure that natural surfaces cannot be exactly described by a small number of parameters, in the way that a perfect sphere could be. In this sense, natural surfaces are complex.

Examples of natural surfaces that can be described as having engravings or markings are to be found on rocks, on water, and on the trunks of trees, for example. There are other types of natural pattern that requires a slightly relaxed use of the term surface. The ground underneath a tree that has just lost its leaves, for example, is neither continuous nor stable as a surface would normally be, but can nevertheless be treated as a surface, although its nature makes it suitable for only some actions such as walking on, and not others such as grasping.

It is possible, although a little odd, to treat the structure of a deciduous tree in winter as having the surface of a sort of ball shape on a stalk. The stalk has slight engravings, the bark, and the ball shape surface is very deeply engraved, to make the branches. It would seem more useful to regard trees, and similar objects, as being three-dimensional structures rather than complex patterns on surfaces.

2 A Computational Approach to Texture

The term, texture, is usually taken to be a simple property of a stimulus. At worst, it is often just equated with a stimulus display that appears to be complex, but this leads to the very obvious difficulty that it is impossible to set a non-arbitrary criterion complexity for a pattern to be regarded as a texture. A pattern of dense, randomly placed lines clearly is a texture, whereas a pattern made up of a single line clearly is not a texture. Texture is sometimes regarded as a property of object surfaces, much like reflectance or specularity, although it would be better to regard it as being a perceptual effects of a surface, like colour or shininess. Even this is inadequate: for natural surfaces that have structure at many scales, it is not obvious what the object is until a task or decision has been chosen. If the object depends on the task, then so does the surface of the object.

2.1 What Are the Uses of Texture In Vision?

There are many benefits that can be obtained from living and seeing in a world of textured objects. As a simple starting point, motion analysis, stereopsis, and even object detection and recognition are enhanced by the presence of patterning across a surface. The patterning provides more measurable information for a visual system to use in these judgements.

The shape of a surface can be computed from the manner in which a pattern varies in an image of that surface. If it is assumed that the pattern is itself stationary across the surface, then any changes in the pattern in an image of the surface have to be due to changes in the angle that the surface makes to the line of sight, and to changes in the distance of the surface from the point of sight.

Some objects can be recognized by their texture. Some classes of object can only be categorized on the basis of statistical properties, rather than geometrical properties. Most natural objects are of this general type. Some examples of the classes of natural object that come into this category will be given below.

Surface properties can be obtained from texture. Simple distinctions between rough and smooth surfaces can be made with some confidence on the basis of texture information. Moreover if the texture has anisotropic patterns, then the surface roughness can be expected to be similarly anisotropic.

2.2 Representations of Texture

For many purposes the domain of texture can be left as a very wide category. Basically, the extraction of depth information across a surface relies either on photometric stereo or on there being dense texture. Likewise, optic flow depends, for its measurement on plenty of luminance contour, and texture surfaces are a useful source of optic flow. Any surface marking or engraving is equally useful for these purposes and there is no point in proscribing any. In these instances, there is no need to generate a representation of the texture itself, merely a representation of the luminance variations that are associated with the texture, and these only as a temporary and intermediate stage before surface shape and orientation or optic flow are represented. Given that explicit representations of texture, per se, are not needed in these areas of visual processing, there is thus no need to give any consideration to how such a representation might be created or for what purposes.

However, there are circumstances where an explicit representation of the form of a texture pattern is useful. When stepping on a rocky surface, it is necessary to know whether the surface will offer sufficient friction to provide a good grip. A representation of the surface roughness can thus be useful. Similarly, the type of rock can be determined to a certain extent by the nature of the surface texture. In these cases, it is necessary to have a description of the surface texture. Very little attention has been given to the problems associated with texture analysis and description.

Most machine vision work has been concerned with either the use of texture and texture gradients for computing surface shape (e.g. Witkin 1981; Blake and Marinos 1990), or with creating algorithms that can compress texture for efficient transmission. Most psychophysical work has been concerned with either the segmentation of images at texture boundaries (e.g. Julesz 1981; Nothdurft 1985a; Landy and Bergen 1991), or with the detection of texture gradients (Nothdurft 1985b; Sagi and Julesz 1987; Nothdurft 1991).

In all of these studies, the usefulness of local orientation has been a major conclusion. For example, the degree of isotropy of texture, locally measured, is a valuable indicator of surface shape (Blake and Marinos 1990). The psychophysical work has concentrated on the question of what image features are used for representing textures. Since the textures used have largely been made up of short line segments, the results have hardly been surprising: orientations and lengths of short line segments. There is some doubt as to what else might be important.

It is far from obvious how a model of texture perception based on such stimuli and results could be applied to natural textures. There are several shortcomings in current approaches to visible texture. There is no information about natural textures. Natural surfaces belong to a small number of different types that can be classified according to how they were made and from what material. There is no information available about differences between surface types and about the natural range of variability of surface forms within a class. There is no clear way in which texture can be described: most propositions are either very global (use the Fourier transform, fractal dimension) or highly local, depending on being able to identify discrete local features, such as short line segments. The former does not make texture gradients explicit; the latter does not readily deal with the continuous patterns of natural texture images.

2.3   Texture and Tasks

It is taken as self-evident that texture is not a physical characteristic of the world, nor a stimulus for the visual system, but a response of the visual system. This means that it is a response to the demands of a particular task and the stimulus. Consider this example. There are several different ways in which the structure of a complex object such as a tree may be visually represented and each form of representation makes explicit different information and hence makes available different actions and decision about the object. If all that is represented is the number of branches, their typical shape and colour, but none of the specific details, then certain types of decision and behaviour are supported and others are not. If the exact structure, the number, individual shapes, colours and layout of all the branches are all represented then certain other types of decision and action are possible. In the first case, the tree would have the same representation as other trees of the same species and hence could be judged as being similar to them. In the second case, it would not have the same representation as any other tree, and would only be similar to itself. With the second representation, on the other hand, a route for climbing up the tree could be calculated, but not with the first, where perhaps only the general difficulty in climbing such a tree could be calculated. The first form of representation will be called texture.

It can be seen that the processing of an image or part of an image as a texture will make available a different set of tasks from those that can be accomplished with other forms of representation. This means that for some tasks and decisions, the image or the appropriate parts of the image need to be treated as texture rather than geometric structure.

2.4   Texture and Scale

Different tasks can be considered that will all use the same part of an image. Each of these will involve the image information in a different way. Even those tasks that require texture information from the same part of the image could require the information to be collected from different spatial scales. Each task is likely to have a characteristic scale which is determined by the scale of the actions to be performed.

From my office window, I can see a rough, craggy hillside. The surface of this hillside has certain structures that are of interest. These are at a fairly wide range of spatial scales (expressed in units of distance on the object, rather than in units of visual angle). From here, I can plan a route for climbing to the summit. In so doing, I am looking to see where there are precipitous rocky cliffs, and where there is some vegetation. For the purpose of considering and planning a journey to the summit, the surface is the hillside, and the natural scale is of the order of a few metres in length. If I make that journey, then I will be concerned with finding places to place my feet, and this will require the surface to be examined at a finer spatial scale, in order to identify areas of a useful size, roughness and orientation and here the appropriate scale will have the order of a few tenths of a metre. If the purpose of that journey is to discover some samples of moss, known to grow in particular environments, then the surface of interest is at a finer scale still, of the order of millimetres.

For any particular purpose, there is a scale which contains the most appropriate information. The exact geometrical structure of the surface at finer scales than those that are most appropriate can be regarded as unnecessary detail. But the statistical structure of the surface at these finer scales cannot. The scale of interest for route planning may be of the order of metres, but judgements about the nature of the surface at this scale (rock, moorland or forest) will be based on its texture – using information at finer scales. Likewise, the scale of interest for my feet is of the order of tenths of a metre, but the roughness of the surface, essential for judgements about grip, is due to the statistical structure of the surface at finer scales. A distinction can be made between the geometrical properties of a surface that are important for knowing where to make an action such as placing a foot, and the statistical properties of the surface that indicate how the surface will react to the proposed action. When stepping onto a rocky ledge, it is important to know where to step, (i.e. where the ledge begins and ends) but it is not important to know where the little rough knobs on the rock are, only that they are there. There are those scales that carry surface geometry that is important in the execution of a task (where the gullys are, where the footholds are), and there are the finer scales that carry statistical information about the nature of the surfaces of interest. Both types of information are important, but in different ways. The second type will be called texture.

2.5   Texture and Control of Processing

Texture is a voluntary representation. All surfaces have markings that could be regarded as texture, but sometimes this is inappropriate and we might not wish to do so. Some surfaces are invariably treated as having texture, whereas others are never treated as such. The markings on the surface of the canvas commonly known as the Mona Lisa, for example, are usually seen as having a structure that is not texture. The markings on the surface of a brick wall, on the other hand, are usually seen as a texture rather than as a structure.

The decision whether to treat a surface as having a texture or not depends to a certain extent on the task that is to be performed, or the task that could be performed. This in turn sets a set of spatial scales at which useful geometric information is available. The decision thus can be seen as a decision about the type of processing that is applied at different spatial scales.

It has already been proposed that human vision is subject to a form of control at very low levels (Watt 1988). The proposal was based on psychophysical evidence about the dynamics of visual processing and the temporal effects of spatial scale. This evidence was found to be further consistent with much of the literature on the subject of visual attention. The main claim is that all spatial scales are or can be processed for the purposes of generating a texture representation, but that the scales at which geometric information is calculated are determined by a coarse to fine spatial scale progression, subject to control. It is therefore useful to be able to note that a logical analysis of texture processing has already lead to precisely this requirement.

2.6   Surfaces and Textures

If texture is a type of visual processing rather than a physical property of the environment, then there is a difficulty in discussing the concept of surface texture. Any image can be analyzed by the visual system as a texture, and once the nature of that processing has been discovered, then there will be no further difficulty in understanding the process. However, in order to discover what the texture processing does, it is necessary to start with some basic idea of what that processing is to achieve, and to achieve with what types of stimuli. There are some types of image for which a texture representation has little function: the texture representation of an image of a kettle is not useful. It follows that an image of a kettle is not a suitable image with which to experiment. On the other hand an image of a piece of tree bark does seem suitable, and a texture representation of such an image could serve several useful functions.

It is therefore proposed to adopt a relaxed terminology where texture is taken to be both the product of a particular form of processing and the type of surface and image for which that processing is notably suitable. The first of these two uses of the term texture could be quite unambiguous, and it is an understanding of that use that is sought. The second use, however, has no real prospect of being unambiguous because there will always be some images whose status as textures or not cannot be agreed.

3   Computational Analysis of Natural Textures

Given the argument that texture perception is inherently linked in with the nature of the tasks that are to be accomplished with that texture, it is necessary to start an analysis of texture processing by considering potential tasks. In this section, a number of different tasks will be described. In each case, it will be shown that at the root of the task is the concept of similarity. Similarity is a relationship between two items that indicates that the two items are not necessarily identical, but that they are not completely different. In this work, a simple definition, taken from elementary geometry will be used. Two items are similar if one can be transformed until it is identical to the other, given only a constrained set of transformations.

3.1   Texture Tasks

The following tasks are all realistically complex. In each case, however, the basic result is a binary decision. This makes the analysis of the tasks simple.

Classification   Images of textures can be obtained from various different types of source. The task is, given a set of images, each belonging to a particular category such as rock, tree bark, foliage, or stone walls, can an algorithm for texture description be devised that will cluster each set together by producing similar descriptions for the members of any one category that are different from the descriptions that are produced from members of the others?

Homogeneity   Very few images of natural texture are completely homogeneous. The next task is, given a set of images, some simple homogeneous textures, and some images containing several different types of texture, such as tree bark and foliage, can the algorithm for texture description be extended so that it is able to differentiate the two sets of images?

Gradients   Most natural textures are seen at an non-perpendicular angle to the line of sight. The consequence of this is that the texture pattern will have a gradient which corresponds to the effects of the local angle of sight and the relative viewing distance of different parts of the surface. The task is, given images of texture surfaces that are grossly planar, can the algorithm for texture description be extended so that it is able to determine the direction and extent of the texture gradient.

Non-planarity   Finally, many natural objects, such as tree trunks are not grossly planar. The task is, given images of textured surfaces that are grossly planar and images that are not, can the algorithm for texture description be extended so that it is able to detect which images are not grossly planar?

In the first task, of classification, two textures that are from the same cause will be similar. Hence two texture that are similar need to be classified together, and two that are not similar must not. Thus, provided we have a good means of assessing texture similarity (against the physical criterion of cause), we have a good means of classification.

In the second task, of homogeneity, the same general argument applies. A texture is homogeneous if different subsamples of the texture are all similar to each other. If there are regions that are not similar to each other, then the texture is not homogeneous. It will be seen that there is a problem in this concerning the area of the texture that should be taken as a subsample.

In the third task, of gradients, the problem is to discover ways in which the texture shows a systematic variation in similarity. A gradient of texture would be where, in most respects a texture is homogeneous across an image, but with respect to a few parameters arising because of viewing arrangements, there is a systematic variation. The concept of similarity between different parts of the texture pattern, subject to some systematic variation is thus at the heart of this task.

In the fourth task, the same general principle applies as did for the third task, although in this case the variation need not be quite so systematic. This task is different from the second task in that the different areas of the texture pattern are all similar in some respects, but show variations in others.

Each of the three last tasks might be accomplished by creating local texture descriptions over small areas of the image and to examine the manner in which these vary across the image. If there exist regions that are completely dissimilar to each other, then the texture is not homogeneous. If there exist regions that are similar, subject to a transformation then they are homogeneous but not flat. If the transformation is graded, then there is a texture gradient, otherwise the texture is non-planar.

The difficulty with this idea is two-fold. First, as will be shown immediately below, the concept of similarity of textures is far from simple. And then, of course the size of the area covered by each local description would seem to be an arbitrary parameter.

The three tasks all involve the concept of similarity of textures. This concept will be elaborated now.

3.2   Similarity of Images

Two images are identical if they have the same intensity value at each pixel – that is, if all the pixel-by-pixel differences between the images are zero. Two images that are completely unrelated might have a distribution of pixel-by-pixel differences that has a mean of zero, but will extend on either side. In these circumstances, the difference cannot be described by the mean signed difference, pixel-by-pixel, but some form of absolute measure of pixel difference must be used. It is usual to use the square of the difference, so that a simple measure of the difference between two images is the sum of the squared numerical differences between corresponding pairs of pixels.

Two images could be identical, subject only to a small transformation of one with respect to the other. It is better to state that the similarity of two images can be measured by transforming (e.g. shifting, rotating or dilating) one of the images to find the transformation at which the sum of the squared numerical differences between corresponding pairs of pixels becomes minimum.

Lastly, the similarity of two images does should not depend in an uninteresting fashion on the grey-level gain of those images. The values in each image should first be normalized numerically so that the difference value obtained does not depend on the range of values in the two images.

When defined like this, the concept of similarity is capable of dealing with many situations, but cannot do so for natural texture: two images of tree bark will not be any more similar than an image of tree bark and an image of a kettle.

3.3   Similarity of Representations

The concept of similarity can be applied, with some small modifications, to representations of images. If we suppose that a description of an image corresponds, at some level, to a list of numbers, then differences in descriptions can be measured in just the same way that differences in images can.

The manner in which images are represented makes explicit certain aspects of those images and not others. If this information is all that is available, then this information also determines the manner in which those images can be judged as similar. For example, take two images that are similar in having the same grey-level histograms. No single pixel value makes explicit the form of the histogram or therefore the similarity between the two images. A description of the images that gives the mean and standard deviation of their grey-level histograms, on the other hand will make the similarity explicit. A description is able to represent certain types of information that are not explicit in the image itself. Images that are not pixel-by-pixel similar may nonetheless be represented by descriptions that are similar, term-by-term.

If the interest lies only in similarity, then a form of description can be used in which the variables that correspond to the transformations that are allowed are simply not present. For example, the description of a triangle can be given in several different ways. A description of the positions of the three corners is not invariant with respect to either size or orientation. A description giving the lengths of the three sides has no angle information explicit and is thus invariant with respect to orientation but not with respect to size. A description of the sizes of the three angles is invariant with respect to both size and orientation. Invariant in these cases means that triangles that are similar subject to changes in the invariant parameter a give identical descriptions.

3.4 Similarity of Textures

The way to determine whether two textures are similar is then to find a transformation of texture images that will cause all image pairs that we would require to be similar to lead to similar representations and would cause to all image pairs that we require to be not similar to give rise to dissimilar representations. The question arises, however, as to what would be the appropriate transformation. This question is equivalent to the question of what the appropriate description of texture should be – i.e. what is the description of texture that is invariant to the types of variation that don't change textures?

4 Texture Descriptions

There are three ways of approaching this question of the appropriate description. The first is simply to use the subjective judgements of people. In its most exhaustive form, the process would involve taking all pairs of textures that are of interest or likely of be of interest and have people judge, numerically for each pair, the perceived similarity. This data then provide a grand look-up table which would serve as the texture transformation. Less exhausting versions of this process could be devised, by using a representative sample of the textures, and then hoping to intuit the nature of the whole space. Such procedures would all serve little use since the goal is to understand the human visual perception of texture, in non-circular terms.

A second approach is similarly empirical, in the sense of being a search for a texture description. In the second approach however, a classification of textures on grounds that are not human perceptual can be used. Many natural textured surfaces belong to discrete classes of object – trees of different species, rocks of different geological type and so on. Given examples of such classes, can a description be found that will behave appropriately? At least half of the problem is constrained by taking this approach.

There is a third approach, however, which is even better. It is plausible that the differences between images of the bark of Scots Pine trees can be explained in terms of the processes that generated those images. There are, of course, imaging variables, such as viewing distance and orientation and lighting factors. There are also, however, variables that relate to the manner in which individual trees grow. Suppose that it is the case that the bark tissue of a tree splits as the bore of the tree trunk expands, but splits in directions and with probabilities that are predefined by anisotropies in the material structure of the tissue. Samples of tree bark will thus be identical in the structures of the materials of which they are made. These structures correspond to the manner in which the tree bark will respond to forces, such as stretching and weathering. Can a description of the texture of a surface be created, therefore, that has as explicit parameters, a representation of the material structures of the surface?

It is a long leap, but a worthwhile one to suggest that a useful visual description of a surface should make explicit the reaction of that surface to forces during the creation of the surface. A slightly shorter journey would be required to get to the proposition that visual descriptions of the texture of surfaces should make explicit the manner in which those surfaces will react (now) to forces. A representation of a rock that makes explicit how much friction will be offered to a walker's boot, and in what directions the greatest friction is available is clearly worth having, if the boot is attached to your leg and the rock is directly in front of you. Such a representation is not far from being a representation of how that rock has reacted to weathering processes.

It is another step to suggest that the main character of natural forces are captured by a simple treatment in terms of vectors. A force such as the wind can be modelled, to a first approximation, as a vector field (direction and magnitude) at any given instant. The long term effects of such a force can then be described by the distribution of such vector fields that are naturally encountered. Perhaps an analogous description of textures in terms of lengths and directions (orientations) would be useful.

5 A Computational Model

We now turn to consider how natural textures can be analyzed by a visual system, and a particular form of computational analysis of texture images will be described. This form has been applied to various types of natural image with some success. It is not sensible to consider an analysis before having decided what the function of the analysis should be and so a simple computational task is assumed.

5.1 A Simple Task

When shown pictures of various different types of texture pattern, such as the bark of trees of a single species, rocks of a single type, or clouds of a single variety, people are very good at being able to say which textures are of the same type, or, in other words, which are similar. Two images of tree bark are very similar, in the sense of having similar causes, but this similarity is not simple to measure directly. In this section, a simple form of texture representation is described. This representation is based on an earlier model for edge-representation.

5.2   An Extension of MIRAGE for Texture Description

The original MIRAGE algorithm (Watt and Morgan 1985) was developed to account for
psychophysical data concerning the precision with which various different aspects of edges
could be judged. The model is a computational model, in that it can be simulated easily
on a computer, but it is also based on a computational-theoretic analysis of the goals
and limitations inherent in edge detection and localization. The MIRAGE model has
been extended several times, incorporating dynamics (Watt 1987, 1988), the use of two-
dimensional primitives to describe zero-bounded blobs (Watt 1990). The latest extension
showed how such primitives could be combined to produce texture descriptions (Watt
1991b).

The computational process begins with filtering the image in parallel with a set of
oriented filters, varying in spatial scale and orientation. The result of this filtering stage
is a set of new images. This step is a standard process to apply, but everything that
follows is novel. The general plan is to break the image down into small pieces.

The resultant filtered images all have a mean value of zero, with roughly the same
distribution of values on either side of zero. If one such filtered image is cut wherever
it crosses a threshold level either side of zero, then a set of blob-shaped segments are
isolated. Each of these can be described by giving its length, width, mass, orientation
and position. The complete filtered image is described by a set of such descriptions. The
image description is a fairly faithful representation of the filtered image and of course is
no more useful for measuring texture similarity than was the original. The next step is to
create from such a image description a two-dimensional histogram showing the distribution
of mass as a function of segment orientation and length. One of these histograms can be
computed for the output of each filter in this way. The complete set of histograms then
constitutes a *texture description*: a four dimensional data structure, plotting mass as a
function of segment orientation, segment length, filter scale and filter orientation.

The texture description contains a vast amount of information. It is obvious that
there will be a close correlation between segment orientation and filter orientation, and
similarly between segment length and filter scale. In practice there appears to be no point
in keeping the dimension of filter orientation, and the histograms from differently oriented
filters at any one spatial scale can be added together without loss of information. It is
also the case that segment orientation and length will often be independent so that the
projections of the histograms onto these axes will be as informative as the full plane.

The resultant version of the texture description is in each case an array of numerical
values, just like an image. The similarity of two such descriptions can thus be calculated
in an exactly analogous fashion, simply by calculating the numerical difference between
corresponding cells, allowing for a translation which corresponds to a rotation and dilation
of the original images.

5.3   Results: 1st Experiment

Experiments with this form of texture description have proved very encouraging. The
texture description just described has been tested with a set of natural texture images
and is capable of performing basic texture classification. The use of oriented filters turns
out to be quite critical in generating the best, that is the most useful, texture descriptions.
The use of an intermediate representational form, the image description with its sets of
parameterized segment descriptions also turns out to be very important.

Original Stimulus

Figure 3: *(a)*

Reconstruction of Original from
Blob Descriptions

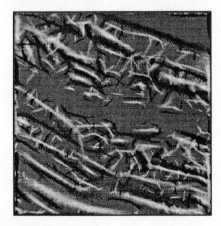

Figure 3: *(d)*

Figure 3: (a)*An example of a natural texture is shown. Each of the stages in the creation of a texture description is also illustrated. A set of filtered images of different orientations, but only at one spatial scale are shown (b). The results (c) of applying a threshold to these to produce blobs are then shown. For each blob, a description can be created. The set of such descriptions is illustrated in the figure (d) by giving an image that is created by reconstructing the blobs just from the descriptions. The orientations and lengths of the blobs in the filtered images can be represented by a pair of histograms giving the distribution of mass as a function of blob length and orientation (e). Note that there are two histograms for orientation and two for length: these correspond to the dark blobs and the light blobs separately (f).*

Filtered Images

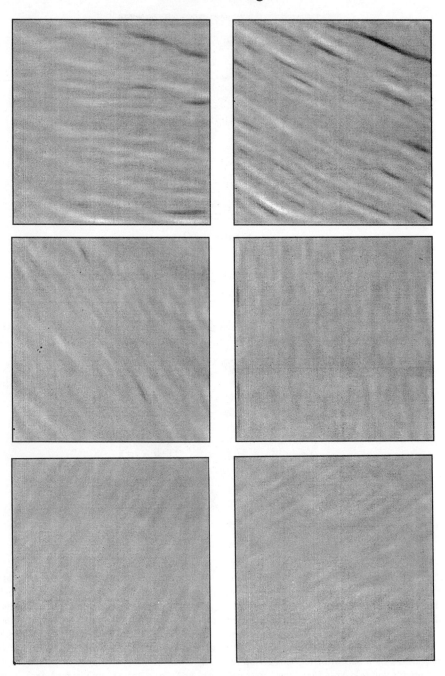

Figure 3: (b)

Blob Images

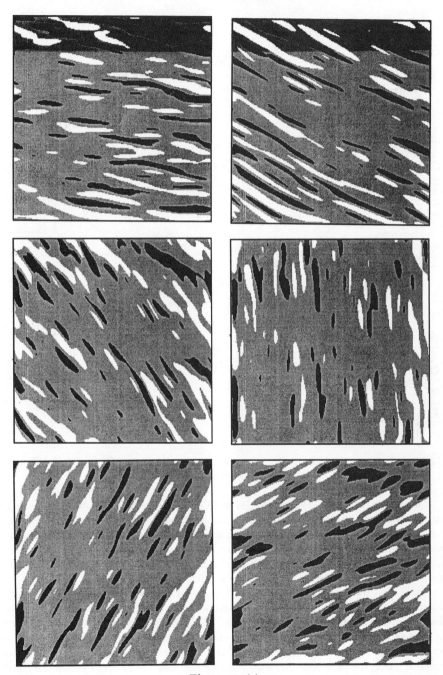

Figure 3: *(c)*

Histograms

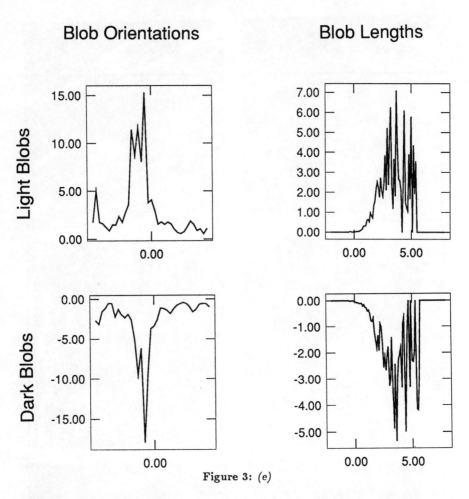

Figure 3: (e)

The basic experiment involved a total of 18 natural images, six chosen each from the categories of tree bark, tree twigs (i.e. trees in winter), and foliage (leaves). For each image, a texture description was created, as described above. Then, for each pair of images, the similarity of the two texture descriptions was calculated. Ideally this will be a high number for each pair involving similar textures, and a low number for all other pairs. Figure 4 shows the ideal form of the results.

The actual results obtained from the 18 images are shown in Figure 5. Each part of the texture description has been treated separately, and so there are four panels. Notice that the main categories can be partly seen in different parts of the texture description.

The results are encouraging, but far from demonstrating success. There are several problems with the experiment, and two of them will now be considered. The first main problem is that the categories that have been selected may not be homogeneous categories. The second main problem is that the actual texture histograms show structure that is

Histograms over Spatial Scale

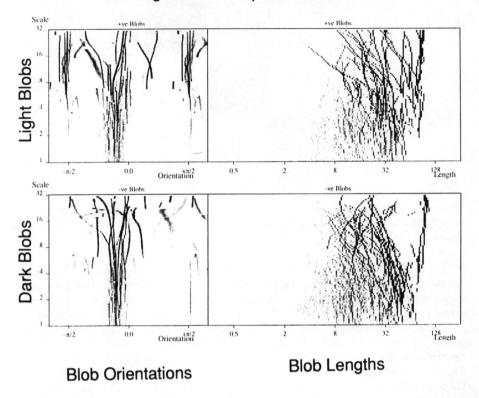

Figure 3: *(f)*

due to individual blobs in the images. They therefore might have too much information about the specific image rather than the general structure of the texture.

5.4   Results: 2nd Experiment

The computational experiment was repeated with more carefully considered stimuli, and with a modification to the texture histograms that will be described in a moment.

A new set of 12 images were obtained. All were images of the bark of trees. These were chosen so that they fell into four groups of three images, with each group of three corresponding to a single species of tree, growing in a single site. Thus each set should be as homogeneous as possible, given only the type of natural variation that is of interest.

Secondly, the texture descriptions, which are two-dimensional arrays of numbers, can be treated as images and processed further. In particular, in an attempt to remove the local and individual blob structures visible in the original descriptions, these were blurred by convolution with a Gaussian of various different standard deviations.

Note that by now requiring the process to discriminate between the bark patterns of different species of tree, rather than between completely different types of texture, the task has been made considerably more difficult.

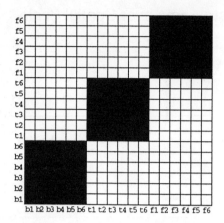

Figure 4: *The 18 images are labelled b1-b6, t1-t6 and f1-f6. If the texture description is suitable for classifying the images according to their type, then the pattern of results should appear as illustrated in this figure.*

The raw results are shown in figure 6. As previously, the results without blurring are not convincing. However, when the histograms are blurred somewhat, then the results, shown in figure 7 are much more encouraging. The blurring is simply applied to an equal extent (measured in terms of the number of bins of the histograms) in each direction.

Finally, it is interesting to examine the effects of the degree of blurring applied. At very small degrees of blurring, the individual blob structures are resolved in the texture descriptions, and they will all be equally dissimilar to each other. At extremely large degrees of blurring, on the other hand, the texture descriptions will all be spread out to the point where they are all equally similar to each other. It follows that there will be an intermediate point where the required pattern of results should be found. The series of results obtained for different degrees of blur are shown in figure 8. These indicate that a degree of blurring of between 4 and 8 bins is appropriate. In the case of the orientation histograms, this corresponds to an equal number of degrees (between 4 and 8).

5.5   Discussion

There are a number of important general aspects of the texture description algorithm that has just been described that should be noted.

Textures have been described as being made up from the random placing of texture elements in an area of an image. The first thing that the algorithm just described does is to split the image up into pieces. These pieces are unlikely to be the same as the texture elements, but that is not perhaps too important. The use of oriented filters appears to be a good choice for the purpose of segmentation. The blobs that are produced as a result of filtering with oriented filters are all simple shapes, and are all generally well-described by the few simple terms of an elongated Gaussian. Figure 9 shows a measure of the discrepancy between the blobs and the best fitting elongated Gaussians for some sample images, as a function of the degree of elongation of the filter. Note that the functions tend to asymptote for elongations of more than about 3 or so.

The switch from the blobs in the filtered image to the histogram is also an important step in the process. This is the step that makes explicit the observation that the placing

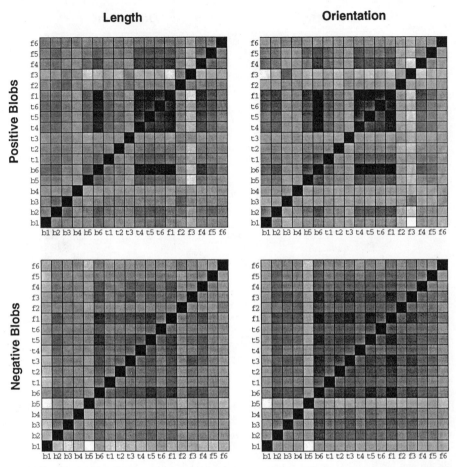

Figure 5: *The 18 images are labelled b1-b6, t1-t6 and f1-f6. The actual measured similarities are shown in this figure. The four panels correspond to the four parts of the texture description. Although there is some hint of the pattern of results from Figure 4, the effect is at best weak.*

of individual elements within a texture is unimportant. It is important to preserve some information about the distribution of blob lengths and blob orientations, beyond just the mean or mode length and orientation. Without this information, the information explicitly represented is too sparse, and too few discriminations can be made.

In the basic implementation of the model, a very fine sampling of all dimensions, orientation, length and spatial scale was used. At this sampling, the tracks of individual blobs can be seen in the histograms, and this is too fine a sampling for optimum performance. The second experiment established that a degree of blurring that corresponds, in the orientation case to a figure of around 4 - 8 degrees, produces consistently better results.

Finally, it does seem likely that the sensitivity of the system to differences in texture could be modulated by adjusting the degree of blurring applied to the histograms (or equivalently, the sampling density). In one limit, no two images are likely to be found to be similar if an extremely fine sampling is used, whereas at the other extreme, all images

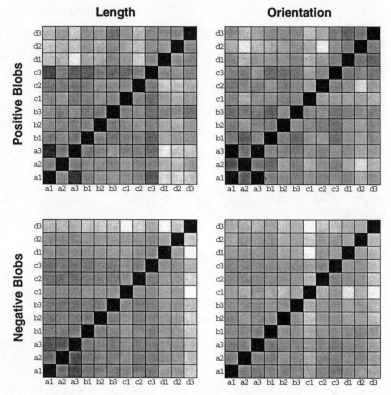

Figure 6: *The 12 images are labelled a1-3, b1-3, c1-3, and d1-3. The actual measured similarities are shown in this figure. As before, the four panels correspond to the four different parts of the texture description. There is no discernable classification.*

will be similar if an extremely coarse sampling rate is used. A continuum of sensitivity might exists the two extremes. It would be interesting to examine the interaction between the degree of similarity between textures and the optimum blurring.

6   Conclusions

6.1   The Present Results

The studies that have been described here have made progress towards understanding the nature of the visual analysis of texture images. Specifically, there has been progress in understanding the nature of texture itself. The use of texture can be shown to be related in many instances to concepts of similarity between textures, and an operational test for similarity has been devised. An algorithm for generating texture descriptions has been described, and some illustrations of its performance have been provided. In the course of this, several new directions for research in this area have been identified.

An initial examination of the logical status of texture has revealed that texture cannot be regarded simply as a feature of surfaces, even natural ones. It is the product of a particular manner of analyzing images of surfaces, particularly natural ones. In these

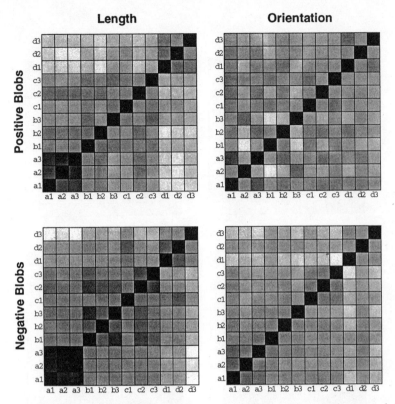

Figure 7: *The 12 images are labelled a1-3, b1-3, c1-3, and d1-3. The actual measured similarities between texture descriptions that have been blurred are shown in this figure. As before, the four panels correspond to the four different parts of the texture description. There is now a clearly discernable classification.*

cases, it is important to make explicit something about the general nature of the surface, rather than its specific details.

The similarity of textures is readily apparent to the human observer but requires an appropriate description of the images involved before it can be measured computationally. The key to this issue is to note that similar textures are made by taking the same population or distribution of texture elements, and randomly dropping them into an image, subject perhaps to some constraints on where they might land. Given this, with the all important random aspect, any process which recovers the elements, whole or in parts, and draws up a description of the appropriate distribution of values on the dimensions of variation between elements, will generate similar descriptions for texture distributions that are generated from the same process.

An algorithm capable of creating useful descriptions of texture and based on human psychophysics, has been described in this paper. Some indication of the performance of the algorithm in classifying natural texture has been given. There are several aspects of the algorithm that require further examination.

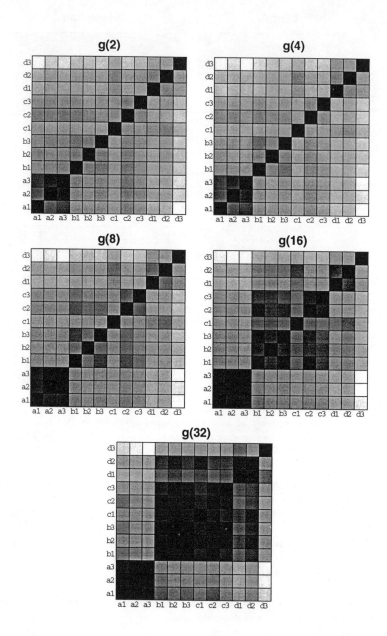

Figure 8: *The effects of increasing blurring of the texture descriptions. The figure shows the measured similarities between the length histograms for dark blobs, at several different degrees of blur.*

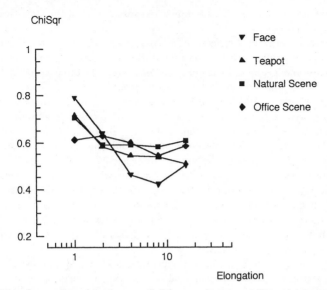

Figure 9: *This figure shows the degree of difference between a filtered image and its reconstruction based on the blob descriptions, as a function of the degree of elongation or orientation selectivity of the filter. Several different images are shown. The function tends to asymptote at an elongation of around 3 or 4.*

6.2 The Next Steps in Studying Texture

The model for texture classification has been described above, with some reasons for the structure that it has. The model proceeds through a filtering stage, producing filtered images as output. By segmenting the filtered images into blobs, and describing each blob as having a mass, length, orientation and position, an image description can be produced. Each such image is described, and a these image descriptions are converted into histograms plotting the mass at each orientation and length. When the process is repeated at a range of spatial scales, a full texture description is created. There arise a number of questions that can be used to refine the texture description, and that can be used to generate psychophysical tests of human texture vision.

It is not clear that the use of simple oriented filters is sufficient to generate the required discrimination at the level of the texture description. It is not difficult to find images of texture that are only marginally discriminable using this mechanism. There are several ways in which the texture description could be enhanced. The blob descriptors themselves could have a curvature term in addition to the present terms. If the problem is that the parts of texture elements are too simple, i.e. too elemental, then an alternative would be to use a more complex shaped filter. The receptive field of so-called end-stopped cells in the mammalian visual cortex would be a suitable candidate for experimentation in this respect.

Then there is the question of whether the descriptions should keep or throw away information about the filters themselves. The proposition that this information be discarded is an interesting one, and one that runs counter to most assumptions that are made in research in this area. It is important to establish the extent to which the original four dimensions are redundant. This requires an examination of the statistical properties of natural texture images.

Another issue to be tackled concerns the types of structures that are found in the texture descriptions themselves and the question of whether such structures can be computed directly. For example, given an image of a tree in winter, the orientation/scale histogram shows oblique blobs, which mean that the predominant orientation changes with changes in spatial scale. Coarser blobs tend to be near to vertical, but finer scale structures tend to be oriented more nearly to horizontal, with a smooth transition between the two. This structure would be very difficult to compute directly, although not impossible.

A related issue concerns how the histograms should be sampled. What are the appropriate densities of samples in the various dimensions? The answer to this question depends partly on the natural scale of the structures within the histograms and partly on the desired discriminability of the histograms. In the limit of only one sample in each dimension, then all textures would be judged similar. In the other limit of an infinite number of samples in each dimension, then perhaps no textures would be judged similar. Once again, more information on the statistics of natural textures is required to provide a principled answer to this question.

A related issue is to look into how these techniques can be applied to images that are not homogeneous. In the present process, the whole image is treated as a single source of a texture, and a texture description is produced for the whole image. A more natural situation would be one where an image was made up with a patchwork of different textures. The key to this is to produce local texture descriptions for small areas of the image and to assess how the structures within these vary across the image. The question is how small such areas can be made, and what effect this has on the most suitable sampling.

References

Blake, A., Marinos, C. (1990) Shape from texture: estimation, isotropy and moments. Artificial Intelligence 45, 323-380

Julesz, B. (1981) Textons, the elements of texture perception, and their interactions. Nature 290, 91-97

Landy, M.S., Bergen, J. (1991) Texture segregation and orientation gradient. Vision Research 31, 679-691

Nothdurft, H.C. (1985a) Orientation sensitivity and texture segmentation in patterns with different line orientations. Vision Research 25, 551-560

Nothdurft, H.C. (1985b) Sensitivity for structure gradient in texture discrimination tasks. Vision Research 25, 1975-1968

Nothdurft, H.C. (1991) Texture segmentation and pop-out from orientation contrast. Vision Research 31, 1073-1078

Sagi, D. and Julesz, B. (1987) Short range limitations on detection of feature differences. Spatial Vision 2, 39-49

Watt, R.J. (1987) Scanning from coarse to fine spatial scales in the human visual system after the onset of a stimulus. J. Opt. Soc. Amer. A/4, 2006-2021

Watt, R.J. (1988) Visual Processing: Computational, Psychophysical and Cognitive Research. Lawrence-Erlbaum Associates, Sussex

Watt, R.J. (1990) The Primal Sketch in Human Vision. In: A. Blake & T. Troscianko (eds.) AI and the Eye. John Wiley

Watt, R.J. (1991) Seeing texture. Current Biology 1, 137-139

Watt, R.J. (1991) Understanding Vision. Academic Press, London

Watt, R.J. and Morgan, M.J. (1985) A theory of the primitive spatial code in human vision. Vision Research 25, 1661-1674

Witkin, A.P. (1981) Recovering surface shape and orientation from texture. Artificial Intelligence 21, 17-47

Segmenting Textures of Curved-Line Elements

David R. Simmons and David H. Foster

Department of Communication and Neuroscience, Keele University

Key words:
Curvature, Orientation, Texture Segmentation, Categorical Perception

Abstract

Performance in a segmentation task was measured for a variety of briefly presented textures composed of curved-line elements. It was found that the threshold increase in curvature required for reliable segmentation performance showed a dependence on background curvature similar to that for the *discrimination* of two curved lines. The form of this threshold dependence for segmentation was robust under changes in texture-element orientation and positional jitter. For textures of vertically oriented elements, variations in segmentation performance with background- and target-element curvature were well predicted by a model in which curved-line elements received a discrete representation in magnitude ("straight", "just curved", "more than just curved") and sign ("curved left", "curved right"). For textures of randomly oriented curved-line elements, a marked asymmetry was found in segmentation performance: targets comprising highly curved lines on a background of straight lines were much easier to segment than targets comprising straight lines on a background of highly curved lines. These data and additional computer simulations are shown to reject some traditional explanations of texture segmentation, including those based on local-luminance and on local-orientation differences. It is argued that the representation of curvature information in textures of curved-line elements is based upon the pre-attentive assignment of one of a few curvature labels to each element, each label associated with magnitude and direction information.

1   General Introduction

The "micropattern" texture, in which a foreground and background region contain a number of copies of different texture elements, has been a popular stimulus for studies of texture segmentation (Caelli and Julesz 1978; Gurnsey and Browse 1987; Landy and Bergen 1991). The "texton" theory of Julesz (Julesz 1981, 1984) explained the pre-attentive segmentation of this class of visual textures by suggesting that foreground and background could be easily distinguished only if the density of a specific set of features – the "textons" – was different in the two regions. The texton theory was later criticized by Nothdurft (1990), who argued that many aspects of texture segmentation apparently explained by texton density differences could be attributed to differences in the luminance or spatial-frequency content of the segregated image regions. One of Julesz's textons, element orientation, proved to be robust under Nothdurft's tests, although Nothdurft (1990) qualified the description of orientation as a texton by noting from his earlier work (Nothdurft 1985b) that local changes in orientation, rather than global differences in region orientation content, provided the most useful information about texture boundaries.

Several models of texture segmentation have emphasized the role of orientation information by modelling the segmentation process as a series of operations performed on the outputs of orientation- and spatial-frequency-tuned filters (Fogel and Sagi 1989; Rubenstein and Sagi 1990; Malik and Perona 1990; Landy and Bergen 1991). None of these models, however, has been able to capture fully all aspects of human performance in texture-segmentation tasks. A particularly problematic class of micropattern textures for such channel-based segmentation models has been the class of textures which comprise simple line figures, such as "L"s, "T"s, "X"s and "+"s. This class of micropattern textures is broadband in both orientation *and* spatial frequency, so complex problems of channel combination must be solved if these textures are to be successfully segmented by multiple-channel models (Landy and Bergen 1991). A further complication has been the existence of performance asymmetries, when the segmentation of a particular micropattern combination depends on which of the pair is foreground or background (Gurnsey and Browse 1987, 1989). These asymmetries have posed problems for segmentation models based on local spatial filtering (Gurnsey and Browse 1989; Malik and Perona 1990; Rubenstein and Sagi 1990).

In psychophysical studies of texture segmentation various quantitative measures of the "segmentability" of textures have been adopted. Some have given demonstrations of textures that "do" or "do not" segment (for example, Nothdurft 1990). Other approaches have included judging whether the two halves of a micropattern texture appear "the same" or "different" (Caelli and Julesz 1978), measuring the effects of different levels of masking on target-region shape discrimination (Nothdurft 1985a; Landy and Bergen 1991), and measuring the effects of masking on target-region localization (Gurnsey and Browse 1987). Few authors have measured thresholds for texture segmentation as a function of differences between target and background elements, the properties of which can be located on a suitable continuum. Such an approach is useful for two reasons. First, segmentation data can be compared directly to discrimination data for individual stimuli parameterized in the same way. Second, differences between target and background are not limited to those that can be characterized by the presence or absence of a particular binary feature, such as a texton, but may assume a range of values, thus allowing a threshold difference to be properly determined. This approach provides information about both the identity and the magnitude of the micropattern differences that are important for texture segmentation.

A spatial parameter that has been neglected in psychophysical studies of texture segmentation is contour curvature, perhaps because of the discouraging evidence of Beck (1973) that line curvature was a poor cue for similarity grouping. Nevertheless, there is agreement that curvature is an important image parameter, particularly for shape discrimination (Attneave 1954; Hoffman and Richards 1984; Richards *et al.* 1986), and there have been many studies of aspects of contour-curvature *discrimination* (Ogilvie and Daicar 1967; Andrews *et al.* 1973; Watt and Andrews 1982; Foster 1983; Watt 1984; Wilson 1985; Ferraro and Foster 1986; Watt 1987; Wilson and Richards 1989; Foster *et al.* 1992). Physiological interest has centred on the question of whether curvature sensitivity is the result of the action of specialized neurones for curvature processing or simply a consequence of orientation sensitivity (Hammond and Andrews 1978; Dobbins *et al.* 1987, 1989; Versavel *et al.* 1990). The question of how curvature-sensitive mechanisms might be constructed from combinations of orientation-sensitive mechanisms has also been addressed in psychophysical studies (Blakemore and Over 1974; Wilson 1985; Wilson and Richards 1989).

Another question pertinent to curvature processing is the possible "discrete" or "categorical" representation of curvature in briefly presented displays (Foster 1983; Ferraro and Foster 1986; Foster and Cook 1989). It has been suggested that in such a representation a curved line is given a discrete label, such as "straight" or "curved", and any further processing of the image has access only to this label information. The implication of this type of representation for texture segmentation is that a given image region could be characterized by the dominant label of the texture elements within it, and segmentation would thus follow automatically. Categorical theories of curvature processing may be related to the "two-process" theories of curvature processing that have been suggested by a number of authors (Watt and Andrews 1982; Wilson and Richards 1989; Versavel et al. 1990).

This study of the segmentation of textures of curved-line elements addressed three questions. First, can psychophysical performance for this class of textures be explained by existing models of texture segmentation? Second, does the processing of such textures demand specialized curvature-sensitive mechanisms? Third, is there evidence of a categorical representation of curvature in the analysis of these textures? Segmentation performance was measured for human subjects who viewed textures of curved-line elements, and performed a target-region shape-discrimination task similar to that used by Nothdurft (1985a). Experiments determined the relationship between curved-line-element texture segmentation and two-curved-line discrimination, the variation in segmentation performance due to texture-element changes that did not affect curvature cues, the importance of the direction of curvature of the texture elements, how curvature-labelling information might be used by segmentation mechanisms, and, finally, the relationship between element curvature and orientation in determining segmentation performance.

2 General Methods

2.1 Stimuli and Apparatus

Each stimulus consisted of a square array of 100 (10×10) curved-line elements. This array subtended 5×5 deg or 7×7 deg of visual angle. Within the array a horizontally or vertically oriented rectangular target patch of eight (4×2) elements was defined. All of the elements within the target patch had the same curvature, but this curvature was different from (either more or less than) that of the elements in the remainder of the array (the "background"; see figure 1). The elements themselves were generated by interpolation between a straight line and a circular arc, and were thus elliptical arcs. Element curvature was defined in terms of a parameter s (the "sag"), which corresponded to the angular distance in minutes of arc visual angle between the mid-point of the chord and the centre of the curve (see Foster 1983). This method of curved-line generation produces a very close approximation to true circular arcs (within 3% for the stimuli used here), and has the theoretical advantage of being directly related to the cue for curved-line discrimination (Foster *et al.* 1992).

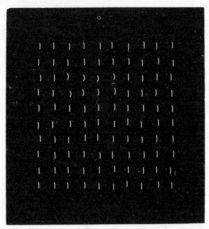

Figure 1: *Example of a curved-line-element texture (see Section 3). The entire array subtended 5×5 deg of visual angle, element chord size was 0.2 deg, background-sag value 0.43 arcmin, target-sag value 2.14 arcmin, and maximum positional jitter ±0.05 deg (±10% of mean element separation). The correct response for this stimulus would be "target horizontal".*

Stimuli were white and appeared superimposed on a uniform white 30×35 deg background with a luminance of approximately 40 cd m^{-2}. At the beginning of each experimental session, the intensity of the stimuli was adjusted by the subject (using a neutral density filter) to be ten times the luminance increment threshold, so that the stimuli were adequately suprathreshold, but not so bright as to produce noticeable afterimages.

Stimuli were presented on the screen of an X–Y display oscilloscope (Hewlett-Packard, Type 1321A) with a P4 sulfide phosphor (decay time approximately 100 μs). Experiments were controlled by a laboratory microcomputer through 12-bit DACs and a vector-graphics true-line generator (Sigma Electronics System QVEC 2150). Each curved-line element was drawn within a patch of 1024×1024 endpoint resolution and total extent 24×24 arcmin on the screen. The screen was viewed binocularly at a distance of 170 cm through a view tunnel and optical system that produced the uniform background field.

The curved-line elements each had a constant chord length of either 0.2 deg or 0.4 deg (the reasons for this choice of values are discussed later), and each consisted of eight concatenated straight-line segments or nine adjacent spots. Over the range of curvatures used, the curved-line elements appeared smooth to the eye (spot size 1.3 arcmin, full width at half height). Stimulus presentation time was nominally 100 ms, which corresponded to five refreshes of the display oscilloscope (refresh rate 50 Hz). In some experiments, stimulus presentation was followed by a 500-ms masking stimulus, which consisted of a sampled superposition of target and background elements.

2.2 Subjects

There were in all 14 subjects, each of whom had good acuity (Snellen 6/5 or better) and either fully corrected or no astigmatism. The subjects were paid volunteers and one of the authors (DRS); ages ranged from 19 to 48 yr.

2.3   Procedure

Subjects fixated a central fixation target and, when ready, initiated a trial by pressing a button on a button-box connected to the computer. After a 40-ms delay the stimulus appeared for 100 ms and was followed by a 100-ms blank field, after which the screen either remained blank or the masking stimulus appeared (see above). Subjects maintained central fixation during the presentation period, after which they indicated the orientation of the target patch (either horizontal or vertical) by pressing one of two buttons on a second button-box. The fixation target reappeared after a short delay, signalling that the next trial could be started.

2.4   Experimental Design and Data Analysis

Details are given in the appropriate sections.

3   A Comparison Between Curved-Line-Element Texture Segmentation and Two-Curved-Line Discrimination

Several studies have suggested that curvature discrimination in briefly presented displays is a categorical process (Foster 1983; Ferraro and Foster 1986; Foster and Cook 1989), in that there is an underlying discrete representation of the curved-line continuum. The interpretation of categorical processing (Wood 1976) was based on the presence of sharp peaks and troughs in the plot of increment threshold against stimulus magnitude, and the correspondence of those extrema with the predictions of separate curved-line labelling experiments (Foster 1983). The purpose of the experiments in this section was to determine if arrays of curved-line elements forming textures were processed in a similar discrete way. Measurements of increment threshold for the segmentation of textures of curved-line elements were thus compared with those for the simple discrimination of two curved lines.

3.1   Methods

3.1.1   Segmentation

Stimuli for the segmentation task were generated as outlined in the General Methods section, with curved-line elements formed from eight concatenated straight-line segments. Target-element sag values were always greater than background-element values. For a background sag of s, $s > 0$, target element sag was $s + ds$, $ds > 0$, and thus all elements curved in the same direction (as in figure 1). Chord sizes were 0.2 deg and 0.4 deg, and seven values of background sag were used (0.0, 0.43, 0.86, 1.28, 1.71, 2.14 and 2.57 arcmin for the 0.2 deg chord elements, and 0.0, 0.86, 1.71, 2.57, 3.43, 4.29 and 5.14 arcmin for the 0.4 deg elements). The field size was 5×5 deg for the 0.2 deg chord stimuli and 7×7 deg for the 0.4 deg chord stimuli. Luminance cues to segmentation were disrupted by giving each curved-line element a small amount of random positional jitter, with a maximum value of 0.05 deg, corresponding to 10% and 7% of the average element separation for the

0.2 deg and 0.4 deg chord sizes respectively. Elements were always vertically oriented. Stimulus presentation time was 100 ms with no poststimulus mask.[1]

Data were collected in blocks of 70 trials, preceded by seven practice trials. In each trial the background-sag value was randomly selected from the seven fixed values, but the target-sag value was chosen by an adaptive routine (PEST; Taylor and Creelman 1967; Hall 1981). In each block there were ten trials for each background-sag value. Chord size was randomized between, but not within, blocks. Subjects took approximately five minutes to complete each block, and normally completed eight blocks of trials in a one-hour session. After 16 blocks of trials had been completed, each of the eight subjects had obtained data for 14 psychometric functions (one for each background-sag and chord-size combination), each function based on 80 trials. Each of these sets of data was fitted by a quadratic function, after transformation of the proportion-correct scores by the inverse of a cumulative-Gaussian function. The threshold value was taken for a performance level of 75% correct. A "bootstrap" procedure (Foster and Bischof 1991) was used to estimate the standard deviation on each threshold value. These standard deviations were used to calculate weighted means and standard deviations when averaging over subjects.

3.1.2 Discrimination

In a parallel series of measurements, sag increment thresholds were determined for the discrimination of two curved lines. The task was a spatial two-alternative forced choice (2AFC): subjects were asked to indicate the position ("left" or "right") of the more curved of the curved lines. The curved lines were presented each side of a central fixation cross, at an eccentricity of 1.4 deg. Stimulus duration was 100 ms. Stimulus presentation was followed, after a 100-ms delay, by a 500-ms presentation of a masking stimulus of five randomly oriented straight lines. Curved lines were similar to the curved-line texture elements used in the segmentation experiments, except that each was formed from 12, rather than eight, concatenated straight-line segments. The minor differences in experimental procedure and methods of curved-line generation in the segmentation and discrimination experiments were caused partly by hardware limitations, and partly by the requirements of consistency with other curvature-discrimination experiments. The curved lines in the discrimination experiments were always vertically oriented and were not positionally jittered. Thresholds were obtained for the same range of reference curvatures and chord sizes as in the segmentation experiments.

3.2 Results

Figure 2a shows sag increment thresholds for texture segmentation and for two-curved-line discrimination as a function of background- or reference-sag value. Each point is a weighted combination of the thresholds from the eight subjects. The peaks and troughs in the two sets of data are almost exactly coincident, although there is an evident increase in threshold values of just under 1 arcmin for the segmentation task. Figure 2b shows corresponding data for the larger chord size of 0.4 deg. The threshold minimum in the discrimination data is again coincident with the minimum in the segmentation data, but segmentation thresholds appear to saturate for background-sag values greater than 2.57

---

[1]A poststimulus mask was used in later experiments (see Sections 5 and 7), for consistency with other studies of texture segmentation. In a control measurement, however, sag increment thresholds for segmentation under the present conditions were found to be independent of whether the mask was used.

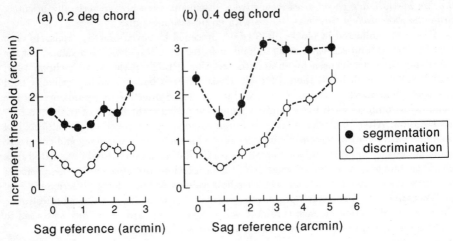

Figure 2: *Sag increment thresholds for curved-line-element texture segmentation (filled circles), and two-curved-line discrimination (unfilled circles), plotted as a function of background or reference sag. Increment threshold corresponds to the increase in sag of the target element or elements (relative to the reference or background elements respectively) required to obtain a performance level of 75% correct. Data points and standard errors are weighted means of thresholds from eight subjects: (a) 0.2 deg chord stimuli, (b) 0.4 deg chord stimuli.*

arcmin. Notice that the peaks in performance (threshold minima) in all four experimental conditions occur at the same value of background or reference sag, namely 0.86 arcmin.

3.3   Discussion

3.3.1   Categorical Segmentation?

Evidence for categorical processing in curved-line discrimination in early vision has been presented elsewhere (Foster 1983; Ferraro and Foster 1986; Foster and Cook 1989). The existence of similar peaks and troughs in the dependence of sag increment threshold on background sag in texture segmentation suggests the possibility of similar categorical processes subserving performance. The lowest thresholds for texture segmentation and curved-line discrimination for both chord sizes occurred at or close to background- or reference-sag values of 0.86 arcmin, which corresponds very closely to the location of one of the peaks in discrimination performance for a slightly different curved-line discrimination task (Foster 1983, Fig 2b). The increase in thresholds on each side of this minimum suggests (Foster 1983) a transition between curved-line categories, the two boundaries of which are close to 0.86 arcmin.

This putative process of segmentation by categorization, which from now on will be called *categorical segmentation*, cannot adequately account for two aspects of the threshold data shown in figure 2. First, if there is only one curved-line category boundary, and it is located at a sag value of about 0.86 arcmin, the segmentation of textures with background sags greater than this value should become increasingly difficult as the probability of the background and target texture elements being assigned to different categories becomes smaller and smaller (Foster 1983; Ferraro and Foster 1986). As shown in the upper graph of figure 2b, sag increment thresholds were approximately constant for background sag values of 2.57 arcmin and larger, and therefore some other segmentation mechanism, perhaps one sensitive to element orientation content, must be involved, at least when background-sag values are large. Second, target-sag values for segmentation threshold were always higher, by about 1 arcmin, than those for discrimination threshold. It has been shown, by means of a computer model based on segmentation by luminance cues (Simmons *et al.* 1991), that this threshold difference is difficult to explain purely in terms of categorical segmentation if only two curved-line labels (for example, "straight" and "curved") were available to the observer; that is, segmentation performance should be as good, if not better, than discrimination performance. If a two-category labelling process did underlie the segmentation of these curved-line-element textures, then there must exist differences between the labelling probabilities for isolated and embedded curved-line elements of a given sag value. A further complicating factor may have been the possible variation in labelling probability with element eccentricity. Some of these issues are considered in more detail later, when the role of categorical processing in segmentation performance is examined more directly (see Section 6).

3.3.2 Element Orientation Content

When there exists a sag difference between two curved lines curved in the same direction and of equal chord lengths, there also exists a difference in their orientation contents (the range of angles turned through by a tangent to the curve moving along the curve). Thus when the chords of all the curved-line elements in a texture are aligned (as in figure 1)[2] orientation-tuned mechanisms should respond differently to target and background regions, and could therefore subserve segmentation performance. (For related material on the role of orientation cues, see Watt 1992.)

It has been shown elsewhere that differences in orientation content are, in fact, a poor cue for isolated curved-line discrimination when presented for long durations (Foster *et al.* 1992). A further argument against differences in orientation as the cue for segmentation is based on the coincidence of the threshold minima in the data shown in figure 2. If orientation differences were the cue for segmentation or discrimination then the positions on the curved-line continuum of any peaks and troughs in performance should scale with the size of the curved lines. Specifically, doubling the chord size from 0.2 deg to 0.4 deg should have resulted in a shift of the threshold minimum to a larger value (approximately twice the size) of the background sag. There was no such shift, suggesting that differences in curved-line orientation content were unlikely to have been the primary cue for either curved-line-element texture segmentation or two-curved-line discrimination in these experiments.

It may be possible to manipulate a texture-segmentation model operating solely on the outputs of orientation- and spatial-frequency-tuned filters so that it accounted for the stability of the positions of the threshold minima in the curved-line-element texture-segmentation data. One possible mechanism for this might be a normalization scheme similar to that described by Gurnsey and Browse (1989). The more parsimonious explanation based on categorical processing of curved-line sag will be discussed later.

---

[2]Note the difference between the orientation of the curved-line element itself, which is taken to be the orientation of its chord and is independent of the sag value, and the orientation *content* of the element, which is dependent on both the orientation of the chord and the sag value of the element; the *range* of orientations contained in the element will not vary with chord orientation.

4   Effects of Positional Jitter and Element Rotation

All of the curved-line textures used in the experiments of Section 3 contained vertically oriented elements which had a maximum positional jitter of ±0.05 deg. The experiments presented in this section investigated to what extent segmentation increment thresholds were independent of element orientation and array regularity.

4.1   Methods

Two sets of experiments were performed in parallel. In the first set, maximum positional jitter was kept the same at ±0.05 deg (±3 arcmin; ±10% of mean element separation), but element orientation was allowed to assume four different values: vertical, horizontal, right-oblique (−45 deg) and left-oblique (+45 deg). The curved-line elements in these textures were always aligned, and orientation was randomized between, but not within, presentation blocks (within-texture randomization of orientation is considered in Section 7). In the second set of experiments, the maximum positional jitter of each element was 0.0 deg, ±0.05 deg, or ±0.10 deg, corresponding to 0%, ±10%, and ±20% of the mean element separation, but element orientation was fixed at either vertical or left-oblique.

As in Section 3, data were collected in blocks of 70 trials, preceded by seven practice trials; in each trial the background-sag value was randomly selected from the seven fixed values, but the target-sag value was chosen by a PEST routine, and in each block there were ten trials at each background-sag value. The eight experimental conditions consisted of all four orientation conditions at the intermediate value of maximum jitter (±0.05 deg) together with vertical and left-oblique orientation conditions at the other two maximum jitter values. Subjects normally completed eight blocks of trials in a one-hour session, one block for each condition in a random order. After 64 blocks of trials had been completed, each of the eight subjects had obtained data for 56 psychometric functions (one for each combination of background sag and experimental condition), each function based on 80 trials. Each of these functions was analysed as described in Section 3, and the threshold values obtained were similarly combined.

Stimulus presentation methods were also the same as in Section 3, except that the curved-line elements were made up of nine spots rather than eight vectors. This modification ensured that the total luminance content of each element was independent of its orientation.

4.2   Results

Figure 3 shows segmentation increment-threshold functions for the four orientation conditions. Threshold values and standard errors are weighted means from eight subjects. The graphs for all conditions except the right-oblique show a minimum threshold value for background-sag values of either 0.43 arcmin or 0.86 arcmin. Thresholds for textures with oblique elements were always higher than those for textures with vertical or horizontal elements. Figure 4a shows the effect of different levels of positional jitter on segmentation increment thresholds for textures of vertically oriented elements. Similar results are shown in figure 4b for textures with left-oblique elements. The thresholds in figure 4b show the expected U-shaped dependence on background sag at all values of positional jitter, although the position of the threshold minimum does vary with jitter, moving from 0.43 arcmin to 0.86 arcmin as jitter increases.

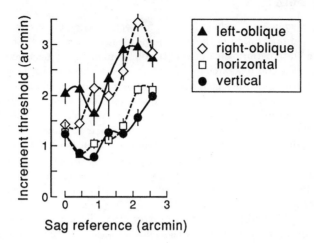

Figure 3: *Sag increment thresholds for curved-line-element texture segmentation plotted as a function of background-sag value for four different element orientations: vertical (filled circles), horizontal (unfilled squares), right-oblique (unfilled diamonds), left-oblique (filled triangles). The chord size was 0.2 deg. The maximum positional jitter of the elements was ±0.05 deg (±10% of mean element separation). Data points and standard errors are weighted means from eight subjects.*

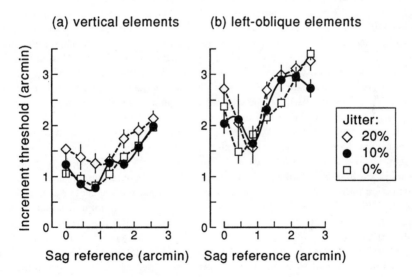

Figure 4: *As figure 3, but for (a) textures of vertical curved-line elements, and (b) textures of left-oblique curved-line elements, at three values of maximum element positional jitter: 0.0 deg (0% of mean element separation; unfilled squares), 0.05 deg (10% of mean element separation; filled circles), and 0.1 deg (20% of mean element separation; unfilled diamonds).*

4.3   Discussion

The form of the increment-threshold function for segmentation did not change with variations in jitter and orientation of the curved-line elements, and in all conditions, except one, there were clear performance peaks (threshold minima) at background-sag values between 0.43 arcmin and 0.86 arcmin. The positions of these peaks, however, did vary with jitter and orientation: threshold minima shifted towards larger values of background sag at larger jitter values, and textures of oblique curved-line elements were more difficult to segment, for a given sag difference, than those of vertical or horizontal curved-line elements. Results for the 10% jitter condition were similar to those from Section 3, suggesting that the different method of curved-line generation (spots rather than vectors) made no significant difference. Thresholds were highest for the largest value of jitter, but the basic shape of the threshold dependence was preserved over all jitter values.

Nothdurft (1990) suggested that the validity of a particular texture-element attribute as a cue for segmentation should be questioned if segmentation by means of this cue could be disrupted by modifications of the texture that did not directly change the cue value. For example, if performance in a segmentation task were based on orientation differences, then random variation in the luminance of the texture elements should not affect the segmentability of the texture. But the fact that image segmentation by orientation differences in the *absence* of global luminance differences is possible, does not necessarily imply that segmentation performance should be the same when luminance cues are present in addition to the orientation cues. Luminance cues themselves could provide alternative valid segmentations of the image, as Nothdurft's own demonstrations have indicated (Nothdurft 1990), although he was careful to reduce the impact of luminance cues as far as possible. Complex questions of cue combination and competition must therefore be addressed before a candidate cue for segmentation is rejected on this basis.

The fact that curvature increment thresholds for segmentation were influenced by positional jitter of the elements does not imply that curvature cues were not used to perform the segmentation. Positional jitter may have introduced additional segmentation cues such as luminance edges that accidentally raised or lowered segmentation thresholds by increasing or reducing the visibility of the texture edges that were already present. The importance of luminance cues to segmentation in textures of curved-line elements, and how these cues are disrupted by element positional jitter, was further investigated in the series of computational control experiments presented in Section 8.

With regard to variations in segmentation performance with element orientation, a number studies have shown that there exists an "oblique effect" for curvature discrimination (Ogilvie and Daicar 1967; Watt and Andrews 1982; Wilson 1985), in that increment thresholds are higher for oblique than for vertical or horizontal curved lines. A reduced segmentation performance with textures of oblique curved-line elements was therefore not surprising. The existence of an oblique effect for curvature discrimination has been used to argue for a representation of curvature information solely in terms of the outputs of orientation-tuned filters (Wilson 1985), but in view of the general decline in discrimination performance along oblique axes it is also possible that more direct representations of curvature could lead to orientation anisotropies, especially if such anisotropies are a consequence of the organization of these representations about horizontal-vertical axes (see Kahn and Foster 1986; Foster 1991, also Foster and Ward 1991b).

The presence of increment-threshold minima at levels of background sag between 0.43 arcmin and 0.86 arcmin for all but one of the eight experimental conditions here is consistent with an explanation based on the categorical segmentation of curved-line textures.

The anomalous threshold dependence found for right-oblique elements may represent a sampling problem in that the category boundary may have been between the background-sag values 0.43 arcmin and 0.86 arcmin, so that no clear minimum was detected. If segmentation performance was determined by categorical processes, then, for certain fixed values of the background curved-line sag, it should be possible to detect rapid changes in discrimination performance as the target-sag value crossed from one putative category to another. This behaviour over large variations in target-sag values was explored as part of the following experiment.

5 Effects of Sag Decrements

The experiments described in Sections 3 and 4 were restricted to measurements on textures of curved-line elements with sag values equal to or greater than those of the background. If a division of the curved-line continuum into discrete categories is valid, then performance should be independent of which category is dominant in the target or the background region, and should depend only on whether these categories are different. It is not clear, however, whether curved lines pointing in opposite directions receive different codings, and how such codings might be related to the magnitude of the curvature. In the following experiment, sag increments were allowed to take negative as well as positive values, resulting in textures in which target-sag values were sometimes less than background-sag values, and some elements curved in opposite directions, as shown in figure 5. This extension to the range of sag values also resulted in the interesting situation of target and background containing elements with precisely the same sag values but of opposite signs. This situation provided a further test for texture segmentation models sensitive only to contour orientation differences, for both target and background elements contained the same range of orientations.

5.1 Methods

The adaptive PEST method for setting target-element sag was replaced by a method of constant stimuli, because of the possibility of the stimuli yielding non-monotonic psychometric functions. A post-stimulus mask was also introduced (see Note 1). Data were obtained from a smaller group of subjects and over a longer period of time than in the experiments of Sections 3 and 4, so that within-subject trends could be more easily followed over the larger range of conditions.

Data were collected in blocks of 250 trials. Each block contained 25 presentations of each of the ten background-sag values (0.0, 0.43, 0.86, 1.28, 1.71, 2.14, 2.57, 3.0, 3.43, 3.86 arcmin). The 20 target-sag values were positive-and-negative values taken from the same set of magnitudes as the background-sag values. These 20 target-sag values for each background-sag value were spread across four separate blocks of trials so that only five target-sag values were used for a given background-sag value in a given block. The distribution of target- and background-sag values within each block was organized such that each of the 20 possible target-sag values was presented approximately the same number of times. Each block took approximately 25 min to complete, and subjects completed two blocks in sessions of one-hour duration. Eight replications of each block by each subject yielded 40 trials at each of the background- and target-sag value combinations.

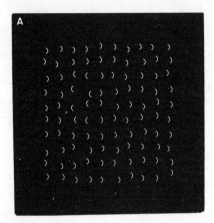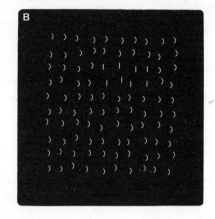

Figure 5: *Examples of curved-line-element textures (Section 5); array and chord sizes are the same as figure 1. (a) Target-sag value −3.86 arcmin, background-sag value 3.86 arcmin. Maximum positional jitter was ±0.1 deg (±20% of the mean element separation). Notice that target and background elements are mirror symmetric. The two subjects scored 80% and 73% correct with this stimulus. (b) Target-sag value 0.0 arcmin, background-sag value 3.86 arcmin. Same positional jitter as (a). Subjects scored 100% and 99% correct with this stimulus.*

Because of the possible very large values of sag increment, a modification was made to the positioning of the curved-lines elements with respect to the underlying texture "matrix". In the experiments described in Sections 3 and 4, the centre of a curved-line element was taken to be the mid-point of its chord. Positional jitter and orientation changes were made in a local coordinate system with the chord mid-point as origin. In the present experiments, which allowed curvature in both directions, this local coordinate origin might have introduced strong luminance cues at the target-region boundary. To reduce such luminance cues, the local coordinate system centre was shifted to the "centre-of-mass" of the curved-lines, which corresponded to a position approximately two-thirds of the distance along the perpendicular bisector of the curved-line element joining the mid-point of the chord to the mid-point of the curved-line element. A randomized positional jitter with maximum values ±0.1 deg (±20% of the mean element separation) was also imposed to disrupt luminance cues further.

5.2   Results

Figure 6 shows, for each subject, the probability of correct performance in the segmentation task as a function of the target-element sag at three representative values of background sag: 0.43, 1.71, and 3.0 arcmin. The piecewise smooth curves are maximum-likelihood linear and quadratic functions (with variable intersections), after transformation by the inverse of a cumulative Gaussian function. Note that the sign of the target-sag value refers to the *relative* direction of background and target elements, so that negative values refer to target elements curving in the *opposite direction* to the background elements (as in figure 5b).

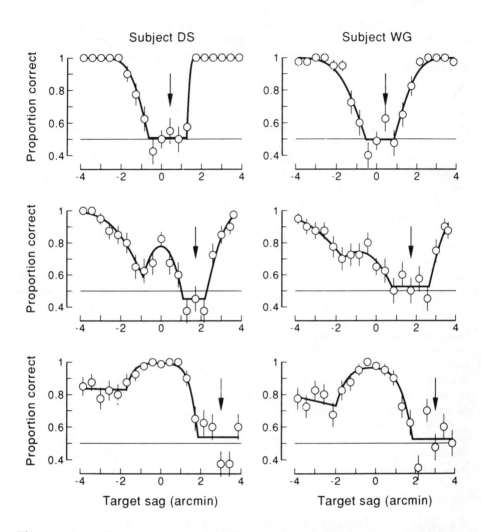

Figure 6: *Proportion-correct performance in the curved-line texture segmentation task for two subjects and three background-sag values plotted as a function of target-sag value. Elements were always vertical. The horizontal line represents chance performance. The vertical arrows indicate the values of the background sag along the continuum. Each data point shows the mean and standard error of 40 trials, assuming a binomial error distribution. The fitted curves are piecewise smooth maximum-likelihood linear and quadratic functions (with variable intersections), after transformation of the proportion-correct data by the inverse of a cumulative Gaussian function. Notice that the sign of the target-sag value refers to the relative direction of background and target elements, so that negative values refer to target elements curving in the opposite direction to the background elements (as in figure 5a). Background-sag values were (top sections) 0.43 arcmin, (middle sections) 1.71 arcmin, (bottom sections) 3.0 arcmin.*

5.3   Discussion

It is evident that segmentation performance does not increase smoothly and monotonically with increase in the difference between target- and background-sag values: there are critical points along the continuum at sag values of approximate magnitudes 0.8 and 1.5–2.0 arcmin where performance accelerates or has an inflexion. These values are remarkably close to the boundary values of 0.87 and 1.81 arcmin reported in the discrimination and categorical labelling of isolated curved lines (Foster 1983).

An explanation of the results in terms of an explicit, categorical representation of curvature in textures of curved-line elements may be formulated as follows, provided that some information about direction of curvature is also included. Suppose (Foster 1983) that there are three curved-line "channels" for a given curved-line orientation: one tuned to straight lines, one to lines that are "just curved", and one to lines that are "more than just curved", and that there are two forms of the second and third channels for the two directions. Suppose also that it is easier to distinguish magnitudes of category labels than their signs, in accordance with the notion of weak sign labels, proposed by Foster (1978) and extended in Kahn and Foster (1986) (see also Foster 1991). When background-sag values are small (0.43 arcmin; figure 6, top), target-sag values in either direction must at least reach the category boundaries, at approximately 1 arcmin and −1 arcmin, before performance is better than chance. When background-sag values are larger (close to 1.71 arcmin; figure 6, middle), target-sag values must be less than 1 arcmin for moderate levels of performance to be obtained, and there is a small *decrease* in performance as the target-sag value passes through the boundary at either −1 arcmin or −2 arcmin. When background-sag values are larger still (3.0 arcmin; figure 6, bottom) performance actually declines at sufficiently large negative values of the target sag, when background and target have the same magnitudes of sag, but different sign. This last result is difficult to interpret in terms of a simple segmentation based on orientation content; for the particular case of the target-sag value of −3.0 arcmin in figure 6 (bottom sections), target and background curved-line elements had identical orientation contents.

Although categorical segmentation appears to provide a plausible explanation for several aspects of performance in the segmentation of textures of curved-line elements, it has not yet been established that the labels assigned in texture segmentation were the same as those used in curved-line discrimination. This issue is addressed in the following section.

6   Target-Region Labelling Experiment

The experiment described in this section made an explicit test of the categorical model of curved-line texture segmentation by requiring subjects to assign curvature labels to the elements within the target region. Labelling performance was then used to predict performance in the 2AFC segmentation task.

6.1   Methods

Methods for the 2AFC segmentation task were the same as for the experiments in Section 5, except that a smaller value of maximum element jitter was used (±0.05 deg; $\pm10\%$ of mean element separation) to enhance performance levels. Stimuli for the labelling experiment were exactly the same as for the 2AFC task, the only difference between the experiments being that the subjects were given a three-button box, and asked to report the curvature of the curved-line elements in the target region instead of the orientation of the segmented patch. Subjects were asked to push the leftmost button if the elements in the target region appeared to be curved to the left, the rightmost button if they appeared to be curved to the right, and the middle button if they appeared to be straight. In view of the results of the previous experiment, and of Foster (1983), it might be argued that a range of five labels should have been offered, but the purpose here was to keep the demands of the tasks as simple as possible. Some disparity between segmentation performance and the predictions of the labelling model was therefore anticipated at large target-sag values. The full set of 2AFC segmentation data was collected first, followed by that for the labelling task. Both subjects were well practiced in curved-line-element texture segmentation.

6.2   Results

The right-hand column of figure 7 shows one of the subject's proportion-correct responses in the 2AFC segmentation task as a function of target-sag value for three background-sag values. The predictions based on the labelling data are shown in the left-hand column. The predictions were based on the assumption that segmentation was successful if and only if target and background regions were labelled differently.

6.3   Discussion

It is clear from figure 7 that there is good qualitative agreement between 2AFC and labelling data for small (0.43 arcmin) and large (3.0 arcmin) values of background sag, and for the negative part of the continuum at the intermediate background-sag value (1.71 arcmin). A deviation between the labelling and 2AFC data occurred in the positive part of the curved-line continuum, where background and target elements were pointing in the same direction. For the 1.71 arcmin background-sag value, segmentation performance was predicted to be at chance levels, yet the 2AFC data showed a clear improvement in performance with increasing background sag.

The main reason for this disparity between predicted and observed performance was mentioned in the Methods section, namely the need for a third curvature-magnitude category. The evidence from the threshold data in Sections 3 and 4 points to one category boundary at a sag value of about 0.86 arcmin. The proportion-correct data in Section 5 suggested that a second category boundary, at about 1.71 arcmin, was also appropriate. Subjects reported that it was often possible to segment targets with elements of very high curvature (sags of about 4 arcmin) from those of intermediate curvature (sag of about 2 arcmin), even though both were clearly given the label "curved".

A potential general source of disagreement between 2AFC segmentation and labelling data may have been that reported category boundaries were sharper than those truly underlying segmentation performance, for in order to label a target region correctly it was necessary only to identify a single element within that region as being different from the background and label that single element correctly, whereas in order to segment that region considerably more elements (at least three) would have to be correctly labelled. Further quantitative comparison of the data would be better achieved within the framework of a more complete model of segmentation (for example, Proesmans and Oosterlinck 1992). Despite these caveats, it should be noted that labelling data give a good qualitative prediction of several aspects of vertical curved-line-element texture-segmentation performance.

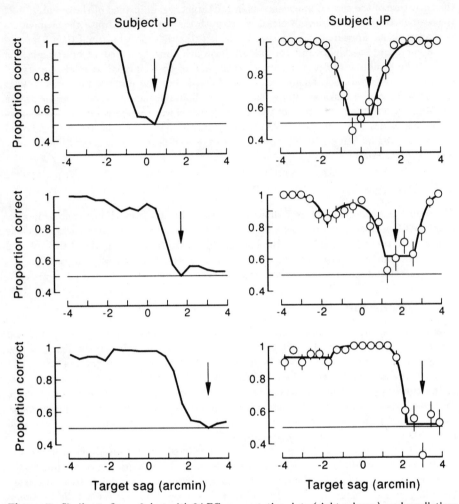

Figure 7: *Similar to figure 6, but with 2AFC segmentation data (right column), and predictions based on labelling data (left column) for one subject (JP). The maximum element jitter was lower than in Section 5 at ±0.05 deg (±10% of the mean element separation). The vertical arrows indicate the values of the background sag along the continuum. Background-sag values were (top sections) 0.43 arcmin, (middle sections) 1.71 arcmin, (bottom sections) 3.0 arcmin.*

The experiments with textures of vertically oriented curved-line elements in Section 5 and this section have indicated the importance of the direction of curvature in the representation of curved lines. The experiments with textures of curved line elements at other orientations (Section 4) showed that sensitivity to curvature differences was reduced at oblique orientations. Physiological studies of curvature-sensitive neurones have indicated that these neurones are sometimes sensitive to both the magnitude and the direction of contour curvature (Hammond and Andrews 1978; Dobbins *et al.* 1987, 1989; Versavel *et al.* 1990). The question that arises is thus: in the segmentation of textures of curved-line elements, to what extent is the representation of curvature independent of the orientations of the elements?

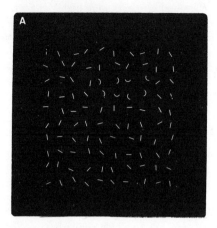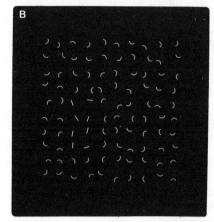

Figure 8: *Example stimuli (Section 7). Array and chord sizes as in figure 1. (a) Target-sag value 0.0 arcmin, background-sag value 3.86 arcmin. Maximum orientation jitter was ±180 deg, maximum positional jitter was ±0.10 deg (20% of the mean element separation). Subjects scored 63% (DS) and 55% (LC) correct with this stimulus. (b) Target-sag value 3.86 arcmin, background-sag value 0.0 arcmin. Other details as in (a). Subjects scored 93% (DS) and 95% (LC) correct with this stimulus.*

7 Textures with Randomly Oriented Curved-Line Elements

With the obvious exception of segmentation by orientation differences, the validity of a particular element attribute as a cue for texture segmentation has been commonly tested with textures of randomly oriented elements (see, for example, Julesz 1981). Measurements of curved-line-element texture segmentation with randomly oriented elements were undertaken to clarify to what extent curvature information is useful for texture segmentation in the absence of orientation cues. Notice that curvature is the space derivative of orientation, and therefore noise in the orientation domain might be expected to interfere with curvature information, leading to reduced performance.

7.1 Methods

Methods were identical to those of Section 5, except that the elements now had a completely randomized orientation (see figure 8).

7.2 Results

Figure 9 shows segmentation performance data for the two subjects at three background-sag values. These background-sag values corresponded to the two limits of the range of sag values used (sag values of 0.0 arcmin and 3.86 arcmin; Figures 9, top and bottom sections) together with one from the centre of the range (1.71 arcmin; figure 9, middle section). There was a clear performance asymmetry. When the target elements were highly curved, and the background elements were straight lines, the target region was easily segmented (target-sag value of 3.86 arcmin; figure 9, top section), but when the target elements were straight lines, and the background elements were highly curved, the target region was much more difficult to segment (target-sag value of 0.0 arcmin; figure 9, bottom section). The two example textures shown in figure 8 illustrate the stimuli that caused this performance asymmetry. Performance levels were considerably lower in these experiments than in those of Section 5: sag increment-threshold values were at least 1 arcmin higher and one subject did not reach threshold performance level for background-sag values of greater than 1.28 arcmin with the range of target-sag values available.

7.3   Discussion

Gurnsey and Browse (1989) argued that asymmetries in visual texture discrimination, in which texture A within texture B is much easier to detect than texture B within texture A, provide *prima facie* evidence against any model of texture discrimination based only on local measurements and comparisons. They proposed that texture-discrimination asymmetries could be explained by assuming that the responses of orientation- and spatial-frequency-selective filters were normalized by the amount to which similarly tuned operators responded elsewhere in the image. As well as adding to the list of asymmetries in texture segmentation (Gurnsey and Browse 1987, 1989), and possibly visual search (Julesz 1981; Beck 1982; Treisman and Souther 1985; Treisman and Gormican 1988; Foster and Ward 1991a, b), the data presented in this section suggest that, if the Gurnsey and Browse (1989) explanation of texture segmentation asymmetries is correct, the responses of *curvature-selective* filters should be similarly normalized across an image.

Results from visual search experiments suggest that visual search for a target that *lacks* a feature among distractors that possess this feature is slower than for a target that possesses that feature among distractors that do not (Treisman and Souther 1985; Treisman 1988; Treisman and Gormican 1988). If curvature is represented in some sort of "feature map" (Treisman 1988), then perhaps the most appropriate labels for the segmentation of textures of randomly oriented curved-line elements are "curved" and "not curved". This type of representation would reduce straightness to being the absence of curvature, and the asymmetry found for this segmentation task would then be consistent with some of those observed in visual search experiments. Such an elaboration of the categorical-labelling model could be related to the Gurnsey and Browse (1989) explanation of asymmetries in segmentation performance in that the meaning of the curved-line label, and possibly also the positions of the category boundaries, might vary if a global normalization of the responses of similarly tuned curvature-selective filters took place.

The fact that the segmentation of textures of randomly oriented curved-line elements was possible provides further evidence that any model of texture segmentation that relies solely on orientation, spatial-frequency, or luminance information is incomplete; curvature information must also be explicitly represented. Yet the increased difficulty that subjects experienced in trying to segment textures of randomly oriented curved-line elements suggests that the representation of curvature is also likely to be directional (that is specifying the orientation and sense of the normal to the chord); therefore the grouping of curved-line elements would be easier when they shared both magnitude and direction of curvature, rather than magnitude of curvature alone.

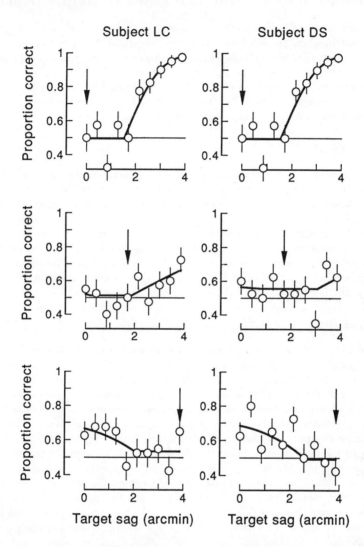

Figure 9: *Proportion-correct performance in the curved-line-element texture segmentation task (Section 7). Data for two subjects are plotted as a function of target-sag value for three representative background-sag conditions. Elements had maximum orientation jitter of 180 deg. Maximum positional jitter was 0.10 deg (20% of the mean element separation). The horizontal line represents chance performance. The vertical arrows indicate the values of the background sag along the continuum. Note that the range is restricted to positive sag values only, as curvature sign has no meaning for randomly oriented stimuli. Each data point shows the mean and standard error of 40 trials, assuming a binomial error distribution. Background-sag values were (top sections) 0.0 arcmin, (middle sections) 1.71 arcmin, (bottom sections) 3.86 arcmin.*

8   The Role of Luminance Cues to Segmentation

The precautions taken in this study against accidental luminance cues to segmentation included matching the curved-line elements for brightness (by generating each of them from equal numbers of vector line-elements or spots); employing positional jitter of the texture elements; and, in some cases, ensuring that the local coordinate system in which element transformations (such as rotations) were performed had its origin at the centre-of-mass of the element. Nevertheless, some additional measurements based on computer simulations were made to control for accidental luminance cues.

8.1   Methods

Textures of curved-line elements were simulated by computer-generated 512×512 arrays of floating-point numbers. They were generated in a similar way to the textures used in the psychophysical experiments in that each curved-line element was constructed from nine adjacent spots with Gaussian intensity profiles. The spot-size was scaled to the correct fraction of the display size for a texture subtending 5×5 deg. Figure 10a shows a scaled example of a computer-simulated curved-line texture, although the number of gray levels has been reduced for ease of reproduction. These texture images were convolved with isotropic Gaussian kernels of unit volume to test for the presence of local variations in the luminance distribution.

8.2   Results

The results of the filtering process were highly dependent on the standard deviation of the Gaussian filter. If the standard deviation was small with respect to the mean element separation, the filter integrated luminance energy only from single curved-line elements, and therefore no differences between target and background elements were obtained. But with filter standard deviations close to half the mean element separation some luminance cues to segmentation became visible. Figure 10a illustrates a curved-line texture with (simulated) background-sag value 0.86 arcmin, and target-sag value 1.86 arcmin (corresponding to 1.7 and 3.2 pixels respectively). There is no positional jitter, and the local coordinate origin is at the centre of the chord (as in the experiments of Sections 3 and 4). After convolution of this image with an isotropic Gaussian filter of standard deviation 30 pixels and unit volume the image shown in figure 10b was obtained. Strong luminance edges are apparent in figure 10b that coincide with the vertical edges of the target region. The amplitude of the simulated luminance difference between the bright and dark patches and the background was approximately 2% of the mean luminance of the image; thus, although the amplitude of the luminance edge was exaggerated in this case by intensity scaling of the computer image, it may still have been detectable in psychophysical experiments.

The texture shown in figure 10c was filtered in exactly the same way as that in figure 10a, and the result is shown in figure 10d. The only difference between the textures in figures 10a and 10c is that the latter contains a small amount of random positional jitter, with a maximum value equal to 10% of the mean element separation (the same jitter level as in the experiments described in Sections 3 and 6, and as in some of those of Section 4). Figure 10d shows that the use of even small amounts of positional jitter was enough to severely disrupt luminance cues to the edge of the target region. Further computational

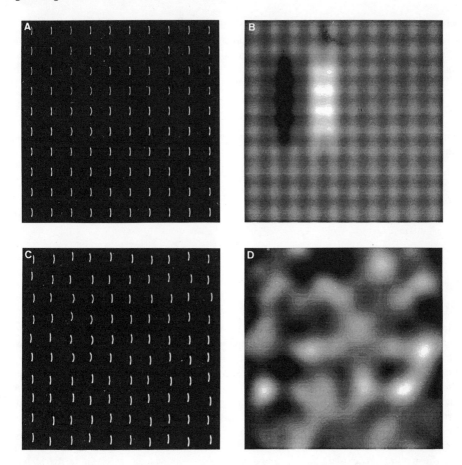

Figure 10: *Illustration of the role of luminance information in assisting target region orientation identification at different levels of positional jitter. (a) Simulated curved-line texture from Section 3. Background sag 0.86 arcmin, target sag 1.86 arcmin, no positional jitter. (b) The same texture filtered with a normalized isotropic Gaussian, standard deviation 1.17 times the mean element separation. The image was re-scaled after filtering. Strong luminance cues to the target region orientation are evident. (c) As (a) except with a small random positional jitter of maximum value ±0.05 deg (±3 arcmin; ±10% of mean element separation. (d) The result of (c) filtered with the same isotropic Gaussian filter as in (b).*

experiments with textures in which the elements had no positional jitter showed that the size of the luminance cue for the location of the edges of the target region was dependent only on the difference in sag between target and background elements. Thus the strength of the luminance cue did not vary with background-sag value for a given sag difference.

8.3 Discussion

The results of these computer simulations suggested that the use of even small amounts of element positional jitter was enough to severely disrupt luminance cues to segementation of the target region. Moreover, in the unjittered textures, the constancy of the strength of the luminance cue with changes in background-sag value suggested that, if this cue had been used to segment the textures used in the experiments of Section 4, then sag increment-thresholds would have been constant with background-sag value, which manifestly was not so.

9   General Discussion

This study addressed three questions concerning the segmentation of textures of curved-line elements:

1. Can psychophysical performance for this class of textures be explained by existing models of texture segmentation?

2. Does the processing of such textures demand specialized curvature-sensitive mechanisms?

3. Is there evidence of a categorical representation of curvature in the analysis of these textures?

9.1   Implications for Models of Texture Segmentation

Many models of texture segmentation have incorporated stages sensitive to the orientation contents of the texture elements. As is emphasized below, this approach is not sufficient to account for the segmentation of curved-line-element textures. Any model must also be able to explain performance asymmetries, where segmentation performance is different for a given micropattern combination, depending on which of the micropatterns forms the target or background. The performance asymmetry found with textures of randomly oriented curved-line elements (Section 7), together with those considered by Gurnsey and Browse (1987, 1989), provide a challenge for comprehensive models of texture segmentation, such as the scheme described by Malik and Perona (1990). Although an attempt has been made to explain performance asymmetries in visual search in terms of local filtering operations (Rubenstein and Sagi 1990), it has been argued strongly by Gurnsey and Browse (1989) that a global normalization of local filter responses was required for a more complete explanation of asymmetry phenomena. Clearly any texture segmentation model that predicts only the strength of the border between two textures (Malik and Perona 1990; Landy and Bergen 1991) is inadequate.

9.2   Necessity of Curvature-Sensitive Mechanisms

The segmentation of textures of randomly oriented curved-line elements (Section 7) can be possible only if curvature-sensitive mechanisms are involved at some stage of the segmentation process. Furthermore, the segmentation of textures in which target and background curved-line elements were aligned but mirror-symmetric (Sections 5 and 6) suggests that these mechanisms must be sensitive to both the magnitude and the direction of curvature. The existence of performance peaks and troughs exhibited in the increment-threshold functions (Sections 3 and 4), and particularly the coincidence of the threshold minima for different sizes of chord (Section 3), provide further evidence that the segmentation of textures of curved-line elements is likely to be performed by specialized curvature-sensitive mechanisms.

9.3   Categorical Segmentation

Categorical segmentation provides a simple explanation of the existence of the peaks and troughs in performance in increment-threshold functions (Sections 3 and 4), and the coincidence of the optimum values of background sag (where sag increment thresholds were minimum) for different sizes of chord (Section 3). The fact that segmentation performance data for textures of vertically oriented curved-line elements (Section 6) can be predicted, albeit partially, by categorical labelling data suggests that the representation of curvature information may take a categorical form.

With regard to the precise nature of the proposed curvature representation, the labelling data (Section 6) suggest that, for a given curved-line orientation, at least five categorical labels are necessary to predict segmentation performance with briefly presented textures of curved-line elements. These labels would correspond to "straight", "just curved" (in two directions), and "more than just curved" (in two directions). This labelling scheme is an elaboration

of that proposed by Foster (1983) to explain performance in a four-alternative forced-choice curvature discrimination task. The boundary between "straight" and "just curved" categories should occur close to a sag value of 0.86 arcmin, and that between "just curved" and "more than just curved" categories should be in the region of 2.0 arcmin, although these values may vary with curved-line orientation and the range of curvatures present within the image. Such a representation scheme is not inconsistent with results from single-cell recording experiments (Hammond and Andrews 1978; Dobbins *et al.* 1987, 1989; Versavel *et al.* 1990).

Acknowledgements — This work was supported under the EC INSIGHT initiative and by the Medical Research Council of Great Britain. We would like to thank Ron Knapper for technical support. We are particularly grateful to Andrew Schofield for his helpful comments on earlier versions of this paper. Preliminary results included in this paper were presented at the 13th European Conference on Visual Perception in Paris, September, 1990 (Simmons and Foster 1990) and at the AVA annual meeting in Manchester, April, 1991 (Simmons *et al.* 1991).

References

Andrews, D.P., Butcher, A.K., Buckley, B.R. (1973) Acuities for spatial arrangement in line figures: human and ideal observers compared. Vision Research 13, 599-620

Attneave, F. (1954) Some informational aspects of visual perception. Psychological Review 61, 183-193

Beck, J. (1973) Similarity grouping of curves. Perceptual and Motor Skills 36, 1331-1341

Beck, J. (1982) Textural segmentation. In: Beck, J. (Ed.) Organization and Representation in Perception. Hillsdale, New Jersey: Lawrence Erlbaum Associates

Blakemore, C., Over, R. (1974) Curvature detectors in human vision? Perception 3, 3-7

Caelli, T., Julesz, B. (1978) On perceptual analyzers underlying visual texture discrimination: Part 1. Biological Cybernetics 28, 167-175

Dobbins, A., Zucker, S.W., Cynader, M.S. (1987) Endstopped neurons in the visual cortex as a substrate for calculating curvature. Nature, London 329, 438-441

Dobbins, A., Zucker, S.W., Cynader, M.S. (1989) Endstopping and curvature. Vision Research 29, 1371-1387

Ferraro, M., Foster, D.H. (1986) Discrete and continuous modes of curved-line discrimination controlled by effective stimulus duration. Spatial Vision 1, 219-230

Fogel, I., Sagi, D. (1989) Gabor filters as texture discriminator. Biological Cybernetics 61, 103-113

Foster, D.H. (1978) Visual comparison of random-dot patterns: evidence concerning a fixed visual association between features and feature-relations. Quarterly Journal of Experimental Psychology 30, 637-654

Foster, D.H. (1983) Visual discrimination, categorical identification, and categorical rating in brief displays of curved lines: Implications for discrete encoding processes. Journal of Experimental Psychology: Human Perception and Performance 9, 785-806

Foster, D.H. (1991) Operating on spatial relations. In: R.J. Watt (Ed.) Pattern Recognition by Man and Machine. Vision and Visual Dysfunction (vol 14), Basingstoke, U.K.: Macmillan, pp 50-68

Foster, D.H., Cook, M.J. (1989) Categorical and noncategorical discrimination of curved lines depends on stimulus duration, not performance level. Perception 18, 519

Foster, D.H., Bischof, W.F. (1991) Thresholds from psychometric functions: superiority of bootstrap to incremental and probit variance estimators. Psychological Bulletin 109, 152-159

Foster, D.H., Ward, P.A. (1991a) Asymmetries in oriented-line detection indicate two orthogonal filters in early vision. Proceedings of the Royal Society, London B/243, 75–81

Foster, D.H., Ward, P.A. (1991b) Horizontal-vertical filters in early vision predict anomalous line-orientation identification frequencies. Proceedings of the Royal Society, London B/243, 83-86

Foster, D.H., Simmons, D.R., Cook, M.J. (1992) The cue for contour-curvature discrimination. (submitted manuscript)

Gurnsey, R., Browse, R.A. (1987) Micropattern properties and presentation conditions influencing visual texture discrimination. Perception and Psychophysics, 41, 239-252

Gurnsey, R., Browse, R.A.(1989) Asymmetries in visual texture discrimination. Spatial Vision 4, 31-44

Hall, J.L. (1981) Hybrid adaptive procedure for estimation of psychometric functions. Journal of the Acoustical Society of America 69, 1763-1769

Hammond, P., Andrews, D.P. (1978) Collinearity tolerance of cells in areas 17 and 18 of the cat's visual cortex: relative sensitivity to straight lines and chevrons. Experimental Brain Research 31, 329-339

Hoffman, D.D., Richards, W.A. (1984) Parts of recognition. Cognition 18, 65-96

Julesz, B. (1981) Textons, the elements of texture perception, and their interactions. Nature 290, 91-97

Julesz, B. (1984) A brief outline of the texton theory of human vision. Trends in Neuroscience 7, 41-45

Kahn, J.I., Foster, D.H. (1986) Horizontal-vertical structure in the visual comparison of rigidly transformed patterns. Journal of Experimental Psychology: Human Perception and Performance 12, 422-433

Landy, M.S., Bergen, J.R. (1991) Texture segregation and orientation gradient. Vision Research 31, 679-691

Malik, J., Perona, P. (1990) Preattentive texture discrimination with early vision mechanisms. Journal of the Optical Society of America A/7, 923-932

Nothdurft, H.C. (1985a) Orientation sensitivity and texture segmentation in patterns with different line orientation. Vision Research 25, 551-560

Nothdurft, H.C. (1985b) Sensitivity for structure gradient in texture discrimination tasks. Vision Research 25, 1957-1968

Nothdurft, H.C. (1990) Texton segregation by associated differences in global and local luminance distribution. Proceedings of the Royal Society, London B/239, 295-320

Ogilvie, J., Daicar, E. (1967) The perception of curvature. Canadian Journal of Psychology 21, 521-525

Proesmans, M., Oosterlinck, A. (1992) The cracking plate and its parallel implementation. This volume, 225-240

Richards, W., Dawson, B., Whittington, D. (1986) Encoding contour shape by curvature extrema. Journal of the Optical Society of America A/3, 1483-1491

Rubenstein, B.S., Sagi, D. (1990) Spatial variability as a limiting factor in texture-discrimination tasks: Implications for performance asymmetries. Journal of the Optical Society of America A/7, 1632-1643

Simmons, D.R., Foster, D.H. (1990) Segmenting textures of curved line elements. Perception 19, 349

Simmons, D.R., Craven, B.J., Foster, D.H. (1991) Categorical models of image segmentation. Opththalmic and Physiological Optics 11, 282

Taylor, M.M., Creelman, C.D. (1967) PEST: Efficient estimates on probability functions. Journal of the Acoustical Society of America 41, 782-787

Treisman, A. (1988) Features and objects: the fourteenth Bartlett Memorial Lecture. Quarterly Journal of Experimental Psychology 40/A, 201-237

Treisman, A., Gormican, S. (1988) Feature analysis in early vision: evidence from search asymmetries. Psychological Review 95, 15-48

Treisman, A., Souther, J. (1985) Search asymmetry: a diagnostic for preattentive processing of separable features. Journal of Experimental Psychology: General 114, 285-310

Versavel, M., Orban, G.A., Lagae, L. (1990) Responses of visual cortical neurons to curved stimuli and chevrons. Vision Research 30, 235-248

Watt, R.J. (1984) Further evidence concerning the analysis of curvature in human foveal vision. Vision Research 24, 251-253

Watt, R.J. (1987) Scanning from coarse to fine spatial scales in the human visual system after the onset of a stimulus. Journal of the Optical Society of America A/4, 2006-2021

Watt, R.J. (1992) The analysis of natural texture patterns. This volume, 298-323

Watt R.J., Andrews, D.P. (1982) Contour curvature analysis: hyperacuities in the discrimination of detailed shape. Vision Research 22, 449-460

Wilson, H.R. (1985) Discrimination of contour curvature: data and theory. Journal of the Optical Society of America A/2, 1191-1198

Wilson, H.R., Richards, W.A. (1989) Mechanisms of contour curvature discrimination. Journal of the Optical Society of America A/6, 106-115

Wood, C.C. (1976) Discriminability, response bias, and phoneme categories in discrimination of voice onset time. Journal of the Acoustical Society of America 60, 1381-1389

Extraction of Shape Features and Experiments on Cue Integration

Jan-Olof Eklundh, Jonas Gårding, Tony Lindeberg, and Fredrik Bergholm

Computational Vision and Active Perception Laboratory, Department of
Numerical Analysis and Computing Science, Royal Institute of Technology

1   Introduction

Although direct information about depth and three-dimensional structure is available
from binocular and dynamical visual cues, static monocular images also provide important
constraints on the structure of the scene.

In our work within the InSight project we have studied mechanisms for estimating
three-dimensional shape from texture and contours. The underlying geometric nature of
the problem has been analyzed. On the basis of this theory two algorithms for estimating
surface orientation have been derived. This work is presented in Section 2.

Static monocular cues will typically not fully determine the shape of the viewed scene.
Hence, they are in human vision *integrated with other available cues,* like binocular disparities. One aim of our work has been to provide a computational basis for the study
of such interactions. The actual work on the problem, which is of an interdisciplinary
nature, has been carried out by J. Frisby and his collaborators in the AIVRU group in
Sheffield. We refer the reader to that chapter for an account.

Most work in computational vision has been concerned with studying what can be
regarded as foveal vision. However, it is well-known that *peripheral cues* in human vision
provide information that is essential for our understanding of the visual world. Moreover,
the integration of central and peripheral cues seems to be of paramount importance.

We have considered how information about *egomotion* can be computationally derived
in the periphery. The aim is to integrate this information with centrally derived motion
cues. That work is described in Section 3.

The notion of *scale* pervades all computational approaches to deriving properties of
imagery. This is true for low-level processing by various operators as well as for shape
and motion computations.

We have studied how *salient image structures* in grey-level images can be derived, as
well as the scales at which these structures can be observed. The approach is based on the
so-called scale-space embedding, that considers information at all scales simultaneously.
Salience is here defined as structure that stands out from its surroundings.

It is shown that the derived representation can be applied to *guide* various low-level operations. The approach and possible applications are described in Section 4.

2 Shape from Texture and Contour

As noted in the introduction, static monocular cues such as texture and contours can provide important information on the three-dimensional structure of the scene. The mechanisms by which this information is conveyed are only partially understood, and our work in the first phase of the InSight project has been aimed both at improving the theoretical understanding of the problem and at developing useful computational models and algorithms.

An additional motivation for this study is to provide a basis for understanding how monocular cues can be integrated with binocular and dynamical visual cues. For example, Porrill *et al.* (1991) have shown evidence that monocular texture cues may be used to calibrate information from binocular disparities.

In this section we will give an overview of three parts of the completed work; a more detailed account can be found elsewhere (Gårding 1990, 1991, 1992a, 1992b). First, we describe the theoretical development, with a strong emphasis on geometry. Then we present the WISP algorithm, which uses the directional statistics of projected surface contours to estimate surface orientation. Finally, we describe the SEALS algorithm. This algorithm is inspired by current theories of independent spatial frequency channels in the mammalian visual cortex, and uses local orientation-selective Gabor filtering to estimate both surface orientation and relative distance locally.

2.1 A Computational Analysis

Most observers would agree that the simple line drawing shown in figure 1 gives a fairly convincing impression of a receding plane covered with circles. Nevertheless, it is far from trivial to determine precisely from which image qualities this interpretation is derived. It could e.g. be the elliptic *shape* of the projected circles, or the *rate of change* of their shape, size or density. It could even be the *vanishing points* determined by the implicit rulings of the grid on which the circles are placed. Common to all these potential cues, however, is the fact that they represent some form of *projective distortion* of the surface pattern.

The use of projective distortion of texture as a cue to three-dimensional properties of surfaces was first considered by Gibson (1950). He defined informally the *gradient of texture density* as a serial change in the retinal image of the cycle length of a repetitive texture, and wrote

> [...] it may be that *the gradient of density is an adequate stimulus for the impression of continuous distance.* (Gibson 1950, p. 67)

Gibson's observations were of a qualitative nature, but during the four decades which have passed since the appearance of Gibson's seminal work, many interesting and useful computational methods for the quantitative recovery of surface orientation from projective distortion have been proposed; see e.g. Gårding (1991) for a review. Furthermore, psychophysical studies (e.g. Frisby and Buckley [this volume]; Cutting and Millard 1984; Todd and Akerstrom 1987) have verified that texture distortion does indeed play an important role in human perception of three-dimensional surfaces.

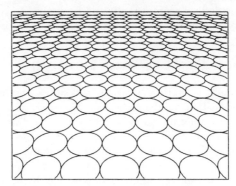

Figure 1: *This image of a slanting plane covered with circles clearly illustrates two important types of projective distortion that can be used to estimate surface shape and orientation. Firstly, the local foreshortening effect causes each circle to appear as an ellipse in the image. Secondly, the varying angle between the surface and the line of sight gives rise to several* texture *gradients; for example, the area of the ellipses decreases towards the top of the image, as does their aspect ratio.*

In the work presented here, we address two fundamental aspects of the computational problem which have often been neglected in the past. Firstly, most of the previously proposed techniques are based on the assumption that the surface is planar. As pointed out by many authors, this assumption can in practice only be justified locally, since real physical surfaces often have significant curvature on a larger scale. However, limiting the size of the analyzed region is not always sufficient. We have shown that even for infinitesimally small surface patches, there is only a very restricted class of texture distortion measures which are independent of surface curvature. Gibson's gradient of texture density, for example, does *not* belong to this class. (Note, however, that it may be perfectly valid to apply piecewise linear approximations of functions on curved surfaces in other cases where there is no explicit dependency on surface curvature; see e.g. Verri *et al.* [this volume].)

Secondly, the possibility of using projective texture distortion as a *direct* cue to surface properties has not been fully exploited in previous work. For example, a local estimate of a suitably chosen texture gradient can be directly used to estimate the surface orientation. This view of texture distortion as a direct cue predominates in the psychophysical literature, but in contrast, most previous work in computational vision has proposed indirect methods where a more or less complete representation of the image pattern is used in a search procedure. A typical example is backprojection, i.e., estimation of surface shape by searching for the surface which optimizes some regularity measure applied to the backprojected image pattern.

Obviously, the question of direct vs. indirect perception pertains to most aspects of vision, and not only to shape from texture. A controversial argument in favor of direct perception was put forth in Gibson's later work (Gibson 1979). Gibson described visual perception in terms of "direct pickup", and postulated that information about the environment is *contained* in the array of ambient light, and not *inferred* from it, so that "computational" mechanisms are in fact superfluous. Although this view seems rather extreme, it is clear that there is a puzzling discrepancy between on one hand the seemingly direct and effortless nature of human perception, and on the other hand the limited success of the sometimes very complex techniques employed in machine vision.

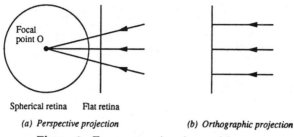

Spherical retina   Flat retina

(a) Perspective projection   *(b) Orthographic projection*

Figure 2: *Two common imaging models.*

The cases where some simple form of "direct pickup" is feasible therefore seem well worth exploring.

In principle, direct perception of shape from texture consists of two relatively independent steps. In the first step, some representation of the local texture distortion is estimated in the retinal array; in the second step this distortion is interpreted in terms of surface orientation and surface curvature. A trivial but crucial observation is that only the first step depends on the nature of the texture, so it is possible to derive general results concerning the interpretation task without making assumptions about the type of texture or the procedures used to analyze it.

The fact that these problems can be treated separately greatly simplifies the analysis. In this subsection we will describe some of the main results of our theoretical analysis of the interpretation problem for the general case of a smooth curved surface viewed in perspective projection. A more detailed account of this work can be found in Gårding (1992a and 1992b). In Sections 2.2 and 2.3 we then describe two complete shape-from-texture algorithms, each using a different way of estimating the projective texture distortion.

Imaging Models

It may be useful to briefly review some of the most commonly used imaging models. In general, a model for image formation determines both *where* the image of some point will appear, and *how bright* it will be (Horn 1986). Here we will only discuss the "where" question, which only depends on the geometry of the projection from surface to retina.

The most common model is *perspective* (or *central*) projection, also known as the "pinhole camera model". In this model, the image of a point P in space is determined as the intersection with the retina of the ray from P to the *focal point O* (figure 2a).

Depending on the shape of the retina we can obtain e.g. *flat* or *spherical* perspective. It is important, however, to realize that the shape of the retina does not in any way influence the information content of the image; given e.g. a spherical perspective image, the corresponding flat perspective image can be obtained by backprojection of the image from the focal point. Nevertheless, the spherical perspective model has certain advantages, in particular that the retina is everywhere perpendicular to the line sight, which eliminates the "position effect" occurring in flat perspective images.

Note that the retina in the perspective model shown in figure 2a is placed in front of the focal point, unlike a real eye or camera where it would of course be located *behind* the lens center. Mathematically, however, it makes no difference which alternative we use.

There are some aspects of image formation which are *not* captured by central perspective. For example, even an ideal lens differs from this projection model in at least

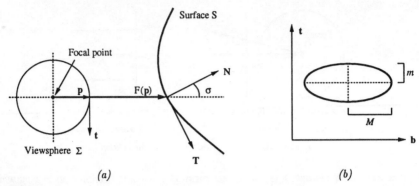

Figure 3: (a) *Local surface geometry and imaging model. The tangent planes to the viewsphere* Σ *at* **p** *and to the surface* S *at* $F(\mathbf{p})$ *are seen edge-on but are indicated by the tangent vectors* **t** *and* **T**. *The image direction* **t** *coincides with the gradient of distance to the surface, and is known as the tilt direction.* (b) *The derivative map* F_* *can be visualized by an image ellipse which corresponds to a unit circle in the surface. The vector* **b** *is the image direction perpendicular to the tilt* **t**.

two respects: firstly, it gathers a finite amount of light, and secondly, only objects within a certain range of distances are imaged sharply.

In order to reduce the complexity of certain mathematical expressions, perspective projection is sometimes approximated by simpler models. In particular, if the field of view is small and the depth range of objects in the scene is negligible, it may be justifiable to assume *orthographic* (i.e., parallel) projection. This is illustrated in figure 2b. Orthographic projection has the convenient property that it leads to a *linear* relationship between the coordinates of a point in a scene and the coordinates of its position in the image. However, it also has some unwanted consequences, e.g. that the size of the image of an object is independent of the distance to the object, contrary to all our visual experience.

Attempts have sometimes been made to define intermediate projection models which account at least qualitatively for perspective effects such as distance scaling, while retaining the mathematical simplicity of orthographic projection. For example, Ohta *et al.* (1981), Aloimonos (1988), and Longuet-Higgins [this volume] have considered various "paraperspective" projections, essentially *local* linear approximations to true perspective. A related type of local linear approximation will be introduced in the next section.

Local Geometry of Perspective Projection

Figure 3a illustrates the basic viewing and surface geometry which will be used in the following. A smooth surface S is mapped by central projection onto a unit viewsphere Σ centered at the focal point.

Consider a small patch in the image. Assuming that this patch is the image of a corresponding patch of a smooth surface in the scene, we have a local differentiable mapping F from the image to the surface. The linear part F_* of F is a 2×2 matrix called the *derivative map*, which can be seen both as a local linear approximation to F and as an exact mapping from the tangent plane of the image (or retina) to the tangent plane of the surface. For a frontoparallel surface, F_* is simply a scaling by the distance, but for a slanted and tilted surface it will contain a shear as well. F_* can be visualized as an ellipse in the image which corresponds to a unit circle in the surface (figure 3b). Elementary matrix calculus shows that the minor and major axes (m, M) of the ellipse are aligned with the eigenvectors of F_*, and that the halflength of each axis is equal to the inverse of the corresponding eigenvalue. It must be emphasized that this ellipse is simply a convenient way of representing the local geometry of perspective projection; it is *not* linked to any assumption about the structure of the actual surface texture.

First-Order Distortion: Foreshortening

Texture gradients have to do with the *rate of change* of the local linear distortion F_*, but it is important to realize that texture gradients are not necessary for slant perception; there is often sufficient information in F_* alone. For example, even a single image ellipse can be interpreted as a circle to produce an estimate of surface orientation. In this context the compression of the circle to an ellipse is called the *foreshortening* effect.

The local surface orientation can be conveniently represented in terms of *slant* σ and *tilt* t (figure 3a). The slant σ is the angle between the surface normal and the line of sight, and it is simply the inverse cosine of the aspect ratio m/M of the ellipse. The tilt direction t is the projection of the surface normal onto the image, and it can only be determined up to sign from the image ellipse, because a circle at any of the two corresponding surface orientations would project to the same ellipse in the image.

In summary, a local estimate of the linear part F_* of the perspective mapping determines the foreshortening of the texture, which in turn determines slant uniquely, and tilt up to sign.

Second-Order Distortion: Texture Gradients

Gibson (1950) was the first to point out the simple and intuitive fact that the rate of change of the appearance of a projected pattern (e.g. from bottom to top of the image in figure 1) can be used to infer properties of the surface shape.

However, a formal analysis of texture gradients turns out to be rather involved. Firstly, the rate-of-change computation must be related to some well-defined property of the projected texture, and there are several reasonable alternatives, e.g. the area of the ellipses, the spacing between the ellipses, the shape of the ellipses, et cetera. Secondly, once a particular definition has been adopted, a non-trivial geometric analysis is required to determine precisely what a given texture gradient says about the shape of the surface.

It turns out that most of the texture gradients that have been considered in the literature can be expressed in terms of some function of the inverse eigenvalues M and m of F_*, i.e., the lengths of the major and minor axes of the ellipse. For example, we have the *major* gradient ∇M, the *minor* gradient ∇m, the *foreshortening* gradient $\nabla \epsilon = \nabla(m/M)$, the *area* gradient $\nabla A = \nabla(mM)$, and the *density* gradient $\nabla \rho = \nabla(1/(mM))$. The major and minor gradients are often referred to as the *perspective* and *compression* gradients.

We will refer collectively to such gradients as *simple distortion gradients*. They are not independent, because by the chain rule the gradient of any function $f(M, m)$ is simply a linear combination of the basis gradients ∇M and ∇m. In practice it makes more sense to consider the *normalized* gradients $(\nabla M)/M$ and $(\nabla m)/m$, since these expressions are free of scale factors depending on the distance to the surface and the absolute size of the surface markings.

Texture gradients represent local properties of a perspective image of a textured surface, and can be measured either by first estimating the linear projective distortion F_* and then estimating the rate of change of its eigenvalues, or by some more direct method.

But what do they tell about surface shape? Previously, Stevens (1981) has answered this question for the case of a *planar* surface. We have extended the analysis to the general case of a curved surface. More precisely, in a coordinate system where the first coordinate coincides with the tilt direction **t**, we have shown that the normalized basis texture gradients are given by the following expressions:

$$\frac{\nabla m}{m} = -\tan\sigma \left(\frac{2 + r\kappa_t/\cos\sigma}{r\tau} \right) \tag{1}$$

$$\frac{\nabla M}{M} = -\tan\sigma \left(\begin{matrix} 1 \\ 0 \end{matrix} \right) \tag{2}$$

r is the distance from the viewer, σ is the slant of the surface, κ_t is the normal curvature of the surface in the tilt direction, and τ is the geodesic torsion, or "twist", of the surface in the tilt direction. To briefly review the meaning of the curvature parameters κ_t and τ, consider the curve Γ formed by the intersection of the surface with the plane NT containing the surface normal **N** and the tangent vector **T**. κ_t is then simply the curvature of Γ, and τ is a measure of the initial rate along Γ by which the surface normal "twists" out of the NT plane.

From (1) and (2) it is straightforward to derive explicit expressions for gradients of any function of m and M. For example, we obtain the normalized *foreshortening gradient*

$$\frac{\nabla\epsilon}{\epsilon} = \frac{\nabla m}{m} - \frac{\nabla M}{M} = -\tan\sigma \left(\frac{1 + r\kappa_t/\cos\sigma}{r\tau} \right) \tag{3}$$

and the normalized *area gradient*

$$\frac{\nabla A}{A} = \frac{\nabla m}{m} + \frac{\nabla M}{M} = -\tan\sigma \left(\frac{3 + r\kappa_t/\cos\sigma}{r\tau} \right) \tag{4}$$

Equations (1)–(4) reveal several interesting facts about texture gradients. Firstly, the minor gradient ∇m depends on the curvature parameters κ_t and τ, whereas the major gradient ∇M is *independent of surface curvature*. This is important because it means that $(\nabla M)/M$ can be used to estimate the local surface orientation, and hence to corroborate the estimate obtained from foreshortening. Furthermore, unlike foreshortening, $(\nabla M)/M$ yields an estimate which has no tilt ambiguity.

Secondly, the direction of any texture gradient which depends on m, such as the foreshortening gradient or the density gradient, is aligned with the tilt direction if and only if the twist τ vanishes, i.e., *if the tilt direction happens to be a principal direction in the surface*. This is of course always true for a planar surface, but for a general curved surface the only distortion gradient guaranteed to be aligned with tilt is $(\nabla M)/M$.

Thirdly, *the complete local second-order shape (i.e. curvature) of S cannot be estimated by distortion gradients*. The reason is that it takes three parameters to specify the surface curvature, e.g. the normal curvatures κ_t, κ_b in two perpendicular directions and the twist τ. The Gaussian curvature, for example, is given by $K = \kappa_t\kappa_b - \tau^2$. However, the basis gradients (1) and (2) are independent of the normal curvature κ_b.

Simple Distortion Gradients: Discussion

The usefulness of any texture distortion measure is determined both by the information it contains about the surface and the conditions that are necessary for its measurement. These two factors are relatively independent.

It is obvious from (1)–(4) that the minor, foreshortening, area, and density gradients all contain similar information about the surface. They differ only in the relative weight of the curvature invariant part. For example, for a *curved surface with known orientation*, any one of these gradients determines the scaled normal curvature $r\kappa_t$ and the scaled twist $r\tau$ (where r is the distance to the surface at that point). For a *planar surface*, it determines slant and tilt, i.e., the surface orientation, uniquely.

The major gradient, on the other hand, is qualitatively different since it is independent of surface curvature. It always determines the surface orientation uniquely.

However, different categories result if the gradients are instead grouped according to the conditions necessary for their measurement. In order to measure the minor, major, or foreshortening gradient, *the local tilt direction must be known*. This is a consequence of the fact that these gradients depend on m and M which are measured in the tilt direction and the perpendicular direction respectively. Hence, to conclude that these gradients determine tilt may appear to be a completely circular argument, but the situation is slightly more involved than that: a "major gradient" computed using an erroneous tilt direction $\hat{t}$ will in general not be aligned with $\hat{t}$, even at a planar point. Hence, there is some possibility of detecting inconsistent assumptions.

On the other hand, the area and density gradients do not have this problem at all. They only depend on the product mM, and it is easily shown that this product is equal to the product of two perpendicular projected lengths measured in *any* direction. As a consequence, the area and density gradients are also "immune" to local anisotropy. Consider e.g. a texture consisting of elongated elliptical texture elements oriented randomly in the surface, as illustrated in figure 4. If these texture elements are erroneously interpreted as isotropic (circular) elements, the estimate of surface orientation from foreshortening will be wrong. The estimates from the major and minor gradients will also be wrong, although typically better than those from foreshortening. The estimate from the area gradient, in contrast, will not be affected *at all* by the incorrect isotropy assumption.

General Texture Gradients

The fact that the simple distortion gradients considered above cannot provide complete information about surface curvature is just a consequence of the restricted way in which these gradients were defined, and must not be interpreted as a property of texture distortion in general. It *is* in fact possible to compute e.g. Gaussian curvature from texture distortion cues, but then the concept of distortion gradients must be generalized.

The simple distortion gradients are defined as the rate of change of some function of the characteristic lengths m and M, everywhere measured relative to the tilt direction which may vary in the image. There are two problems with this definition. Firstly, in order to measure these gradients the tilt must in principle be known, and secondly, the information contained in the rate of change of the tilt direction is discarded.

An alternative and less problematic approach is to measure the rate of change in some direction in the image of projected length in some *fixed* direction w. It is a non-trivial fact that when w coincides with the tilt or the perpendicular direction, this measure is equivalent to the corresponding simple distortion gradient. However, a gradient can be computed for projected lengths in *any* direction, not just t and b. This way we can obtain information about the surface shape which cannot be provided by any distortion gradient.

A particularly useful example is the derivative in some direction of the projected length measured in *the same* direction. This derivative could e.g. be estimated in a given direction by measuring the rate of change of the distances between the intersections of a reference

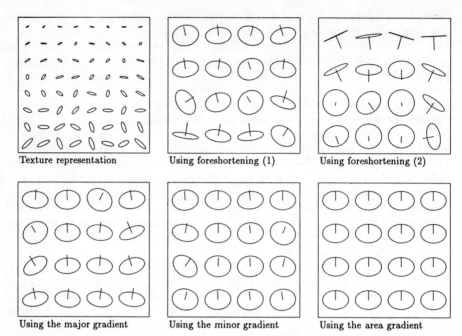

Figure 4: *If an anisotropic texture is interpreted under the incorrect assumption that it is isotropic, the resulting error in the estimated surface orientation depends on the texture distortion measure used. The error is large for foreshortening, smaller but significant for the major and minor gradient, and zero for the area gradient.*

line with projected surface contours. We have shown that the normalized directional derivative computed this way in the direction $\mathbf{w} = \alpha\mathbf{t} + \beta\mathbf{b}$ (where $\mathbf{t}$ is the tilt direction and $\mathbf{b}$ is its perpendicular) is given by

$$- \alpha \tan \sigma \left(2 + \frac{r(\alpha^2 \kappa_t + 2\alpha\beta\tau \cos\sigma + \beta^2 \kappa_b \cos^2\sigma)}{\cos\sigma \, (\alpha^2 + \beta^2 \cos^2\sigma)} \right) \tag{5}$$

which at a planar point simplifies to $-2\alpha \tan \sigma$.

Note that $\alpha = 0$ in the direction perpendicular to the tilt, so that this derivative vanishes. This observation is in keeping with Stevens' (1981) suggestion that the tilt of a *planar* surface can be computed as the direction perpendicular to the direction of least variability in the image, and we now see that this suggestion is in principle valid for curved surfaces as well. However, this direction is not necessarily unique; for a non-convex surface with sufficiently large negative curvatures the derivative may vanish in other directions as well.

It is also worth noting that the normalized directional derivative (5) can be measured even for textures that only exhibit variation in a single direction (such as wood grain), whereas there is no obvious way to measure neither first-order distortion (foreshortening) nor distortion gradients for such textures. For a planar surface, it suffices to measure this derivative in two perpendicular directions in order to determine the surface normal uniquely.

Direct Estimation of Shape from Texture: Conclusions

Several psychophysical studies have already been carried out in order to determine the influence of various texture distortion cues. The typical stimuli are synthetic textures consisting of independent texture elements of uniform intensity on a background of a different uniform intensity. Assuming that the visual system estimates texture distortion by using the shape of the texture elements in each local neighborhood to compute the minor or major characteristic lengths or area etc., the simple distortion gradients can be manipulated by changing the pattern of texture elements. However, when manipulating one particular aspect of the element shapes, it is quite hard to prove that one is not inadvertently at the same time also changing some other texture distortion feature which the visual system might use.

To guide further psychophysical research, it would be useful to have an explicit hypothesis about the computational mechanisms behind human perception of shape from texture. Although the formulation of such a hypothesis is obviously a very difficult task, we believe that the theoretical analysis we have outlined here is a valuable tool which can provide specific and precisely formulated predictions.

For example, assuming that surface orientation is estimated from area or density gradients, we obtain the predictions that for a general view of a curved surface, perceived slant will be systematically wrong due to the factor κ_t in the first component of (4), and perceived tilt will be systematically wrong due to the factor τ in the second component of (4). Furthermore, this assumption leads to the prediction that local texture anisotropy will have no influence at all on perceived surface orientation (see figure 4). The degree to which these predictions are verified or falsified by psychophysical experiments can be taken as an indication of the relative importance of area and density gradients.

Stevens (1981) has proposed a simple model where foreshortening m/M is used to compute local surface orientation, and the major axis M is used to compute depth up to a scale factor. This scheme is certainly reasonable on theoretical grounds, since these quantities are independent of surface curvature and "almost" independent of each other. However, this model will produce internal inconsistencies in the case of anisotropic or spatially inhomogeneous texture, and unless a mechanism for resolving such conflicts is available the model is incomplete. Furthermore, in viewing conditions where the depth range of the surface is small compared to the viewing conditions, the model predicts that foreshortening alone will account for perceived surface orientation. Hence, an elongated shape would always be perceived as a slanted isotropic shape, and this is clearly an oversimplification.

Todd and Akerstrom (1987) described a series of experiments using ellipsoid surfaces of varying eccentricity. In one of their experiments (#3), they compared two surfaces differing only in the constant elongation of the projected texture elements. Hence all simple distortion gradients were equal, but nevertheless there was a dramatic difference in the perceived surface shape: when the texture elements were isotropic the surface was perceived as flat. This effectively proves that simple distortion gradients alone cannot account for human perception of shape from texture.

In summary, it appears that we have to look for more complex models, taking into account more than one texture distortion cue. Furthermore, it is quite possible that other measures than foreshortening and simple distortion gradients must be considered, since neither of the latter can provide complete information about surface curvature. Given a model containing several components it is natural to examine the relative weights. Such studies have already been done within the InSight project by Frisby and Buckley [this volume], but the problem has many degrees of freedom and much work remains to be done.

2.2   WISP: Shape from Texture and Contour by Weak Isotropy

We have proposed an approach, called WISP, which uses the foreshortening effect to estimate surface orientation (modulo tilt reversal as explained earlier). This approach is not based on the local linear approximation F_* of the perspective mapping; it uses the full perspective projection and can thus be applied globally as well as locally in an image. WISP operates on contours derived from the image, and utilizes the assumption that the corresponding surface contours lack directional bias (in a precise sense to be defined in a moment) to recover the surface orientation.

WISP is related to previous theories on shape from foreshortening by Witkin (1981), Brady and Yuille (1984), Kanatani (1984), and Blake and Marinos (1990). It extends these theories in several ways, most notably to include perspective projection and to provide an efficient estimate based directly on simple image statistics. A more detailed description of WISP is available elsewhere (Gårding 1990, 1991).

Consider the distribution of arc length with respect to tangent direction α for a planar piecewise smooth curve. The sign of the tangent direction is immaterial, and α therefore belongs to the interval $0, \pi)$. The fact that α and $\alpha + \pi$ represent the same direction means that conventional statistical concepts can lead to paradoxes; for example, the average of $1°$ and $179°$ should be $0°$ and not $90°$. A standard technique (Mardia 1972) which eliminates these problems is to map α onto the point $(\cos 2\alpha, \sin 2\alpha)$ on the unit circle.[1]

In this representation, the distribution of arc length with respect to tangent direction is a weight distribution $F(\alpha)$ on the unit circle. The most important characteristic of such a distribution is its center of mass (C, S), defined by

$$C = \int_0^\pi \cos 2\alpha \, dF(\alpha) \qquad \text{and} \qquad S = \int_0^\pi \sin 2\alpha \, dF(\alpha) \tag{6}$$

where the distribution F may be continuous, atomic, or a mixture of both. It turns out to be more practical to express the center of mass in polar coordinates $(Q, 2\psi)$:

$$Q = \sqrt{C^2 + S^2} \qquad \text{and} \qquad \psi = \frac{1}{2} \arctan(S, C) \tag{7}$$

Hence, Q is the distance from the center of the circle to the center of mass and 2ψ is the corresponding direction.

The interpretation of Q and ψ is straightforward: ψ is the *mean direction*, and Q is a measure of anisotropy, i.e. how much the distribution is biased towards any particular direction. The quantity $1 - Q$ is often referred to as the *circular variance*. In the case when $Q = 0$, the mean direction ψ is undefined. The interpretation is that there is no single dominant direction. We call such distributions *weakly isotropic*. It is worth noting that this concept is independent of any particular interpretation of the distribution—it can e.g. be applied to both probability distributions and finite samples from such distributions.

Weakly isotropic contours seem to be "simple" in the sense that there is no directional bias. For example, we have shown that all contours which have rotational symmetry of order three or higher, such as all regular polygons, are weakly isotropic. This leads to the following principle (the weak isotropy heuristic): *If the image pattern (contour or*

---

[1]The doubling of the angle is necessary to match the period π of α to the period 2π of the unit circle.

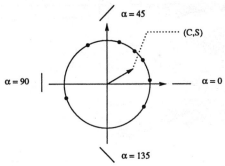

Figure 5: *An atomic weight distribution on the circle and its center of mass.*

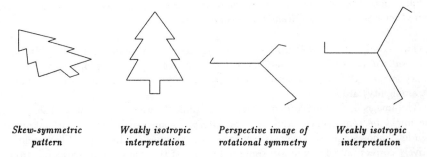

| Skew-symmetric pattern | Weakly isotropic interpretation | Perspective image of rotational symmetry | Weakly isotropic interpretation |

Figure 6: *The weak isotropy algorithm interprets skewed and rotational symmetries as oriented real symmetries.*

texture) can be the projection of a weakly isotropic pattern, then we assume that this is actually the case. The interesting fact is now that it is possible to devise an algorithm that in a simple and robust manner recovers the orientation of the plane that makes the backprojected pattern weakly isotropic. If this orientation is given in terms of slant and tilt, (σ, τ), one is looking for the solution to the non-linear equations

$$
\begin{aligned}
C(\sigma, \tau) &= 0 \\
S(\sigma, \tau) &= 0
\end{aligned}
\tag{8}
$$

where (C, S) is the center of mass of the directional distribution of the backprojected contours. The solution can be found using e.g. a Newton-Raphson technique, but more interestingly we have shown that an approximate solution can be found directly from the directional statistics (Q', ψ') of the *image* contours, ignoring completely all other aspects of their shape. This approximation is given by

$$
\begin{aligned}
\tau &= \psi' \pm \pi/2 \ (\mathrm{mod}\ 2\pi) \\
\sigma &= \arctan(\tan(A^{-1}(Q'))\cos(\nu/2))
\end{aligned}
$$

(9)

(10)

where A is a precomputed function defined in terms of elliptic integrals, and ν is the visual angle subtended by the image pattern. Note the decoupling of the variables. In certain simple cases this initial approximation is exact, and in the general case it is typically within a few degrees from the exact solution to (8).

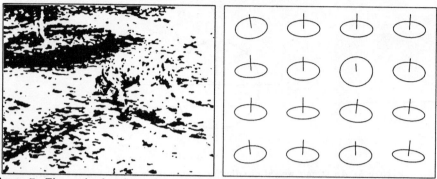

Figure 7: *The result of applying the weak isotropy algorithm locally in an image. Note that the interpretation of the ground plane seems more or less correct, and that the window containing most of the dog yields a significantly different estimate.*

Two examples of how WISP interprets symmetries are shown in figure 6. In figure 7 the results obtained by applying WISP locally in rectangular windows of a well-known binary image are shown. In both figures only one of WISP's two solutions (related by the tilt ambiguity) are shown.

The image in figure 7 presents significant difficulties even to human observers, and one might be tempted to think that no interpretation can be made without invoking high-level knowledge. However, the surface orientation estimates from WISP seem qualitatively correct, and the window containing most of the dog clearly stands out from the background.

In psychophysical experiments conducted at AIVRU, Frisby and Buckley [this volume] have found about equal weighting for texture and stereo cues under natural viewing conditions. Hence, it is natural to ask how well WISP can match the performance of a good stereo algorithm. To this end, a series of experiments were carried out using wallpapers mounted on a movable platform. A stereo image pair was captured and the true surface orientation estimated from the output of the PMF stereo algorithm (Pollard *et al.* 1985). This was repeated for four different surface orientations. WISP was then applied to the left image of each pair, producing two mutually exclusive estimates of surface orientation. The most correct estimate of the two was selected manually. For the highest slant value the edge of the wallpaper was visible in the image, but that region of the contour map was masked out before applying WISP.

One of the results is shown in figure 8. The estimate from WISP closely follows the estimate from stereo, and as expected the error decreases with increasing slant. However, in interpreting these results it is important to keep in mind that the success of WISP ultimately depends on how well the surface texture matches the weak isotropy-assumption, whereas the estimate from stereo is independent of such assumptions.

2.3   SEALS: Shape Estimation by Analysis of Local Spectrum

As an alternative to WISP, we have developed an algorithm which avoids all contour or feature detection and operates directly on the local image intensities. This algorithm, called SEALS, can be seen as direct implementation of the computational principles outlined in Section 2.1. It is also inspired by current theories of the early stages of the visual pathway in biological vision, and bases its computations on the outputs of a set of

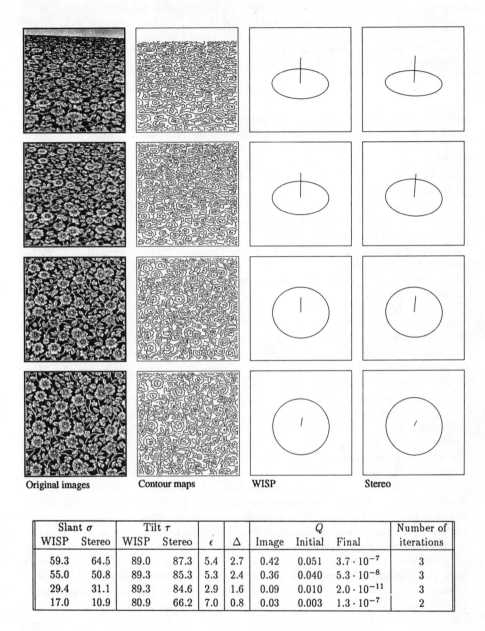

Original images   Contour maps   WISP   Stereo

| Slant σ | | Tilt τ | | | | Q | | | Number of |
WISP	Stereo	WISP	Stereo	ϵ	Δ	Image	Initial	Final	iterations
59.3	64.5	89.0	87.3	5.4	2.7	0.42	0.051	$3.7 \cdot 10^{-7}$	3
55.0	50.8	89.3	85.3	5.3	2.4	0.36	0.040	$5.3 \cdot 10^{-8}$	3
29.4	31.1	89.3	84.6	2.9	1.6	0.09	0.010	$2.0 \cdot 10^{-11}$	3
17.0	10.9	80.9	66.2	7.0	0.8	0.03	0.003	$1.3 \cdot 10^{-7}$	2

Figure 8: *Results of WISP with a wallpaper texture. ϵ is the angle between the estimated surface normal and the true surface normal as measured by stereo correspondence. Δ is the angle between the initial and final estimates of the surface normal. σ, τ, ϵ and Δ are given in degrees.*

orientation-selective filters of a shape similar to that of simple cell receptive fields in the mammalian visual cortex (Jones and Palmer 1987).

The approach is based on a proposition (proven in Gårding 1991) which states that the second moment matrix μ_Σ of the spectrogram of the image intensities is related to the spectrogram μ_S of the reflectance of the surface pattern by the simple relation

$$\mu_\Sigma = F_*^T \mu_S F_* \qquad (11)$$

under the simplifying assumption that the image intensity at a point is directly proportional to the reflectance at the corresponding point in the surface.

If μ_S is known and μ_Σ can be measured, the eigenvectors and inverse eigenvalues (m, M) of F_* can be recovered by factoring (11). Slant and tilt up to sign can then be computed from the ratio m/M and the direction of the eigenvectors, and various texture gradients can be computed from the spatial rate of change of m and M. In practice it is unrealistic to assume complete knowledge of μ_S, but the weaker assumption that the surface reflectance pattern is (weakly) isotropic, i.e., that μ_S is a factor ξ times the identity matrix, may often be justifiable.[2]

Then the eigenvectors of μ_Σ and F_* will be the same, and the eigenvalues of F_* will be equal to the square root of the eigenvalues of μ_Σ divided by ξ. The scale factor ξ is normally unknown, but this is of no consequence since both foreshortening and normalized texture gradients are independent of scale factors in F_*, as shown in Section 2.1.

In our implementation the image spectrogram is sampled by convolving the image with a set of complex 2-D Gabor filters, tuned to a range of spatial frequencies and orientations. The second moment matrix μ_Σ is estimated from the samples by a simple rectangle-rule approximation of the defining integrals, and the scaled characteristic values (m, M) are then computed as explained above. Finally, the local values of m and M are pooled in Gaussian windows to provide local estimates of foreshortening m/M as well as the normalized minor gradient (1), major gradient (2), and area gradient (4).

Figure 9 illustrates the results obtained with a synthetic image of a curved (cylindrical) surface covered by a random isotropic reflectance pattern. The surface is viewed in perspective with a visual angle of approximately 59° across the diagonal of the image. The surface orientation is visualized by a dish with an attached needle parallel to the surface normal, viewed in parallel projection.[3]

The top row shows the original image and the estimated local distortion F_* on a superimposed 8 × 8 grid. The middle row shows the true surface orientation, and the estimates from foreshortening. The bottom row shows estimates from the major, minor, and area gradients. The latter estimates were computed from (2), (1), and (4) under the assumption that the curvature is negligible, i.e., $\kappa_t = \tau = 0$.

There are several things worth pointing out in these results. Firstly, the estimates from all three distortion gradients are significantly less stable than the estimates from foreshortening, which was to be expected since the former are based on the spatial derivatives of an estimated quantity. Note, however, that they are clearly good enough to resolve the tilt ambiguity in the estimates from foreshortening. Secondly, this surface actually

---

[2]This assumption is an instance of the general weak isotropy concept defined in the previous section. In WISP it was applied to a weight distribution defined on curves; here it is applied to a weight distribution in the spatial frequency plane.

[3]The reason for using parallel projection when visualizing the surface normal is to simplify comparisons between estimates; with this convention the shape of each projected dish specifies the orientation regardless of the position of the dish in the image.

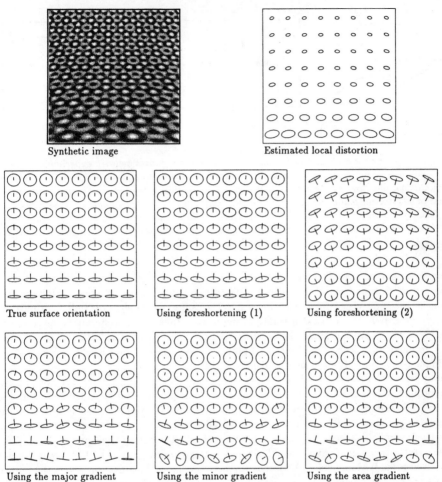

Figure 9: *Estimation of local surface orientation from first- and second-order projective distortion for a cylindrical surface.*

has a significant curvature, making the assumption $\kappa_t = \tau = 0$ invalid. Hence, the estimates from the minor and area gradients will be *biased*, as indicated by (1) and (4). This bias will lead to underestimation of the slant, since $\kappa_t < 0$ for a convex surface, and this effect is quite apparent in figure 9. The bias will also cause a systematic tilt error due to the fact that $\tau \neq 0$ except along the central vertical line in the image, but this error is much lower in magnitude and is quite hard to observe. In contrast, the estimates from the major gradient (2) are not biased at all, since this gradient is independent of surface curvature.

3   Egomotion From Peripheral Vision Cues

Humans have a field of view exceeding 180°, and many animals have an even larger field of view (Carpenter 1988). As in biological vision, peripheral machine vision may well turn

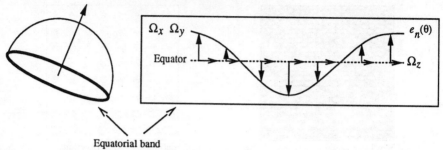

Figure 10: *Equator-normal and equator-parallel flows for rotation with superimposed trans-lational velocities sideways.* Ω_x *and* Ω_y *influence equator-normal flow, whereas* Ω_z *influences equator-parallel flow, as indicated in the figure. The vertical arrows in the diagram indicate the size of equator-normal components for certain longitudes and the endpoints of these vectors are joined by a solid curve indicating the equator-normal components for every longitude* $[0, 2\pi]$. *The horizontal arrows show some equator-parallel components for certain longitudes along the equator.*

out to be a vital ingredient for the task of estimating egomotion. The ability to estimate ones own motion from a sequence of images is of primary importance for locomotion, and in machine vision for navigating a moving camera through some indoor or outdoor environment. In this context it is interesting to note that certain cells with wide receptive fields in cortex are specialized for summing peripheral motion data (Orban [this volume]).

In particular, from a mathematical point of view, egorotation seems to be reasonably simple to calculate from velocities along that band of points that lies at 90 degrees angle away from the optical axis. Using a spherical image surface with the optical axis (z-axis) pointing towards the "north pole" N, the band in question could be called the *equatorial band*, subtending, say, 20 degrees in latitudal direction. *Decomposing* the velocities on the equator (or close to it) into components parallel and normal to the equator, we obtain *two* velocity fields (see figure 10) as a function of longitude along the equator. We call the two vector fields *equator-normal flow* and *equator-parallel flow*. The latter was studied by Nelson and Aloimonos (1988), who furthermore show that rotation about the optical axis ($=\Omega_z$) can be calculated robustly from equator-parallel flow.

We will use the following notation:

(x, y, z)	space coordinates
(V_x, V_y, V_z)	translational velocities
$(\Omega_x, \Omega_y, \Omega_z)$	rotational velocities about x, y, z-axes
dist	distance from projection center $(0, 0, 0)$
θ, φ	longitudes, latitudes on image sphere
e_n	velocity of equator-normal flow

The equator-normal flow (velocity), (12), is essentially a sinusoidal for pure rotation, and translation sideways, see figure 10. This is intuitively clear. When rotating our head, points at one side go out of view, while new points at the other side come into view. Measuring the phase and amplitude of a sine wave in a one–dimensional signal (12) would be a trivial task, were it not for the translational velocity V_z forwards, which causes an almost arbitrary function (V_z divided by distance) to become superimposed on the sinusoidal. The velocity of equator-normal flow is

$$e_n(\theta) = \frac{V_z}{\text{dist}(\theta)} + \sqrt{\Omega_x^2 + \Omega_y^2} \cdot \sin(\theta - \psi) \tag{12}$$

where $\psi = \arctan(\Omega_y/\Omega_x)$ is the phase and $\sqrt{\Omega_x^2 + \Omega_y^2}$ is the amplitude of the sine wave. Equivalently,

$$\begin{cases} \text{Amplitude} \cdot \cos\psi = \Omega_x \\ \text{Amplitude} \cdot \sin\psi = \Omega_y \end{cases} \tag{13}$$

3.1 Experiments

In a series of experiments, Bergholm and Liu (1992), on synthetic *image* data depicting indoor scenes, we have tried to estimate the phase and amplitude of the sinusoidal in (12). The phase and amplitude are linked to rotational velocities Ω_x, Ω_y by (13). The velocity field before decomposition was obtained from using a local image operator that locates interest points (corners, strong curves in smooth contours, centers of small blobs, endpoints of thin lines etc.), see Förstner and Gülch (1987). This operator collects statistics (sample variances and covariances, which form a 2×2 matrix) on the vector field of image intensity gradients, locally, in small image regions (e.g. 7×7 windows). These points were tracked and matched in the image sequence, exploiting a simple technique by Andersson *et al.* (1989). Let $\bar{p}_i^{(j)}$ be the jth Förstner point at time t. The velocity vector (u, v), at $t = i - 0.5$, is calculated making use of four consecutive image frames:

$$\begin{pmatrix} u \\ v \end{pmatrix} = \frac{1}{4} \left(\bar{p}_{i+1}^{(j)} + \bar{p}_i^{(j)} - \bar{p}_{i-1}^{(j)} - \bar{p}_{i-2}^{(j)} \right) \tag{14}$$

After having obtained the equator-normal flow, the approach by Bergholm (1990), Section 4, was used for calculating phase and amplitude. This is done in three steps. First, the equator-normal flow is median filtered to remove crude matching errors (or the influence of thin objects close to the camera), then we obtain phase and amplitude from the one-dimensional Discrete Fourier Transform (DFT), and afterwards the amplitude is further noise corrected.

Our synthetic image data contain errors of two kinds. There are errors of 0-3 pixels in point positions (due to the Förstner operator (Förstner and Gülch 1987)), as well as crude matching errors of 5-20 pixels. The results indicate that the rotation parameters can be estimated with an accuracy of about 10% in this kind of data. Pixel sizes were of the order $0.1° \times 0.1°$ and the rotation Ω_y was 3° per second with V_z of similar magnitude, sampled at 10 frames/s for 2 s. Furthermore, experiments suggest that the *direction* of rotation $= \arctan(\Omega_y/\Omega_x)$ can be estimated with higher accuracy than the *magnitude* of rotation $\sqrt{\Omega_x^2 + \Omega_y^2}$. The method seems to degrade gracefully when confronted with noisier input data. Experiments performed on purely synthetic velocity data (not originating from synthetic images) were presented earlier in Bergholm (1990) and suggested that rotation can be robustly calculated in noisy data, as long as the function V_z/dist behaves in a fairly random way, or varies slowly.

3.2 Ongoing and Future Work

Currently, we are investigating the task of combining peripheral and central motion cues, where we by "central" mean a cone subtending an angle corresponding to a standard camera field of view; $40° - 50°$, centered around the optical axis.

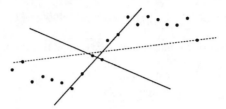

Figure 11: *Illustration of the basic scale problem involved when computing gradients as a basis for edge detection. The lines show the effects of computing derivative approximations from noisy data (here represented as a set of dots) using a central difference operator with a varying step size. More sophisticated approaches exist, but they will face similar problems.*

We derotate (assuming no Ω_z) the velocity fields on the equator exploiting our peripheral vision algorithm. The derotated vector field is fed into an algorithm for estimating Focus of Expansion (FoE) by Gårding (1991) which gives us the direction of egotranslation. This algorithm was originally intended for calculation of vanishing points, but can with a slight modification be exploited also for estimating its dynamic counterpart, FoE.

The acquisition of images with a field of view of 180° is a rather tricky technical problem under investigation. It is not so easy to mimic the human eye, where equatorial flows are produced by refraction of light in the (bulging) cornea and the lens onto the "equator" of the spherical-like retina. Total reflection is avoided since the lens in the human eye is surrounded by fluid.

4   The Scale-Space Primal Sketch

4.1   Detecting Salient Image Structures and their Scales

In order to *extract* information from image data it is necessary to interact with the data using some operators. Some of the most fundamental problems in computer vision concern how to determine *where* in the image such operators should be applied and at *what scale(s)* the information should be extracted. If these issues are not properly dealt with, then the problem of interpreting the output from image operators can be very hard.

As an example of this, consider the task of detecting edges. For simplicity, assume that the gradient is approximated by a central difference operator. It is well-known that in this case the selection of step size leads to a trade-off problem. A small step size gives a small truncation error, but the noise sensitivity can be severe. Conversely, a large step size will in general reduce the noise sensitivity, but at the cost of an increased truncation error. In the worst case, we may even miss the slope of interest and get meaningless results, if the difference quotient approximating the gradient is formed over a wider distance than the size of the object in the image (see figure 11).

Then, how should we be able to cope with these problems? To determine *in advance* where and at what scale(s) to apply image operators seems very much like a chicken-and-the-egg problem. For example, if the task is to do object detection or recognition, we obviously have to apply some operators to the image data in order to get any information about the object(s) we are to investigate. However, how do we determine the scale of an object and where to search for it before knowing what kind of object we are studying and before knowing where it is located?

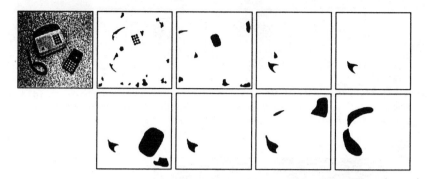

Figure 12: *The 50 most significant dark blobs extracted from a telephone and calculator image. Each such blob is an individual entity existing at an individual level of scale.*

Figure 13: *Boundaries of the most significant blobs from figure 12 using different thresholds on significance (set in gaps in the sequence of significance values).*

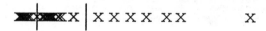

Figure 14: *Significance values of the different blobs from the telephone and calculator image marked with "x" along a logarithmic scale. The vertical lines indicate the manually selected thresholds used in figure 13.*

In our work we have demonstrated that such scale and region determination actually can be performed computationally from raw image data by low-level processing and without access to any prior information about what is in the scene. We have focused on one particular aspect of image structure, namely blob-like image regions which are brighter or darker than the background and stand out from the surrounding, and developed a representation called *the scale-space primal sketch* (Lindeberg 1991; Lindeberg and Eklundh 1990, 1991, 1992), from which a ranking of events in order of significance can be generated. An example of the type of information that can be extracted from this representation is shown in figure 12. We display an image of a telephone and a calculator together with the 50 most significant dark blobs extracted from that image (using the blob extraction scheme to be described below). Each such blob is associated with a scale value suggesting a scale level at which the later stage processing could be performed.

An illustration of the significance values of the different blobs is given in figure 13, where the boundaries of the blobs are displayed using different thresholds on significance, which have been set in gaps in the sequence of significance values, and in figure 14, where the distribution of the significance values has been indicated.

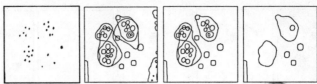

Figure 15: *Multi-scale dot pattern grouping with the scale-space primal sketch. (a) Original grey-level image. (b-d) Boundaries of the most significant dark blobs extracted by the algorithm using different threshold on significance.*

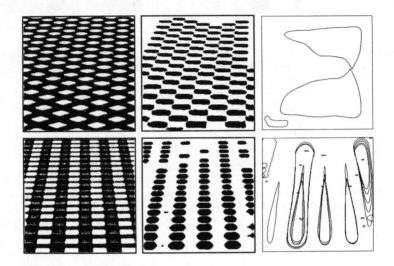

Figure 16: *Blob detection in texture data with the scale-space primal sketch. Left column: Two images of a wall-paper from different views. Middle and right columns: the 100 and 125 most significant dark blobs respectively. (The middle column displays a number of blob support regions, while the right column gives the boundaries of the remaining blobs.)*

Figures 15-16 show the results of applying the scale-space primal sketch to some other types of image data; for dot pattern grouping and for blob detection in texture data.

The ranking on significance is obtained by studying the stability over scale of certain (4D) objects, called *scale-space blobs*, which constitute the basic primitives in the scale-space primal sketch. These objects are defined from geometrical constructions in the *scale-space representation* that one obtains by eroding the grey-level image using the diffusion equation, see e.g. Koenderink and van Doorn (1984). As significance measure we have taken the *volume of the scale-space blobs*, which basically comprises the *spatial extent* of the blob, the *contrast* of the signal and the *lifetime* of this structure in scale-space. This approach is related to an observation by Witkin (1983) about a correspondence between perceptual salience and lifetime in scale-space. However, we do not use the scale-space lifetime alone, since we have noticed that blobs due to noise can survive for a long time in scale-space, if they are located in regions with slowly varying grey-level intensity.

A scale-space blob will in general exist over some scale interval in scale-space. In fact, it is defined as a family of (3D) objects called *grey-level blobs*, each of those existing at

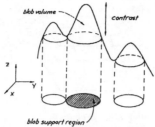

Figure 17: *Grey-level blob definition for bright blobs of a two-dimensional signal. In two dimensions a grey-level blob is generically given by a pair consisting of one local extremum and one saddle point, denoted delimiting saddle point.*

one *single* level of scale. To select a scale level to represent a scale-space blob we take the scale at which the grey-level blob volume assumes its *maximum* volume as function of scale. The intention behind this selection is that is should reflect the scale at which the blob response is "maximally strong". As spatial representative of a scale-space blob we take the support region of the grey-level blob at this representative scale, which is the projection of that grey-level blob onto the spatial plane.

So far we have no firm theoretical support for the proposed methodology for extracting image structures from grey-level data. It is based on a number of postulates expressing that the four-dimensional scale-space blob volume is an appropriate significance measure and that the scale-space blobs can be represented at single scales as described above. However, based on these assumptions and the scale-space theory, several theoretical results have been derived. Moreover, by experiments on different types of images we have demonstrated that this methodology gives intuitively reasonable results in a large number of different situations. It should be stressed that the regions obtained in this way should be seen just as indicators signalling that *"there might be something there of approximately that size — now some other module should take a closer look".* In other words, they should be regarded as coarse spatial descriptors that need to be verified in some way and whose localization must be treated as very approximate. Later, in Section 4.3, we will develop how the localization of the blob boundary can be improved, using the local scale and region information associated with each blob for guiding a local edge detection scheme.

First, however, we will describe in more detail how the scale-space primal sketch is defined from a grey-level image, a section that can be skipped by the hasty reader.

4.2 Construction and Theory of the Scale-Space Primal Sketch

Although the scale-space theory provides a canonical framework for dealing with image structures, which naturally occur at different scales, it gives no information about what scale(s) are appropriate for further analysis or what structures in the image should be treated as significant. The scale-space primal sketch is a multi-scale representation of grey-level shape aimed at providing cues for such decisions, by *making explicit* blobs in scale-space as well as the relations between blobs at different levels of scale.

It is constructed by first defining one type of blobs, called *grey-level blobs*, at *all* levels of scale. The definition of this concept should be obvious from figure 17. Basically, a grey-level blob is a *local extremum with extent*. The extent is determined by the level curve through a specific saddle point, called *delimiting saddle point*, constructed by a water-shed analogy similar to the blob hierarchy described by Koenderink *et al.* in this book.

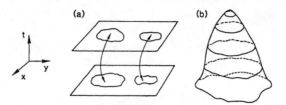

Figure 18: *(a) By linking similar grey-level blobs at adjacent levels of scale we obtain (b) scale-space blobs, which are objects with extent both in space, scale and grey-level. (In this figure the grey-level coordinate has been omitted. The slices illustrate the support regions of the grey-level blobs.)*

Obviously, this blob definition will be highly noise sensitive when treated at a single level of scale only. Therefore, the blob extraction is embedded in a multi-scale framework. These grey-level blobs are then *linked across scales* into four-dimensional objects called *scale-space blobs* (see figure 18), having extent in space, grey-level and scale t. The extent of the scale-space blobs in the scale direction is delimited by *bifurcations* between critical points, or equivalently, by bifurcations between blobs. These events also define *hierarchical relations* between scale-space blobs at different scales.

To summarize, the scale-space primal sketch can be seen as a tree-like data structure with the scale-space blobs as vertices and the bifurcation events as arcs between those (see figure 19). The representation is obtained in a completely bottom-up data-driven manner, without relying on any specific parameters or error criteria. Its definition is expressed for grey-level images, but the approach is valid for any bounded function and can be extended to any (finite) number of dimensions. It can therefore, as we will see in Section 4.3, be used also for deriving properties of e.g. spatial derivatives.

Figure 20 displays some experimental results of extracting grey-level blobs from a grey-level image at different scales. We see that at fine levels of scale mainly small blobs, due to noise and surface texture, are detected. When scale increases, the noise blobs disappear gradually, although much faster in regions near steep gradients (for example, near the calculator). Notable in this context is that blobs due to noise can survive for a long time in scale-space if located in regions with slowly varying grey-level intensity (like in a smooth background). This shows that scale-space lifetime *alone* is not appropriate as a significance measure, since then the significance of such blobs due to noise would be substantially overestimated. We note that the buttons on the telephone keyboard manifest themselves as individual blobs after a small amount of smoothing and at coarser levels of scale they merge into one unit (the keyboard). We can also observe that some other dark details in the image, the calculator, the cord and the receiver, appear as single blobs at coarser level of scale. The idea behind the suggested linking of features across scales is to enable determination of which of these blobs should be regarded as significant as well as what scales(s) are appropriate for treating those.

The proposed significance measure, the scale-space blob volume, depends upon the actual parameterization of the four coordinate axes in the scale-space representation. In order to enable comparisons of significance between structures at different scales, the coordinates obviously need to be (re)parameterized in such a way that structures at different scales will be treated in a uniform manner. We have shown that natural requirements on a transformed scale parameter, *effective scale* τ, imply that there is in principle only one reasonable way to define it, namely by

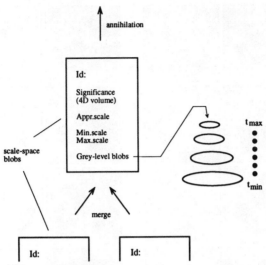

Figure 19: *The scale-space primal sketch can be seen as a tree-like multi-scale representation of blobs with the scale-space blobs as basic primitives (nodes) and the relations (bifurcations) between scale-space blobs at different levels of scale as branches.*

$$\tau(t) = \log\left(\frac{p_0}{p(t)}\right) \qquad (15)$$

where $p(t)$ is the expected density of local extrema at scale t in the scale-space representation of a reference signal and p_0 is a constant. From estimates of how the density of local extrema can be expected to vary with scale in white noise signals, we have proved (Lindeberg 1992) that for continuous signals this function will essentially be a logarithm, while for discrete signals it will be approximately logarithmic at coarse scales and approximately linear at fine scales. It turns out that the volumes of the grey-level blobs must be transformed in a similar manner. That normalization is based on simulation results accumulated from evolution properties of grey-level blobs extracted from random noise signals.

An important aspect to consider in this context is that realistic image data from sensors are *discrete*. For example, if the results from the continuous theory would be directly applied and the scale-space lifetime of a scale-space blob would be measured by $\log t_D - \log t_A$, where t_A and t_D denote the appearance and disappearance scales of the scale-space blobs respectively, then a structure existing in discrete data at scale value zero would be assigned an infinite lifetime. Therefore, the implementation of the scale-space primal sketch is based on a scale-space concept especially developed for discrete signals (Lindeberg 1990, 1992a), and the reparametrization of the coordinates in the scale-space representation is adapted to this fact. For example, the effective scale parameter for discrete signals, which is given by (15), increases linearly with scale at fine scales and is equal to zero when the scale parameter is equal to zero.

Further theoretical properties of the representation are developed in (Lindeberg 1992d), where a precise definition of the scale-space blob concept is given and it is shown that, with respect to these blob definitions based on local extrema, there are basically

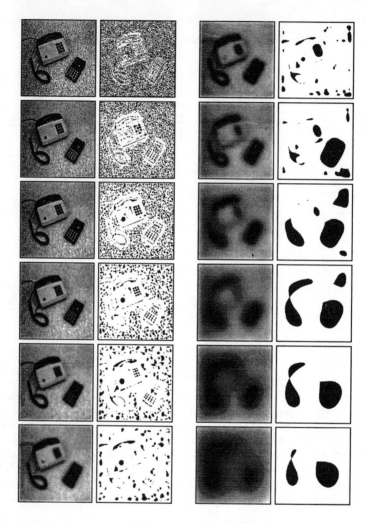

Figure 20: *Grey-level and dark grey-level blob images of a telephone and calculator image at scale levels $t = 0, 1, 2, 4, 8, 16, 32, 64, 128, 256, 512$ and 1024 (from top left to bottom right).*

possible types of (2D) blob events that can occur when the scale parameter increases; a blob may be *annihilated*, two blobs may *merge* into one, one blob may *split* into two or, a blob may be *created*. We have also in more detail analysed the behaviour of critical points and blobs in scale-space, derived generally valid expressions for their drift velocities, and studied the qualitative behaviour of critical points and blobs in a number of characteristic situations. In this way, we can analytically describe how the blobs displayed in figure 20 behave over scales.

4.3 Guiding Early Visual Processing with Qualitative Scale and Region Information

A principle that we argue for is that the *qualitative* spatial and scale descriptors, which are obtained from the scale-space primal sketch, can be useful for *guiding* other processes in early vision and will *simplify* their tasks (Lindeberg 1991, 1992c).

As an example of this, we describe an integration of the scale-space primal sketch with *edge detection*. Figure 21 illustrates the principle. The main idea is to use to blob information to guide a *regional* edge detection scheme working at an *adaptively determined level of scale*. We have selected the seven most significant dark blobs from another telephone and calculator image and sorted them in scale order. In the left column we show a significant blob as extracted from the scale-space primal sketch. The middle column displays edges detected at the scale given by the scale-space blob. (We define edges as the ridges of the gradient magnitude map. It should be noted that we do not use any threshold on the gradient magnitude.) A matching step is performed between the edges and the blobs. The edges matched to blob are marked in black, while the other edges are grey. Then, the edges detected at coarse scales are localized to finer scales using the edge focusing method developed by Bergholm (1987). The right column shows such localized (accumulated) edges at successively finer scales, just before a new blob is considered. The "final result" of this procedure is shown in the lower right corner, where we have localized all the edges to a (pre-selected) fine scale.

Observe that with this method, which we call *blob-initiated edge focusing*, we obtain edges that are more meaningful entities than the output that is obtained from an ordinary edge detector. Note also that the matching between the blob and the edges at a coarse scale has already induced a coarse grouping of edge pixels into higher order units. We explicitly have the relation between edge pixels and the blobs, and know that the edges so obtained are related to the selected dark blobs from the image. Therefore we believe that there is a strong potential in using these relations to delimit the search space for higher order interpretations as e.g. model matching, abstractions of edge descriptors etc. However, there is still more work to be done in order to explore these latter suggestions.

4.4 Focus-of-Attention

More generally, the scale-space primal sketch can serve as a primitive mechanism for focus-of-attention. To illustrate this, we briefly describe some experiments (Brunnström *et al.* 1992), where we have been using this method for guiding the focus-of-attention of a *head-eye system* applied to a specific test problem of *classifying junctions*. In this context, the scale-space primal sketch was used for two main purposes;

- for generating hypotheses about the existence of objects (or facets of objects) in the scene and for selecting scales for computing junction candidates, and

- for finding candidate junctions in curvature data and regions of interest around those, called *curvature blobs*, which are used for acquiring new image data with higher resolution as well as for providing context information necessary for the junction classification procedure.

Figures 22-24 illustrate some of the main processing steps. Figure 22 shows an overview image of a scene under study together with the most significant dark and bright blobs extracted from the grey-level image. Each such region constitutes a hypothesis about the existence of an object, a facet of an object, or an illumination phenomenon in the scene.

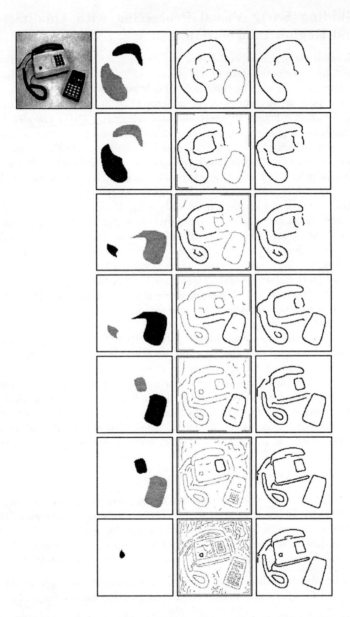

Figure 21: *Illustration of the composed blob-edge focusing procedure for the telephone and calculator image. The left column shows the active blob hypothesis. Its blob support region has been marked with black. The middle column shows the edge image at the level of scale given by the previous blob. The matching edge segments have been drawn black while the other edge pixels are grey. The right column shows the result after focusing, just before a new blob is considered. The image in the lower right corner displays "the final result", i.e. the edges that are related to the dark blobs in the image.*

Figure 22: *(a) Overview image of a scene under study. (b-c) Boundaries of the 20 most significant dark and bright blobs respectively extracted by the scale-space primal sketch.*

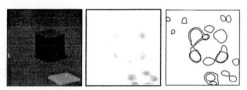

Figure 23: *Zooming in to a region of interest given by a blob from the previous processing step. (a) A window around the region of interest, set from the location and the size of the blob. (b) The (rescaled) level curve curvature computed at the scale of the blob. (c) The boundaries of the 20 most significant curvature blobs obtained from blob detection in the curvature data.*

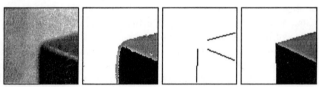

Figure 24: *Zooming in to a junction candidate given by a curvature blob. (a) Maximum window size for the focusing procedure set from the size of the curvature blob. (b) Backprojected peaks from the intensity histogram. (c) Lines computed from the backprojected peaks from the directional histogram. (d) Schematic illustration of the classification result in which a simple junction model has been adjusted to the data. (This junction candidate was classified as a 3-junction).*

In figure 23 we have simulated foveation by redirecting the cameras of the head-eye system towards one of the dark blobs corresponding to the central toy block and zooming in to that structure. From a new image with increased resolution junction candidates were then extracted in a window around the interesting blob. The interest operator, essentially the curvature of level curves (Koenderink and Richards 1998), was tuned to the scale of the blob determining the current region of interest. Then, the scale-space primal sketch was applied to that curvature data generating a number of blobs, called *curvature blobs*, corresponding to regions of strong response from the interest operator.

In figure 24 we have zoomed in further to one of those and initiated a classification procedure (described in Brunnström *et al.* 1990) tuned to the size of the curvature blob. This junction candidate was classified as a 3-junction, based on three peaks, stable to variations in window size, detected in the grey-level and the directional histograms respectively.

An important aspect of this integration of the analysis with a head-eye system is that the algorithm can control the image acquisition and e.g. *acquire new image data* with increased resolution as to *clearly resolve* the structures under study and simplify its classification tasks. Of course, this leads to several issues concerning control strategies of the reasoning that need to be addressed. Nevertheless, such a system, which is *looking at the world*, can zoom in to interesting details and avoid the rather artificial loss of information that occurs if the analysis is restricted to pre-recorded images of overview nature. It can moreover use active control of camera parameters for acquiring additional cues about the physical nature of structures.

References

Aloimonos, J. (1988) Shape from texture. Biological Cybernetics, vol. 58, pp. 345-360

Andersson, M., Eklundh, J.O., Kakimoto, J. (1989) Rigid body motion analysis from the correspondence of interest points. In Proc. 6th Scandinavian Conference on Image Analysis, (Oulo, Finland), pp. 786-791, June 1989

Bergholm, F. (1987) Edge focusing. IEEE Trans. Pattern Anal. and Machine Intell., vol. 9, pp. 726-741, November 1987

Bergholm, F. (1990) Decomposition theory and transformations of visual direction. In Proc. 3rd Int. Conf. on Computer Vision, (Osaka, Japan), pp. 85-90, December 1990

Bergholm, F., Liu, S. (1992) Time path and displacement estimation with applications to peripheral vision. tech. rep., Dept. of Numerical Analysis and Computing Science, Royal Institute of Technology, Stockholm (In preparation)

Blake, A., Marinos, C. (1990) Shape from texture: estimation, isotropy and moments. J. of Artificial Intelligence, vol. 45, pp. 323-380

Brady, M., Yuille, A. (1984) An extremum principle for shape from contour. IEEE Trans. Pattern Anal. and Machine Intell., vol. 6, pp. 288-301

Brunnström, K., Eklundh, J.O., Lindeberg, T. (1990) Scale and resolution in active analysis of local image structure. Image and Vision Computing, vol. 8, pp. 289-296, November 1990

Brunnström, K., Lindeberg, T., Eklundh, J.O. (1992) Active detection and classification of junctions by foveation with a head-eye system guided by the scale-space primal sketch. In: G. Sandini(ed.) Proc. 2nd European Conf. on Computer Vision, vol. 588 of Lecture Notes in Computer Science, Springer-Verlag, May 1992 (Santa Margherita Ligure, Italy)

Carpenter, R.H.S. (1988) Movements of the Eyes. Pion Limited, London, second ed.

Cutting, J.E., Millard, R.T. (1984) Three gradients and the perception of flat and curved surfaces. J. of Experimental Psychology: General, vol. 113(2), pp. 198-216

Förstner, W., Gülch, E. (1987) Detection and precise location of distinct points, corners and centres of circular features. In Proc. Intercommission Workshop of the Int. Soc. for Photogrammetry and Remote Sensing, (Interlaken, Switzerland)

Gårding, J. (1990) Shape from texture and contour by weak isotropy. In Proc. 10th Int. Conf. on Pattern Recognition (A), (Atlantic City, NJ), pp. 324-330, June 1990

Gårding, J. (1991) Shape from surface markings. PhD thesis, Dept. of Numerical Analysis and Computing Science, Royal Institute of Technology, Stockholm, May 1991

Gårding, J. (1992a) Shape from texture for smooth curved surfaces. In: G. Sandini (ed.) Proc. 2nd European Conf. on Computer Vision, vol. 588 of Lecture Notes in Computer Science, Springer-Verlag, May 1992 (Santa Margherita Ligure, Italy)

Gårding, J. (1992b) Shape from texture for smooth curved surfaces in perspective projection. Tech. Rep. TRITA-NA-P9203, Dept. of Numerical Analysis and Computing Science, Royal Institute of Technology, Stockholm, February 1992 (Submitted)

Gibson, J. (1950) The Perception of the Visual World. Houghton Mifflin, Boston

Gibson, J. (1979) The Ecological Approach to Visual Perception. Houghton Mifflin, Boston

Horn, B.K.P. (1986) Robot Vision. MIT Press, Cambridge, Mass.

Jones, J., Palmer, L. (1987) An evaluation of the two-dimensional Gabor filter model of simple receptive fields in cat striate cortex. J. of Neurophysiology, vol. 58, pp. 1233-1258

Kanatani, K. (1984) Detection of surface orientation and motion from texture by a stereological technique. J. of Artificial Intelligence, vol. 23, pp. 213-237

Koenderink, J.J., Richards, W. (1988) Two-dimensional curvature operators. Journal of the Optical Society of America, vol. 5:7, pp. 1136-1141

Koenderink, J.J., van Doorn, A.J. (1984) The structure of images. Biological Cybernetics, vol. 50, pp. 363-370

Lindeberg, T. (1990) Scale-space for discrete signals. IEEE Trans. Pattern Anal. and Machine Intell., vol. 12, pp. 234-254

Lindeberg, T. (1991) Discrete scale space theory and the scale space primal sketch. PhD thesis, Dept. of Numerical Analysis and Computing Science, Royal Institute of Technology, Stockholm, May 1991

Lindeberg, T. (1992a) Discrete derivative approximations with scale-space properties: A basis for low-level feature extraction. Tech. Rep. TRITA-NA-P9212, Dept. of Numerical Analysis and Computing Science, Royal Institute of Technology, Stockholm, April 1992 (Submitted)

Lindeberg, T. (1992b) Guiding early visual processing with qualitative scale and region information. Submitted

Lindeberg, T. (1992c) Scale-space behaviour of local extrema and blobs. J. of Mathematical Imaging and Vision (To appear)

Lindeberg, T. (1992d) Effective scale: a natural unit for measuring scale-space lifetime. IEEE Trans. Pattern Anal. and Machine Intell. (To appear)

Lindeberg, T., Eklundh, J.O. (1990) Scale detection and region extraction from a scale-space primal sketch. In Proc. 3rd Int. Conf. on Computer Vision, (Osaka, Japan), pp. 416-426, December 1990

Lindeberg, T., Eklundh, J.O. (1991) On the computation of a scale-space primal sketch. Journal of Visual Communication and Image Representation, vol. 2, pp. 55-78, March 1991

Lindeberg, T., Eklundh, J.O. (1992) The scale-space primal sketch: construction and experiments. Image and Vision Computing, vol. 10, pp. 3-18, January 1992

Mardia, K.V. (1972) Statistics of Directional Data. Academic Press, London

Nelson, R.C., Aloimonos, J. (1988) Finding motion parameters from spherical motion fields (or the advantage of having eyes in the back of your head). Biological Cybernetics, vol. 58, pp. 261-273

Ohta, Y., Maenobu, K., Sakai, T. (1981) Obtaining surface orientation from texels under perspective projection. In Proc. 7th Int. Joint Conf. on Artificial Intelligence, (Vancouver, B.C., Canada), pp. 746-751, August 1981

Pollard, S.B, Mayhew, J.E.W., Frisby, J.P. (1985) PMF: A stereo correspondence algorithm using a disparity gradient limit. Perception, vol. 14, pp. 449-470

J. Porrill, J. Gårding, J.O. Eklundh, J.P. Frisby, D. Buckley, S. Pollard, Mayhew, J., Spivey, E. (1991) Using shape-from-texture to calibrate stereo. Perception, vol. 20, no. 1, p. 90

Stevens, K.A. (1981) The information content of texture gradients. Biological Cybernetics, vol. 42, pp. 95-105

Todd, J.T., Akerstrom, R.A. (1987) Perception of three-dimensional form from patterns of optical texture. J. of Experimental Psychology: Human Perception and Performance, vol. 13, no. 2, pp. 242-255

Witkin, A.P. (1981) Recovering surface shape and orientation from texture. J. of Artificial Intelligence, vol. 17, pp. 17-45

Witkin, A.P (1983) Scale-space filtering. In Proc. 8th Int. Joint Conf. Art. Intell., Karlsruhe, West Germany, pp. 1019-1022, August 1983

Appendix I: Addresses

Jan-Johan Koenderink, Astrid Kappers and Andrea van Doorn
Rijksuniversiteit te Utrecht
Faculteit der Natuur- en Sterrenkunde
Buys Ballot Laboratorium
Princetonplein 5
Postbus 80 000
3508 TA Utrecht
The Netherlands

Guy A. Orban
Katholieke Universiteit te Leuven
Faculteit Geneeskunde
Laboratorium voor Neuro- en Psychofysiologie
Campus Gasthuisberg
Herestraat 49
B-3000 Leuven
Belgium

Klaus-Peter Hoffmann
Ruhr-Universität Bochum
Fakultät für Biologie
Allgemeine Zoologie und Neurobiologie
N.D. 7, Postfach 10 21 48
D-4630 Bochum 1
Federal Republic of Germany

Alessandro Verri, Marco Straforini and Vincent Torre
Università di Genova
Dipartimento di Fisica
Via Dodecaneso 33
I-16 146 Genova
Italy

Rachid Deriche, Olivier Faugeras, Gérard Giraudon, Théo Papadopoulo,
Régis Vaillant and Thierry Viéville
Institut National de Recherche en Informatique et en Automatique
Unité de Recherche – Sophia Antipolis
2004 Route des Lucioles
B.P. 109
F-06561 Valbonne – Cédex
France

Michel Demazure and Jean-Pierre Henry
Ecole Polytechnique
Centre de Mathématiques
U.R.A. au C.N.R.S. no. 169
F-91 128 Palaiseau – Cédex
France

Michel Merle
Université de Nice – Sophia Antipolis
Laboratoire de Mathématiques
U.R.A. au C.N.R.S no. 168
F-06108 Nice – Cédex 2
France

Bernard Mourrain
Institut Nationale de Recherche en Informatique
et en Automatique – Sophia Antipolis
Project SAFIR
2004 Route des Lucioles
F-06565 Valbonne
France

H. Christopher Longuet-Higgins
University of Sussex
Laboratory of Experimental Psychology
Brighton BN1 9QC
United Kingdom
and
University of Oxford
Department of Engineering Science
Oxford OX1 3PJ
United Kingdom

Hans-Hellmut Nagel
Institut für Algorithmen und Kognitive Systeme
Fakultät für Informatik der Universität Karlsruhe (T.H.)
and
Fraunhofer Institut für Informations- und Datenverarbeitung II TB
Fraunhoferstrasse 1
D-7500 Karlsruhe
Federal Republic of Germany

Marc Proesmans and André Oosterlinck
Katholieke Universiteit te Leuven
E.S.A.T. M.I.2
Kardinaal Mercierlaan 94
B-3001 Heverlee
Belgium

Brian J. Rogers
University of Oxford
Department of Experimental Psychology
OX1 3UD Oxford
United Kingdom

John P. Frisby and David Buckley
University of Sheffield
A.I. Vision Research Unit
P.O. Box 603
Western Bank
S10 2TN Sheffield
United Kingdom

Roger J. Watt
University of Stirling
Department of Psychology
FK9 4LA Stirling
Scotland
United Kingdom

David R. Simmons and David H. Foster
University of Keele
Department of Communication and Neuroscience
ST5 5BG Keele
Staffordshire
United Kingdom

Jan-Olof Eklundh, Jonas Gårding, Tony Lindeberg and Frederik Bergholm
Royal Institute for Technology
Department of Numerical Analysis and Computing Science
Computational Vision and Active Perception Laboratory
S-100 44 Stockholm
Sweden

Appendix II: Curricula Vitae

Fredrik Bergholm received the MSc degree in mathematics, computer science, and economics in 1981 from Stockholm University, Sweden. From 1979 to 1984 he was employed at the Industrial Institute of Social and Economic Research (I.U.I.) in Stockholm. In 1984 he joined the Department of Numerical Analysis and Computing Science at the Royal Institute of Technology in Stockholm, and received the PhD degree in computing science in 1989. He is currently a research associate at the Computational Vision and Active Perception Laboratory (C.V.A.P.) at the Royal Institute of Technology. His primary fields of research are edge detection, multi-scale analysis, and motion analysis.

David Buckley holds a postdoctoral research post in the Artificial Intelligence Research Unit at the University of Sheffield. He received his BSc in Psychology from the University of Birmingham and his PhD in visual psychophysics from the University of Sheffield. His research interests are centred on the how the human visual system combines stereo and texture information.

Rachid Deriche was born in Thénia, Algeria, on June 5, 1954. He received the *Diplôme d'Ingénieur* from the *Ecole Nationale Supérieure des Telecommunications de Paris* in 1979 and the PhD degree from the University of Dauphine, Paris XI in 1982.
He is currently a Research Director at the National Research Institute in Computer Science and Control Theory (I.N.R.I.A.), Sophia Antipolis, France, where he works in the Computer Vision and Robotics Group. His major areas of research interest are in computer vision with an emphasis on low-level, stereo and motion analysis.

Jan-Olof Eklundh first did research in mathematics at Stockholm University (BA 63, PhD 70), working on problems in functional analysis. In 1969 he joined the newly formed Laboratory for Image Analysis at the National Defense Research Institute, F.O.A., Stockholm (from 1978, Linköping) becoming head of the laboratory in 1974. He remained with F.O.A. until 1982, but spent 1977-79 at the Computer Vision Laboratory, University of Maryland, besides some other extended visit to laboratories in the U.S. In 1980 he became Research Director and Head of the Department of Computer Science at F.O.A. The same year he received the degree of Dr. Tech. at the Royal Institute of Technology, K.T.H., on work in algorithm analysis and complexity. In 1982 he joined the Department of Numerical Analysis and Computing Science at K.T.H. as an associate professor. He there founded the Computer Vision and Active Perception Laboratory, now employing about twenty fulltime researchers and research students. In 1986 he became a professor in computer science. He has served on several conference committees, including the International Conference on Computer Vision and the European Conference on Computer Vision. He is and has been on the editorial boards of several well-known journals. Between 1984 and

1988 he was chairman of the Swedish Society of Image Analysis. He is also an associate member of the recently formed Swedish Research Council for Engineering Science.

Olivier Faugeras is Research Director at I.N.R.I.A. (National Research Institute in Computer Science and Control Theory) where he leads the Computer Vision and Robotics Group. His research interests include the application of Mathematics to Computer Vision, Robotics, Shape Representation, Computational Geometry, and the architectures for Artificial Vision systems as well as the links between artificial and biological vision.
He is an Associate Professor of Applied Mathematics at the Ecole Polytechnique in Palaiseau, France, where he teaches Computer Science, Computer Vision, and Computational Geometry.
He is an Associate Editor of several international scientific Journals including the International Journal of Robotics Research, Pattern Recognition Letters, Signal Processing, Robotics and Autonomous Systems, and Vision Research. He is Editor-in-Chief of the International Journal of Computer Vision. He has served as Associate Editor for I.E.E.E. P.A.M.I. from 1987 to 1990 and is a senior member of I.E.E.E.
In April 1989 he received the "Insititut de France - Fondation Fiat" prize from the French Science academy for his work in Vision and Robotics.

David H. Foster is Professor of Theoretical and Applied Vision Sciences in the Research Department of Communication and Neuroscience, Keele University. He received his BSc, PhD, and DSc degrees in physics and biophysics from Imperial College, London University, in 1966, 1970, and 1982 respectively. He was elected Fellow of the Institute of Physics and of the Institute of Mathematics and its Applications in 1981. His research interests cover fundamental, technical, and clinical aspects of human visual information processing, concentrating particularly on the mechanisms underlying spatial perception, pattern recognition and discrimination, and colour vision. He also has a long-standing interest in the development of mathematical and statistical techniques in visual psychophysics. He is a founder and Co-Editor-in-Chief of the journal "Spatial Vision" and has been Chairman of the Applied Vision Association since 1986.

John P. Frisby is Professor of Psychology and Chairman of the Arificial Intelligence Vision Research Unit at the University of Sheffield. The research goals of this Unit are to divise methods and fast hardware for using stereo and motion image data for visual object recognition and guidance of autonomous vehicles fitted with steerable 'eye/head' camera rigs controlled by neural net architectures. He received his BA degree in Psychology from the University of Cambridge and his PhD in visual psychophysics from the University of Sheffield. His main research interests are in developing and testing computational models of human stereoscopic vision.

Jonas Gårding received the MSc degree in engineering physics and applied mathematics from the Royal Institute of Technology, Stockholm, in 1985. He subsequently joined the Department of Numerical Analysis and Computing Science at the same institute, and in 1991 he received the PhD degree with the dissertation "Shape from Surface Markings". He is currently a research associate at the Computational Vision and Active Perception laboratory (C.V.A.P.) at the Royal Institute of Technology. His research interests include most aspects of human and machine vision, in particular estimation of surface shape from monocular and binocular visual cues, and techniques for fusion of data from multiple cues.

Gerard Giroudon was born on March 14 1954 in St. Gaudens (France). He received the Applied Mathematic PhD degree from the University of Nice in 1979. He has worked with IBM R&D Laboratory (La Gaude - France) from 1976-1979, then with Sintra-Alcatel (Paris - France) from 1980-1984 before joining the National Research Institute in Computer Science and Control Theory (I.N.R.I.A.) at Sophia Antipolis (France) in 1984. He is currently a Research Director at I.N.R.I.A. where he works in the Image Understanding Group. His main areas of research interest are image segmentation, modeling and knowledge representation and parallel architecture.

Jean-Pierre Henry (Thesis, Paris, 1979) graduated as engineer at Ecole Polytechnique in 1967. He is currently Chargé de Recherche au C.N.R.S., France, and member of the "Unité associée au C.N.R.S n° 169", Centre de Mathématiques de l'Ecole Polytechnique, a State Research Center founded by Laurent Schwartz in 1965. He studied singularity theory with Frederic Pham and Heisuke Hironaka, and published on that subject, especially with Michel Merle. More recently he extended his interests to effectivity in algebraic geometry, Complexity theory, Symbolic Algebra and algorithms, and wrote with M. Merle and Guillermo Moreno several implementations of efficient algebraic computations in Singularity Theory. He has also worked on applications of real analytic geometry to robotics and computer vision.

Klaus-Peter Hoffmann is Professor of Zoology and Neurobiology and Dean of the Faculty of Biology at the Ruhr-University in Bochum, Germany. He received his PhD degree in Natural Science from the Ludwig-Maximillians-University in Munich in 1970 and his habilitation from the same University in 1974. Dr. Hoffmann is member of the editorial boards of several leading neuroscience journals. His main research interest is the comparative neurobiology of cortical and subcortical interactions in the organisation of visuo-motor behaviour. This includes work on organotypic cell cultures up to the analysis of brain functions in trained monkeys. His work has contributed to the understanding of the parallel pathways in the visual and visuo-motor system.

Astrid M.L. Kappers graduated in physics from Utrecht University in 1984. She received her PhD from the University of Technology in Eindhoven, where she participated in research on speech analysis and recognition. Since 1989 she is connected with the Utrecht Biophysics Research Institute, where she works on various topics in visual psychophysics and haptic perception.

Jan-Johan Koenderink (PhD Utrecht, 1967) is currently professor of biophysics at Utrecht University and scientific director of the Utrecht Biophysics Research Institute (U.B.I.). He has worked on topics in theoretical brain science, experimental psychophysics of vision, touch, and hearing, modelling the phenomenology of perception (*e.g.,* geometrical aspects of image processing and machine vision). He also works on the physics of the visual world, *e.g.,* geometrical aspects (perspective, optic flow), photometry, optical charaterization of materials.

Tony Lindeberg received his MSc degree in engineering physics and applied mathematics from the Royal Institute of Technology, Stockholm, Sweden in 1987. After his studies he was first with the research group in Computer Fluid Dynamics at the Aeronautical Research Institute of Sweden. Then he joined the Department of Numerical Analysis and Computing Science at the Royal Institute of Technology, from which he in 1991 received his

PhD degree in computing science with the dissertation "Discrete Scale-Space Theory and the Scale-Space Primal Sketch". He is currently a research associate at the Computational Vision and Active Perception Laboratory (C.V.A.P.) at the Royal Institute of Technology. His primary research interests in computer vision are scale-space methods, description of shape and focus-of-attention.

H. Christopher Longuet-Higgins FRS is Professor Emeritus at Sussex University where he works in the Centre for Research on Perception and Cognition associated with the Laboratory of Experimental Psychology.

Professor of Theoretical Chemistry at Cambridge until 1967, he has spent the last 25 years in the cognitive sciences, contributing to such diverse areas as language acquisition, speech recognition, the perception of music and the geometry and kinematics of vision; much of this work is documented in his book *Mental Processes: Studies in Cognitive Science* (M.I.T. Press: 1987). In collaboration with the Robotics Group at Oxford University he has devoted much recent attention to the computational problems involved in the interpretation of image sequences.

Michel Merle is Professor of Mathematics at the "Université de Nice-Sophia Antipolis" (U.N.S.A.), Nice, France, and member of the "Unité associée au C.N.R.S n° 168". His research is focused on the study of local singularities of analytic mappings, a subject widely investigated by Thom (Catastrophe Theory), Mather, Hironaka, Arnold ...

He is interested in theoretical aspects of computer vision and effective problems in computer algebra.

Bernard Mourrain is working at I.N.R.I.A. (Sophia Antipolis) since October 1991 as researcher in the team SAFIR, concerned with symbolic computations. Born in 1964, he integrated the "ENS" of Cachan in 1984, passed the "Agregation" of Mathematics in 1987 and began a PhD (1988) in the team of M. Demazure (his advisor) at the center of Mathematics of the Ecole Polytechnique. He defended his thesis on September 1991. The principal subject of his work is related to computational invariant theory, more precisely algebraic methods and algorithms which deals with intrinsics quantities or directly with geometric objects like points, lines, conics, matrices, ...

Hans-Hellmut Nagel received the Diplom degree in Physics from the Universität Heidelberg in 1960 and the Doctor degree in Physics from the Universität Bonn, Federal Republic of Germany, in 1964.

After 18 months as a Visiting Scientist at M.I.T., he worked on the automatic analysis of bubble chamber film at the Deutsche Elektronen-Synchrotron at Hamburg as well as at the Physikalische Institut der Universität Bonn from 1996 through 1971. In Fall 1971 he became o. Professor für Informatik (Computer Science) at the Universität Hamburg. Since 1983 he has been Director of the Fraunhofer-Institut für Informations- und Datenverarbeitung at Karlsruhe, Federal Republic of Germany, in a joint appointment as o. Professor at the Fakultät für Informatik der Universität Karlsruhe. In addition to his primary interest in the evaluation of image sequences and associated questions in Computer Vison, A.I., Knowledge Representation and Pattern Recognition, his interests include the implementation and use of higher level programming languages for the realization of image analysis as well as the design of corresponding system architectures.

Dr. Nagel is a member of editorial boards and advisory boards of various international journals in the field of Computer Vision, A.I., and Pattern Recognition. He is a Senior

Member of I.E.E.E., member of Gesellschaft für Informatik as well as of the Pattern Recognition Society, among others.

André Oosterlinck received his PhD from the Catholic University of Leuven in biological image processing in 1977 after having worked at a number of laboratories, including the Jet Propulsion Lab.
He is vice-president for the Exact Sciences and member of the board of directors of the university of Leuven. He is also director of the E.S.A.T.-division of the department of Electrical Engineering and research manager of the M.I.2 (Machine Intelligence & Imaging) unit of E.S.A.T. M.I.2 consists of over 50 researchers active in digital signal processing.
André Oosterlinck is also founder of an image processing company for automatic visual inspection.
His work has been widely published in more than two hundred international publications. His special research interest is image processing, coding, robot vision, in particular the development of special hardware for this purpose.

Guy A. Orban is currently professor of Neurophysiology at the Katholieke Universiteit Leuven Medical School in Leuven, Belgium and scientific director of the Neuro-and Psychophysiological laboratory, an interdisciplinary centre involved in neuroscientific, perceptual and computational studies in vision. He holds an MD and PhD (Neurophysiology) as well as an MSc in applied mathematics. He is interested in all disciplines which help understand the relationship between the visual cortex and visual perception and memory: single cell physiology, chemical neuroantomy, positron emission tomography, animal and human psychophysics, lesions studies, neural networks and brain modelling. His current work is centred on optic flow, kinetic shapes and contours, and mechanisms of simple discriminations. He is the coordinator of the Insight I and II B.R. projects.

Théodore Papadopoulo was born in Sophia (Bulgaria) on February 4, 1966. He finished the Ecole Normale Superieure de Paris in 1990 and received the aggregation in mathematics in 1989.
He is currently a PhD Student since 1990 at I.N.R.I.A. (National Research Institute in Computer Science and Control Theory), Sophia Antipolis, France, where he works in the Computer Vision and Robotics Group on the structure and motion detection of rigid curves.

Marc Proesmans obtained a degree in Electrical Engineering in 1987 at the Katholieke Universiteit Leuven, after which he joined the E.S.A.T. Robot Vision group at the same university. His early research concerned edge detection, as well as optical flow. His current topics of interest comprise parallel processing in image analysis and anisotropic diffusion models for segmentation.

Brian Rogers currently teaches and researches in the Department of Experimental Psychology at the University of Oxford. He received his BSc and PhD degrees from the University of Bristol in 1969 and 1976 respectively. His research over the past fifteen years has been concerned with the perception and representation of 3-D surfaces and their layout from optic flow and disparity information using psychophysical methods of investigation on human observers. He is presently writing a joint book on 'Binocular Vision and Stereopsis' with Ian Howard as well as a monograph based on his own research into motion parallax and binocular stereopsis.

David R. Simmons is a Research Fellow in the Research Department of Communication and Neuroscience at Keele University. He received a BSc in Physics from Imperial College, London University, in 1985 before commencing his research training in the Oxford University Laboratory of Physiology. He is currently completing his D. Phil. thesis on the spatiotemporal properties of human stereoscopic mechanisms. His research has focused on the relationships between different visual psychophysical tasks, in particular how data from discrimination experiments may provide a suitable metric for understanding more complex tasks, such as image segmentation.

Marco Straforini was born in Genova, on June 19, 1960. He received the PhD degree in Theoretical Physics from the University of Genova, Italy, in 1989. Since 1989 he has been working with Prof. Vincent Torre at the Department of Physics of the University of Genova on problems concerning edge detection and tri-dimensional reconstruction. During the past few years he has also been working on simulation and analysis of the process of electrical transduction in photoreceptors.

Vincent Torre was born in Johannesburg, South Africa, on July 24, 1950. He received the PhD Degree in theoretical Physics from the University of Genova, Italy, in 1973. From 1974 to 1978 he worked on the electrophysiology of retinal cells at the Laboratory of Neurophysiology C.N.R., Pisa, Italy. From 1979 to 1981 he worked on the mechanisms of phototransduction at the Physiological Laboratory, Cambridge U.K., under the supervision of Sir Alan Hodgkin. Since 1991 he has been Full Professor in Cybernetics and Information Theory. He is interested in Biophysics and information processing in man and machine.

Andrea J. van Doorn graduated in physics and mathematics in 1971 from Utrecht University. She has participated in research on vision at Groningen University and is now connected with the Utrecht Biophysics Research Institute, where she works on various topics in visual psychophysics and modeling of visual functions in humans. Her research interests include visual psychophysics, models of the visual system, differential geometry, and image processing and interpretation.

Alessandro Verri was born in Genova, Italy, on October 31, 1960. He received the PhD degree in Theoretical Physics from the University of Genova, Italy, in 1983. From 1983 to 1985 he worked with Prof. V. Torre at the Department of Physics, University of Genova, on problems concerning edge detection and motion analysis in machine vision. His scientific interests range from mathematical to computational aspects of information theory, theory of dynamical systems and bifurcations, and signal theory. He has been studying various aspects of the processing of visual information in both machine and biological systems and he is currently working on the modeling of motion perception, the understanding of complex motion, the interpretation of 3D scenes, and object recognition.

Thierry Viéville is a Researcher at I.N.R.I.A. (National Research Institute in Computer Science and Control Theory), where he works in the Computer Vision and Robotics group. His research interests include the application of Mathematics to Computer Vision, Robotics, Motion Analysis, Active Vision, and the Architectures for Dynamic Vision as well as the links between artificial and biological vision.

He is an Associate Professor of the Ecole Supérieure des Sciences de l'Informatique, where he teaches Computer Vision and Symbolic Methods in Robotics and Vision.
He participates in different European projects on Active Vision and Vision Machines Architectures.

Roger Watt is Professor of Psychology at Stirling University, Scotland. He did a PhD with D.P. Andrews at Keele University which was concerned with comparing human sensitivity to line curvature with that of a notional ideal processor. His work with M.J. Morgan was some of the earliest work to examine psychophysically the various computational edge detection theories available. His interest has continued to be concerned with the interrelationship between psychophysical work and computational theory in human vision. His current interests concern the use of the rigourous psychophysical techniques in the study of natural visual tasks, with all their inherent experiments that are more usually studied. He leads a large interdisciplinary laboratory approaching the study of human vision from several different directions.